BRITISH POLITICS

BRITISH POLITICS

Continuities and Change

THIRD EDITION

DENNIS KAVANAGH

OXFORD UNIVERSITY PRESS

Oxford University Press, Great Clarendon Street, Oxford OX2 6DP
Oxford New York
Athens Auckland Bangkok Bogota Bombay
Buenos Aires Calcutta Cape Town Dar es Salaam
Delhi Florence Hong Kong Istanbul Karachi
Kuala Lumpur Madras Madrid Melbourne
Mexico City Nairobi Paris Singapore
Taipei Tokyo Toronto
and associated companies in
Berlin Ibadan

Oxford is a trade mark of Oxford University Press

Published in the United States by
Oxford University Press Inc., New York

First published 1996
Reprinted in paperback 1997

British Library Cataloguing in Publication Data
Data available

Library of Congress Cataloging in Publication Data
Kavanagh, Dennis.
British politics: continuities and change/Dennis Kavanagh.—3rd ed.
Includes bibliographical references.
1. Great Britain—Politics and government. I. Title.
JN231.K3 1996 320.941—dc20 96-7399
ISBN 0-19-878167-9 (pbk)
ISBN 0-19-878168-7

Typeset by J&L Composition Ltd, Filey, North Yorkshire
Printed in Great Britain on acid-free paper by
Butler & Tanner Ltd., Frome, Somerset

To Catherine

PUBLISHER'S PREFACE: ABOUT THIS BOOK

This new edition builds upon the values already established in the earlier editions. There are two core objectives: to introduce the student to the formal study of political systems; and to give an account of British political institutions and their performance. These objectives have shaped the improvements to the third edition.

The distinguishing features of *British Politics* (3rd edition) are:

Clear and authoritative coverage

1. Change and continuity remains the unifying theme of the book. The text seeks to explain both the new directions and the continuities within British politics.
2. Coverage has been fully revised and updated to take account of the changes within the main political parties, the ongoing debates regarding Britain's membership of the European Union, the search for a settlement in Ireland, the reform of the public services, and the changes in British political culture.
3. New chapters have been provided on the mass media and Europe.

Accessible and engaging features for the student

Of particular appeal to students will be the book's range of learning aids:

Reader's guides. These provide the reader with a guide to the coverage of each chapter. This feature aids effective learning.

Boxed examples. The text now contains more than eighty such examples which provide general examples, historical notes, and further analysis.

Highlighted quotations. These provide an insight into the thoughts of some of the historical and contemporary personalities that make up British politics.

Chapter summaries. In a bullet-point format these summaries are an invaluable learning and revision aid.

Chronology sections. Each chapter includes this feature which place the chapter's contents within a historical context by highlighting key dates.

End of chapter questions. Two types are provided: essay/discussion questions and research exercises.

Further reading. This feature highlights some of the key works in each area.

AUTHOR'S PREFACE

BRITISH politics has altered a great deal over the past two decades. It is appropriate, therefore, that the formal study of the political system has also developed. The now substantial literature, much of it of very high quality, has been transformed by different approaches (whether they be quantitative, institutional, behavioural, pluralist, or Marxist), contrasting evaluations (supportive or critical), and the introduction of new topics. The most striking feature is the more critical stance towards political institutions and performance. Events during recent years, particularly the record of government since 1979, have intensified challenges to many long-established ideas about British politics. If one danger is that our perceptions are shaped by knowledge that is obsolete, a different one is that we may too readily assume the 'permanence' of the new Thatcherite settlement. In *British Politics: Continuities and Change* I have attempted to separate the new directions from continuities in British politics.

Much has changed since the second edition was completed in October 1989. A book on British politics written during the 1980s inevitably bore the heavy impress of Mrs Thatcher and the ideas and policies associated with her. In the late 1990s it is possible to achieve some perspective on the 1980s and what remains of the Thatcher settlement. Some changes strike one immediately. These include the altered state of the Labour and Conservative parties, the failure of the centre party advance, the growing importance of Britain's membership of the European Union and the instability this has caused for the Conservative party, the reforms of the Civil Service and public services, and the change in the political culture and terms of political debate.

Compared with the second edition, chapters on Political Culture and on Policy-making have been omitted and some of the material has been incorporated elsewhere in the book. New chapters have been added on the Mass Media and Europe.

The format of this third edition is also much altered. Readers will note the addition of numerous summaries, diagrams, tables, and exercises, to make the book more useful for teachers and students. These 'aids' are entirely the work of Dr Sue Pryce and I am much indebted to her. As ever I am heavily indebted to my secretary at Nottingham, April Pidgeon, who coped with my dictation and handwriting with unfailing patience and cheerfulness.

D.K.

Nottingham
September 1995

CONTENTS

DETAILED CONTENTS

FIGURES

BOXES

TABLES

1 UNDERSTANDING BRITISH POLITICS: THE BRITISH MODEL

Reader's guide

Britain has long been regarded as a model of stable government. It has achieved a balance between order and liberty, authority and accountability, and, by a gradual evolutionary process, has been transformed from a feudal kingdom into a modern representative nation state. It is argued that an island position, a tradition of rule by consulation and consent, flexible and adaptable institutions, and the sequential order in which demands for change have arisen are some of the factors which account for Britain's stability. Since the late 1950s doubts have been raised about the British model. The post-war era has been characterized by the politics of decline in which Britain has been forced to come to terms with the reality of relative economic decline and its implications for internal political stability and social harmony, as well as external prestige and influence. By the 1970s the continued failure of political élites to deliver economic prosperity to the citizens raised questions of legitimacy, and led to a widening of the ideological divide, as political parties rejected the former consensus about policies and institutions in their search for radical solutions to persistent economic policy failure.

This chapter suggests reasons why the student of politics might find the British model worthy of study. It discusses the historical development and social and geographical characteristics which have helped to shape Britain's modern liberal democracy and explores the kind of pressures that seem to be transforming Britain from a model of good politics to a model of what to avoid.

MANY of Britain's political institutions and practices have been established for a long time. Viewed in international and historical perspective, the British political experience has been widely admired, although national pride has not always taken the extreme form shown by the *Edinburgh Review* in 1807: 'All civilised Governments may be divided into free and reactionary: or more accurately . . . into the Government of England and other European Governments' (cited in Halevy 1924: 147). This chapter tries to explain why in several respects Britain has been viewed as a political success story.

Since the eighteenth century, Britain has been regarded variously as a model of constitutionalism (Montesquieu), representative government (de Tocqueville, Bagehot), or political balance (the authors of the American constitution). Sociologists, anthropologists, and historians have often admired 'the British way of life'. Even Marx and Engels (no friends of British capitalism or its political system) in the nineteenth century used Britain—as the first and most advanced industrial society—to plot the future course of social and economic development of other countries.

Eighteenth-century observers admired the way in which the gradual consolidation of the 1689 constitutional settlement seemed to provide a political balance, one in which government was secure but also accountable to Parliament, ruling according to laws and conventions, and in which civil liberties were better protected than on the Continent. After 1776 the American rebels fought to retain what they regarded as the best features of the British system. The instability unleashed by the French Revolution in 1789 created a political model against which Britain could be viewed favourably.

From that historic event France and Britain have often been contrasted as models of political development, the former alternating between political extremes of despotic and

BOX 1.1. THE BILL OF RIGHTS, 1689

The **1689 Bill of Rights** confirmed the constitutional changes introduced by the **1688 Glorious Revolution**. It brought a definitive end to monarchy by divine right and its replacement by constitutional monarchy.

In 1685 James II, Charles II's brother and heir, succeeded to the throne, despite attempts by some Parliamentarians to exclude him on grounds of his Catholicism (Exclusion crisis 1676–81). James, by use of his prerogative powers, and supported by a purged and compliant judiciary, introduced positive discrimination in favour of Catholics. In a bid to secure these changes James resorted to more drastic measures, including sacking JPs and remodelling borough charters. In 1688 the birth of a Catholic male heir dashed hopes that James's Catholic rule would be a temporary interval before he was succeeded by his Protestant daughter, Mary, and her husband the Protestant Dutch ruler, William of Orange. This prompted parliamentary and political leaders to invite William and Mary to England to defend the Protestant religion and ancient liberties of Englishmen. William invaded Britain. James fled to France. The revolution was bloodless, except in Catholic Ireland, from where James attempted to regain his throne and was defeated by William at the Battle of the Boyne, 1690.

The Glorious Revolution permanently altered the relationship between the Crown and Parliament. It also created a perpetual blot on Anglo-Irish relations that still informs the conflict in Northern Ireland 300 years later.

weak government, the latter moderate and balanced, the one unstable, the other stable. The experience in the twentieth century of so many European states with variants of authoritarian and totalitarian rule made them unattractive models for other countries to follow. One has only to consider the chequered democratic records in the twentieth century of Italy, Germany, Spain, and Japan to be aware of the lack of competition. Students have therefore turned to Britain, in large part in default, in search of the efficient secret of stable democracy.

Britain is one of perhaps a score of mature (say, fifty years or more) liberal democratic states in the world today. These are regimes which permit both a high degree of popular participation in politics and the opportunities for institutionalized opposition to the government of the day. They are found predominantly in Anglo-American societies and Western Europe. At the close of the twentieth century the main rival to the liberal democratic model—the single-party regime with a centrally planned economy—has been discredited. A consequence of the collapse of the USSR and the spread of competitive elections to East European states has been a sharp increase in the number of democracies.

The case of Britain

Notwithstanding present discontents, Britain is an interesting political system to study. Certain features merit particular attention.

1. Since the late seventeenth century the

BOX 1.2. BALANCED GOVERNMENT

The executive, exercising wide-ranging powers, is balanced by a legislature, composed of elected representatives, to which the executive is accountable for the exercise of power.

In reality there is a precarious balance, depending upon political actors recognizing that there are 'rules of the game' (i.e. a constitution) and abiding by them.

The government, by definition, is supported by a majority of the House of Commons. Coupled with the doctrine of **parliamentary sovereignty**, this means that the executive can pass any law.

Liberty, and checks upon authority, rest upon:

- institutionalized Opposition
- accountability of ministers to Parliament
- tradition of government by consent
- vigilant and free media
- free and regular elections

political system has managed to *balance the twin elements of governmental authority and limits on that authority*. The exercise of political authority fuses the ideals of strong executive rule with a recognition of the need for power to be subject to constraints. The limits may be in the form of widely understood 'rules of the game', such as the rights of the Opposition in Parliament to criticize government, the accountability of ministers to Parliament and ultimately to the electorate, a general sense of self-restraint among ministers, and various political, social, and economic checks and balances. There is a balance between the principles of order and liberty, and between effective government and accountable government.

2. A second feature is the *legitimacy of its political institutions*. Legitimacy is the recognition by the vast majority of people of the government's authority and a duty to obey its rules. That in turn depends on popular belief, for example, that power is not abused but exercised in line with established traditions and laws and that the rulers have come to power by established procedures. As a rule, the greater the legitimacy of the regime the less the need for government to use coercion. Compared with many other countries, the low level of political violence in Britain in the twentieth century is striking. Obedience to a regime rests on a mixture of motives and sanctions: habit, law-abidingness, loyalty to nation and state, self-interest, and fear of coercion. A regime's ability to command obedience also rests in part on its monopoly of coercion in society through the armed forces, police, and courts. Yet we will see that force has been lightly used in the British system, although Ireland—as usual—is the great exception to the statement.

3. A third feature is that Britain has enjoyed *democratic stability* since 1945, achieved against a background of relative decline, both economic and diplomatic. Many other West European states have achieved stable democracy on the back of

BOX 1.3. CHARACTERISTIC FEATURES OF A NATION STATE

A state is an organization which controls the population occupying a known territory. Its characteristic features are:

- the ability to impose rule throughout recognized **sovereign territory**: it acknowledges no internal or external superior authority; it has its own nationalized, paid armed forces to deter/repel external threats and police forces to enforce internal law and order
- a **centralized, secularized authority**
- funding from **taxation**
- administration by a central apparatus of paid officials — a **bureaucracy**
- the **allegiance/loyalty** of citizens to a self-conscious 'national' political institution, the state
- **representative government**; an accepted feature of a 'modern' nation state is that the nation is represented in the institutions of state (i.e. representative democracy).

'miracles' of rapid economic growth. Historian Sidney Pollard has noted, in *The Wasting of the British Economy*, that the only 'British economic miracle' was its absence (1981: 6). Indeed, Britain has been a case-study in economic decline, spawning a massive literature—both scholarly and popular—as well as numerous university courses on this topic.

There have been upheavals and violence. There was a bitter civil war between 1640 and 1649, ending with the execution of Charles I, Jacobite rebellions in 1715 and 1745, threats of insurrection in 1830, 1848, and 1914, and guerrilla battles in Ireland in 1918–21. By luck and good judgement, however, the system survived and violence has not been part of British life (except in Northern Ireland).

4. A fourth reason for interest lies in Britain's *pattern of political development*, over a thousand years. Many of the crucial stages of forming a state and nation in Britain were solved long ago. Central authority in England was established by Anglo-Saxon monarchs in the tenth century and consolidated by the Normans after 1066. The Tudors established a strong monarchy in the sixteenth century. The secular and religious claims of the Roman Catholic Church were rebuffed when King Henry VIII broke the ties with Rome in 1534, and set up a national Church of England, subordinate to the State. The constitutional settlement at the end of the seventeenth century provided for a sharing of power between Parliament and the monarchy. A constitutional revolution was achieved peacefully in 1688 when James II's attempts to assert monarchical power were resisted and he fled: henceforth, Britain enjoyed a constitutional (or limited) monarchy, and sovereignty lay in Parliament. Britain's borders were consolidated as England formally absorbed Wales in 1536, united with Scotland in 1707, and enjoyed uneasy dominance in Ireland until 1922.

Continuity and change

In spite of several important changes occurring over time, the frequent use of such terms as 'gradualist', 'evolutionary', and 'traditional' to describe the development of British politics reflects this continuity of key elements of the system. Such pre-democratic institutions as the monarchy and the House of Lords as well as the common law and hereditary peerage have been retained and adapted rather than abolished. The outer shell of the constitutional settlement of 1689 has been preserved largely intact even though, over the years, the political system has been democratized, major shifts in the balance of power between particular institutions have occurred, and the role of government has been transformed.

' . . . the British Constitution has continued in connected outward sameness, but hidden inner change, for many ages . . . an ancient and ever-altering constitution is like an old man who still wears with attached fondness clothes in the fashion of his youth: what you see of him is the same, what you do not see is wholly altered'.

Walter Bagehot, *The English Constitution* (London: Fontana, 1963), 1.

One reason why absolute monarchs in Europe in the seventeenth century required large standing armies was to defend their extensive borders. Yet the same forces could be (and were) used by rulers to coerce taxation and extract taxes from the population, and to overcome challenges to their authority from rivals. This cycle of *coercion–extraction* was important in assisting the process of political centralization and state-building in many European states in the seventeenth century (Finer 1975*a*). In Britain, by contrast, the attempts of the Stuarts to follow this path were defeated. The pretensions to absolute power of Charles I were rebuffed in the Civil War (1640–9), as were those of James II in 1688. By then it was firmly established that the consent of Parliament was required for levying taxation and maintaining an army. No powerful central bureaucracy developed, as happened in much of Europe. The military and police functions were left to local Justices of the Peace—men who saw themselves as members of the local community rather than as agents of central government power. There was substantial freedom of speech, association, and of press, an independent judiciary, and trial by jury.

Britain's *island position* lowered the geopolitical threat from other countries. The English Channel provided a twenty-mile barrier between England's south coast and the European mainland. Naval strength became the key to the country's security and was hardly suitable for coercing the population. If Britain developed as a 'low-profile' state in the eighteenth century, in comparison with France, Prussia, or Russia, then its island location was an important part of the explanation.

Once more it is worth emphasizing how this continuity of basic features of the political system contrasts with the experience of many other states. The present boundaries and constitutions of many West and East European states date only from 1945 or later. Many were occupied between 1939 and 1945 by another power, such as the USSR or Nazi Germany, or suffered military defeat (France, Italy, Germany) and then had their political institutions redrawn subject to the approval of the victorious powers. The British political system did not collapse, as Russia's had in 1917, Germany's in 1945, or France's in 1940 and again in 1958. Continuity and adaptation, rather than the creation of a completely new system at one point in time, have been the main themes of Britain's political development. They are a tribute to the flexibility and effectiveness of its ruling politicians and institutions.

Some historians have argued that the 'load' or set of problems on a state is affected by the timing and intensity of conflicts, on such basic issues as securing its borders, forging a sense of national identity, and establishing the authority of central government. If these challenges have to be faced more or less *simultaneously*, then important questions of authority, legitimacy, and participation coincide and impose a heavy 'load' on the system. This has been the fate of many 'new' states which achieved independence after 1945. But the British have been able to face many of these problems *sequentially*, thus allowing élites time to cope with or solve one problem before confronting another. Britain provides a *model of low-cost political modernization* in which the achievement of central authority and national identity preceded the democratization of the political system. The records and political lessons to be extracted from countries such as Russia, Italy, France, or Germany in the twentieth century have appeared less attractive in comparison.

Representative democracy

Britain's passage to representative democracy was also drawn-out. As noted, the claims of an absolute monarchy were checked more effectively and earlier than on the Continent. It was during the early nineteenth century that the Cabinet, rather than the monarchy, emerged as the effective executive, and the principle of its accountability to the Commons was established. The idea of responsible parliamentary government was the kernel of the development of representative democracy in Britain. Once Parliament had established its claims to control government, the next stage was to make it representative of the population.

Although the House of Commons might take account of outside public opinion, it was in no sense a democratically elected body: the electorate for the House of Commons amounted to less than 5 per cent of the adult male population before 1832. A series of Acts of Parliament from 1832 down to 1969 extended the right to vote until today virtually all men and women over the age of 18 can vote. And within Parliament the supremacy of the popularly elected Commons over the unelected House of Lords was gradually established, decisively so in 1911 when a Parliament Act was passed which abolished the veto power of the Lords, and allowed it only

BOX 1.4. THE LIBERAL VIEW OF THE CONSTITUTION

According to the classic **liberal view of the constitution**, the House of Commons is the supreme political institution with power to make and unmake governments, pass any law, and resolve great political issues of the day. The monarchy and the House of Lords have only limited political significance, and this view takes little or no account of the role of political parties, the Civil Service, the media, pressure groups, or public opinion.

It is associated with the writings of A. V. Dicey and Walter Bagehot, who observed Parliament during the 'golden age' (1832–67) between the two Reform Acts, before the rise of disciplined mass parties, when the House of Commons really did reign supreme.

Democratic chain of authority according to the liberal view:

Citizens (after 1832, etc.)

↓

House of Commons

↓

Cabinet and ministers

↓

Civil and military servants

to delay by two years legislation passed by the Commons. The growth of the Civil Service in the twentieth century followed the extension of the suffrage and the emergence of organized political parties; senior civil servants saw themselves as impartial servants of the State, serving the government of the day, regardless of party. In essence, the history of responsible parliamentary government encapsulates what is called a 'liberal view' of the Constitution. The view is inadequate in some respects, but its depiction of the steady shift in power from the monarchy and Lords to the Commons since 1832 is broadly correct.

The major exception or qualification (though an important one) to this rather benign account concerns Ireland. In the nineteenth century, a largely Catholic Ireland struggled for independence from the United Kingdom. This demand was finally met in 1921, but six largely Protestant counties in the north were allowed to remain British. Many Irish nationalists, in the north and south, never accepted this, and a clause in the Irish constitution still looks forward to the six counties' eventual incorporation in a united Ireland. The Unionist politicians in the north were determined to resist such an outcome and threatened an armed uprising if they were forced into an independent Ireland. They succeeded and a separate north and south coexisted uneasily. But Catholic protests in the north broke out in 1968, and since 1969 Irish and Ulster paramilitary forces have waged a bloody conflict. In 1972 the British government imposed direct rule in Ulster and stationed the army there to maintain a semblance of order. 'The Irish Problem', as the English have revealingly termed it, has combined divisions over religion (Protestant versus Catholic), nationality (Irish versus British), and statehood (Ireland versus Britain), and is a good example of the intractability posed by simultaneous conflict in several spheres. Britain ruled the six counties of Northern Ireland, but for the large Catholic minority it has been a case of 'Governing without Consensus' (Rose 1971). For long the Unionists rejected any initiatives that involved talks with a 'foreign' government in Dublin, or the IRA and the nationalists insisted on Dublin's participation. By 1995, at last, the violence had ceased and talks about the future political arrangement had opened.

Social structure

Successive waves of immigration from the time of the Romans, Celts, Saxons, and Danes down to the Irish and Afro-Caribbeans ensure that the British are not ethnically homogeneous. But many of the social differences are rather weak and some of the differences do not translate into political disagreements.

- The country is highly industrialized, with less than 2 per cent of the working population engaged in agriculture. It is also heavily urbanized, with nearly two-thirds of the population concentrated in six large conurbations.
- Less than 5 per cent of the population is non-white, mainly immigrants from the new Commonwealth states and their offspring, and most of these are now British-born.
- Some two-thirds of people admitting a religious allegiance are Church of England, a

much higher proportion than Nonconformist (15 per cent) or Catholic (10 per cent). Comparative surveys show that religious feeling is weak; a notable exception is the strong religious attachment of Muslims.

- Nearly five-sixths of the total British population reside in England.

In a nutshell, when we look at race, language, religion, and nationality, we find that the social differences which might express themselves in politics are limited. *Britain is overwhelmingly urban, white, Anglican, and English.* Claims about social homogeneity and the importance of social class derive from the relative weakness of other cleavages. Britain is very different from the ethnic melting pot of the United States.

Britain did not always have such a simple social structure. The validity of the preceding statements dates only from the withdrawal of the twenty-six Irish counties from the United Kingdom in 1922. At a stroke this simplified the social structure, removing an overwhelmingly Catholic, rural, and Irish population, strikingly different from that of the mainland. In 1851, even following the years of famine and heavy emigration, the Irish still amounted to 24 per cent of the UK population (compared to 61 per cent who were English). With such a differentiated population the pressures upon the British political system to change would have been immense. There would almost certainly have been a multi-party system—eighty Irish Nationalist MPs sat in the House of Commons between 1880 and 1918. There would probably have been regular coalition governments, federalism, a more territorial style of politics, and pressures to adopt a written constitution and a more proportional electoral system. All this is speculative but it is reasonable to assume that Ireland's exit from the United Kingdom provided an important force for the continuity of Britain's political institutions and values. Ireland is a very different society from mainland Britain, economically poorer, more rural, and with a culture which is shaped by a powerful Catholic church.

Geography

Notwithstanding British membership of the European Community since 1 January 1973, most Britons still do not think of themselves as Europeans. Although Britain has become a key member of the North Atlantic and West European security alliance, surveys show that the British are less 'European' and more resistant to further moves to European integration (in the sense of agreeing to cede more power to the institutions) than the populations of most other member states. One reason for the weaker attachment among the elderly in Britain is that most middle-aged citizens in the other member states have had personal experience of their country being militarily defeated or occupied. In Britain's case the outcome of the two great wars of this century has helped to keep its people psychologically apart from their West European neighbours.

We have noted that the natural frontiers of Great Britain (England, Scotland, and Wales), by helping the country to withstand military invasion since 1066, have been important in providing for the continuity of the regime. In many other countries military defeat has often been a catalyst for revolution and/or displacement of a regime. In Britain, victory only seemed to vindicate essential features of the political system. The preservation of its territorial integrity for such a long period has also probably promoted a strong sense of national identity, allowed the political institutions to become deeply rooted, and perhaps encouraged a politically relaxed society. Such institutions as the monarchy and the House of Lords, the common law, and the absence of a written constitution or a proportional electoral system make Britain 'different' from most other West European states and, some might claim, may explain why it is a 'reluctant' member of the European Union. Yet a number of political and economic changes have been inspired by foreign example: adoption of economic planning in 1961 from France, introduction of the ombudsman (or Parliamentary Commissioner) in 1965 from Sweden, and legislation to outlaw discrimination on the grounds of race and gender from the United States.

Post-war decline

The belief that the British political system is a success story has become less fashionable recently. For at least the last thirty years there has been a sustained—if not necessarily coherent—critique of the political system. The origins of this critique lie in the problems that Britain—and particularly its economy—has faced in the post-war world, clear policy failures, and concern about the quality of political democracy.

Dissatisfaction is not new. In the last quarter of the nineteenth century informed observers noted how Germany and the United States were overtaking Britain in terms of industrial performance. At the beginning of the twentieth century, for example, critics campaigned for 'National Efficiency', just as in the 1960s their successors campaigned for 'Modernization' or 'Remodernization'. The thematic similarity of the two campaigns was remarkable and in both cases transcended party lines. For example, the main complaints of both were addressed to Britain's economic weakness, international decline, and the alleged amateurism and lack of expertise among administrative and economic élites. Critics have divided into two groups, the 'Jeremiahs', who see the trends as part of a long-term decline, and the 'Pragmatists', who claim that changes in personnel, policies, and institutions may bring improvements (Gwyn 1980, Weiner 1981, Barnett 1988).

Domestic politics in post-war Britain has been dominated by problems concerning the economy. International diagnoses of the 'British disease' and domestic criticism of successive governments have usually fastened on economic failures, but the failure in the post-war period has been relative, not absolute. The rate of economic growth for the period 1945–95 has matched that of any similar period over the past two centuries. And in these past fifty years the living standards of the population have more than doubled. But the faster rates of economic growth in neigh-

BOX 1.5. POST-WAR CONSENSUS

There is a high level of agreement across political parties and governing élites about the the substance of public policy.

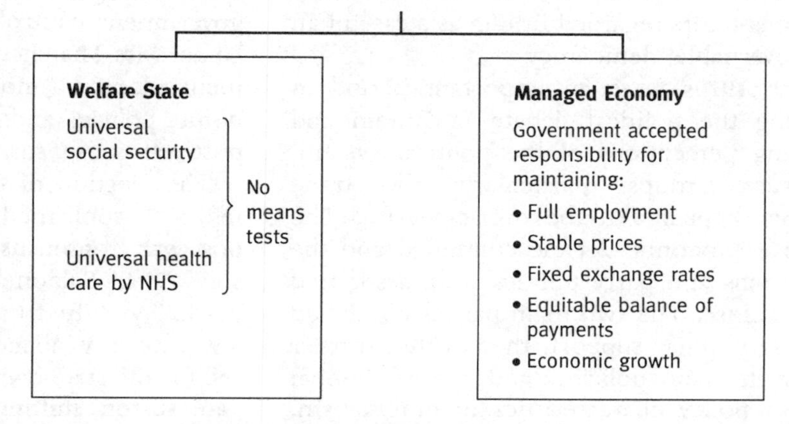

A large measure of consensus has also existed in respect of foreign policy:

- **special relationship with the USA, institutionalized in NATO**
- **ambivalence towards Europe**
- **possession of independent nuclear capability**
- **acceptance of the end of the empire**

bouring states have meant that their citizens' living standards as well as the quality of their public services overtook Britain's.

In 1945 the first majority Labour government was formed. It delivered full employment, created the National Health Service and an ambitious social security programme, placed the utilities of gas, electricity, rail, coal, and steel under public ownership, and began the process of giving colonies independence. These policies established the main outlines of British policy over the following thirty years. When the Conservative party returned to office in 1951 it accepted many of them. It presided over a steady improvement in living standards and, as the party of prosperity, won three elections consecutively in 1951, 1955, and 1959, the first time a party had done this in over a century. Labour, tied to an image of wartime austerity and economic controls, and a shrinking working class, was forced to rethink its policies. It returned to office in 1964, pledged to reverse the country's relative economic decline. It failed. Its National Plan for the economy was abandoned, the pound was devalued in 1967, and economic pressures forced it to abandon many of its social programmes.

A new Conservative government led by Ted Heath between 1970 and 1974 started out determined to reduce government intervention in the economy and encourage the free market. By 1972 it had executed a U-turn, and introduced statutory controls on incomes and policies and powers to intervene in industry. It lost an election in 1974 forced in large part by the coal miners' strike against the pay policy. The successor Labour government (1974–9) was no more successful. Its anti-inflation pay policies led to conflicts with the unions and a 'winter of discontent' which left the policy in ruins. By 1979 some commentators regarded Britain as a case of an 'ungovernable' democracy.

The 1970s were an important period in shifting the political debate in Britain and altering perceptions of the political system. Pressure groups, particularly the trade unions, appeared to be over-powerful. The relative economic decline continued and the politicians and party politics were associated with failure. The two main parties continued to lose popular support, they drifted further apart in their policies, and the traditional British policy characteristics of pragmatism, trust, and compromise appeared to wane.

Not surprisingly, political rivals differed in their analyses of what had gone wrong, and there was a sharpening of the ideological debate in the late 1970s and 1980s. Right-wing Conservatives talked of 'a crisis of social democracy' or socialism, consisting of excessive government intervention in the economy, irresponsible trade union power, and excessive levels of personal taxation and public expenditure. These policies, it was argued, reduced incentives, weakened the free-enterprise system, undermined thrift and individual self-help, and damaged the authority of the government. Socialists talked of a 'crisis of capitalism', of a system that failed to generate economic growth or allocate resources efficiently or fairly, and produced mass unemployment.

Until the mid-1970s debates within and between the parties and the thrust of government policies were resolved in favour of a mix of 'conservative socialism', or so-called consensus politics. Thereafter, the groups which gained influence in the two main parties sought to break away from the consensus policies which had prevailed since the end of the war, and to provide their own radical substitutes for the 'failed' consensus. The Labour party moved to the Left, opposing Britain's membership of the European Community and possession of nuclear weapons and pressing for more public ownership and government control over the private sector. Under Mrs Thatcher the Conservative government adopted a more free-market set of economic policies and explicitly condemned the post-war consensus.

The election of the Thatcher government in 1979 confirmed the breakdown of the post-war consensus (Kavanagh 1990, Addison 1994, Seldon 1994). The government gradually shifted the direction of public policy, explicitly abandoning full employment, selling off state-owned industries to the private sector, shifting the balance of the tax burden from direct to indirect taxes, imposing wide-ranging legal controls on the operation of trade unions and ending the post-war practice of consulting with them, and generally encouraging market rather than public-sector solutions to problems. A succession of electoral defeats forced Labour also to break with many of its old policies and effectively to abandon socialism in the 1990s.

There has been a sharp decline in Britain's international standing since 1945. At that time Britain was widely regarded as the third of the three great powers, largely because it had been on the winning side of the war and potential rivals, like Japan, Germany, and France, had been devastated. Britain was also the only state to be involved in 'three overlapping circles', Churchill's term for Britain's links with the Empire/Commonwealth, the United States, and Western Europe. Hopes of maintaining such widespread responsibilities were unrealistic. The population is less than a quarter of those of the two post-war superpowers and the country is smaller than Texas or California. It was only a matter of time before other countries restored their economies and overtook Britain. However, if Britain was precluded from being a superpower like the United States or the former Soviet Union, it remains the world's eleventh largest state in terms of population and ranks seventh in terms of total economic wealth or Gross National Product (GNP).

The post-war period has seen a rapid readjustment to the status of a medium-sized power. The dissolution of the British Empire began in 1947, with the granting of independence to India, Pakistan, and Burma. Since then, more than thirty other colonies have achieved statehood, and most have joined the Commonwealth. Britain has shed a number of overseas responsibilities and its armed forces have been withdrawn from various countries. It agreed to hand Hong Kong back to China in 1997, though British possession of Gibraltar and the Falklands creates continuing tensions with Spain and Argentina, respectively, which have claims on those two territories. The dispatch of the naval task force to recapture the Falklands from Argentina in 1982 evoked memories of Britain's imperial past.

Foreign policy, in contrast to domestic policy, is an area in which government is dependent on the actions or non-actions of other governments and agencies. Such outside events as Egypt's seizure of the Suez Canal in 1956 (prompting the ultimately abortive Franco-British occupation of the Canal), General de Gaulle's veto of British application for membership of the EEC in 1963 and 1967, the quadrupling of Arab oil prices in 1973–4, the International Monetary Fund's (IMF) 'rescue' of the pound in 1976, and the Argentine

BOX 1.6. BRITISH POLITICS TODAY

Continuity and change, institutions adapting to accommodate new demands and shifting political pressures, is a continuous process.

Some contemporary (1990s) issues/pressures include:

- What kind of relationship is appropriate between the state and citizen in the post-welfare state? As the costs of welfare rise, will people in work be willing to pay higher taxes to pay for benefits, and will there be more pressure for people to take care of themselves in sickness and retirement?
- What is the appropriate role of the state in the economy in the post-industrial age?
- The traditional role of the state has been that of setting rules for society: maintaining internal law and order, and protecting citizens from external attack. How will rising crime figures and increased vulnerability to terrorist attack affect the traditional role of the state?
- By a gradual evolutionary process divine-right monarchy was transformed into constitutional monarchy. Will the current questions about the succession of Prince Charles bring a further reduction in the constitutional role of the monarch or even the end of the monarchy?
- Whither the British nation state if it accepts a single European currency?
- What will be the outcome of the next European Union Inter-governmental conference (IGC) scheduled for 1996? Will Britain accept further integration, and, if so, how will this affect the British state and political institutions?

seizure of the Falklands in 1982 all had major impacts on British domestic politics. British governments are heavily involved in concerting economic, foreign, and defence policies with such bodies as the EU, IMF, UN, Organization for Economic Cooperation and Development (OECD), and NATO, and so on. Membership of the EU involves a formal limit on Britain's sovereignty as the treaties impose obligations on Britain's defence, trade, and taxation policies. It may be argued, however, that the obligations of the treaties only recognize the actual limits on an independent British economic or defence policy.

In the immediate post-war years, foreign policy issues were, in general, matters of agreement between the parties. However, the various applications for membership of the EC, the proposed rearmament of Germany in the 1950s, the invasion of Suez, the retreat from Empire in the 1950s and 1960s, Britain's possession of nuclear weapons, and moves in the 1990s to further integration in Europe have divided one or the other of the two main parties.

Conclusion

Political learning may take one of two forms. One is the *acquisition of knowledge* for its own sake. Studies of comparative government, for example, may observe how broadly similar functions are carried out differently across societies (for example, a largely state-run health service in Britain or a largely private insurance-based system in the USA) or how similar institutions operate differently in different periods and contexts. A second is to study with a view to *learning lessons*, for example, borrowing admired features from another society or seeking to avoid its mistakes. Britain has attracted both types of study and there is a rich store of explanatory writing about its 'success'. Political scientists have looked at the British experience, the manner and sequence in which its major problems were confronted, its institutions and procedures, and sought lessons which could be applied to other states. The two-party system, the deferential political culture, the gradual historical development, the Government's power to dissolve the House of Commons, or the figurehead role of the monarchy have all been suggested as mechanisms for promoting a state's political stability and effectiveness, *à la* Britain.

In politics, however, there are no permanent solutions, and many of the once-praised features of British politics have been subject to criticism in recent years. There are always pressures to change or to adapt, from international forces, economic trends, interest groups, and public opinion, to name but a few. When a country's political institutions and practices have been as moulded by the past as Britain's have been, it may be difficult to accommodate the forces for change. The long-established institutions may become too entrenched to adapt. For at least three decades now, attempts to promote social, political, and economic change or modernization have been a central part of the political agenda in Britain. Critics of the social class divisions, the education system, industry, the civil service, and the political institutions claim that these are burdened by too much traditionalism. A growing body of opinion now regards parts of the political system as a barrier to desirable social and economic development. Few member states of the European Union now look to Britain for democratic guidance. Indeed, the opposite is often the case, as supporters of electoral reform, a Bill of Rights, devolution, or a written constitution look to the Continent for inspiration. It is telling that so few leaders in the East European states who drew up their new democratic structures, following the collapse of the USSR, looked to Britain's political institutions, least of all its centralized majoritarian system.

Summary

- Students have traditionally looked to Britain as a model of a mature stable liberal democracy which since the seventeenth century at least, has accommodated political change without bloody revolution and has avoided the extremes of ideology of both left and right.
- The political system has achieved a balance between authority and liberty: government is strong and effective but accountable.
- With the notable exception of Northern Ireland, legitimacy of the regime is widely accepted and coercion has played a minimal role in ensuring political obedience. Rebellions and violence have been limited and rare.
- Political modernization has been a gradual sequential process and this may account for the way in which élites and the system have responded successfully to change.
- By an incremental process the power of the monarch and hereditary nobles was gradually ceded to the House of Commons, elected on universal adult (18 years and over) suffrage. Government became the function of a cabinet, primarily composed from and answerable to the Commons and thence the electorate.
- Ireland has been a major exception to this story of peaceful development of democracy and general recognition of the government's right to rule. In Ireland, until 1922, and in Northern Ireland since that time, questions have been raised about national identity, statehood, and religion, and the resort to terrorism and state coercion have been commonplace.
- Britain is overwhelmingly urban, white, Anglican, and English. This substantiates the view that British society is homogeneous and has resulted in the economic category of class becoming the most politically significant social division.
- Britain's island position has had an impact on the development of the relationship between the state and its citizens and continues to inform Britain's relationship with the EU and international organizations. Insularity partly accounts for Britain's rather different institutional development that continues to set it apart from its European partners.

- In the immediate post-war period a wide measure of agreement existed between the main political parties in respect of both policies and institutional arrangements. It was accepted that the state was responsible for the welfare and economic well-being of its citizens.

- The post-war period has also been characterized by the political problems associated with the management of a weak economy. In the 1970s relative economic decline and clashes with the trade unions undermined governments.

- By the 1970s and 1980s the post-war consensus began to break down. Some sections of society became more willing to resort to political disobedience and more willing to reject the traditional channels and negotiated settlements in favour of direct action.

- Political parties have been forced to explore more radical options in their search for solutions to persistent problems.

- There are some signs in the 1990s of a new consensus emerging which recognizes that there are limits to the state's ability to deliver economic and welfare benefits.

- Britain, once the ruler of an empire upon which the sun never set, has had to come to terms with the realities of imperial overstretch. The thrust for independence in former colonies and discrediting of the ideological basis of colonialism has forced Britain to withdraw from its former empire and accept the role of second-rank power.

- Emerging democracies now seem less willing to emulate the British model.

CHRONOLOGY

1066	Normans begin to consolidate a centralized state
1215	Magna Carta: documentary recognition of consent; the king is not above the law
1485	Beginning of the Tudor dynasty; national identity begins to develop
1534	Henry VIII breaks ties with Rome and becomes Supreme Head of Church of England
1536	Wales incorporated into England
1640	Civil War
1649	Republican experiment under Cromwell
1660	Restoration of the monarchy
1681	Exclusion crisis
1688	Glorious Revolution
1689	Bill of Rights: constitutional monarchy
1707	Act of Union between Great Britan and Scotland
1801	Union of Great Britain and Ireland, formation of United Kingdom
1832	Reform Act: beginning of democratization of the state
1854	Northcote–Trevelyan Report creates modern civil service bureaucracy
1867	Reform Act extends suffrage
1906	Suffragettes become active
1911	Parliament Act establishes House of Commons supremacy
1918	Franchise extended to all men over 21 years and women over 30 years
1921	Irish Free State created; Northern Ireland remains British
1926	Women get vote on equal terms with men
1945	Welfare State: government is responsible for social and economic well-being of citizens
1949	NATO institutionalizes special relationship with the United States
1952	Elizabeth II succeeds to the throne
1965	Office of ombudsman created
1970	Ushers in decade of economic problems and doubts about British political institutions
1972	Stormont suspended, direct rule of Northern Ireland begins
1973	Britain joins EC
1974	February 'Who Governs Britain?' election
1978	'Winter of Discontent'
1979	Margaret Thatcher elected
1986	Britain signs Single European Act
1992	Britain signs Treaty of European Union (Maastricht)

ESSAY/DISCUSSION TOPICS

1. Account for Britain's success in achieving political modernization without widespread political upheaval or violent revolution.
2. The political modernization of Britain is a success story which has lessons for any country in the process of transformation to modern nation statehood. Discuss.
3. Why were questions of legitimacy on the political agenda in Britain in the 1970s?
4. Is Britain a modern nation state?
5. Why is economic policy of fundamental political importance?

RESEARCH EXERCISES

1. Rebellion and political violence, though uncharacteristic, are not absent from the British political experience. Briefly discuss *three* twentieth-century examples in which the legitimacy of the state has been challenged.

2. Explain the term 'sovereignty'. Select three examples from the last twenty-five years which demonstrate that limits exist to British sovereignty.

FURTHER READING

The literature on British politics is enormous. Most of the topics have now acquired a substantial literature of their own. At the end of each chapter will be found a selection of books and articles which are up to date and useful. For more recent and research-based material the main journals should be read. Particularly important are *Parliamentary Affairs*, *Public Adminstration*, and *Political Quarterly*. *Political Studies* and *The British Journal of Political Science* also have useful material as well as articles on subjects other than British government. Three new journals, *Politics Review*, *Contemporary Record*, and *Talking Politics*, have been established; they produce brief articles which are aimed at sixth formers and first-year degree students.

On decline see A. Gamble, *Britain in Decline*, 3rd edn. (London: Macmillan, 1990). A good guide to the debate about the post-war consensus is A. Seldon, 'Consensus: A Debate too Long?', *Parliamentary Affairs*, 47 (1994) 501–15. On foreign policy see P. Byrd (ed.), *British Foreign Policy under Thatcher* (Oxford: Philip Allan, 1988).

Other general works which may be useful are:

Bagehot, Walter, *The English Constitution* (London: Fontana, 1963).
Barnett, C., *The Audit of Britain* (London: Macmillan, 1986).
Childs, David, *Britain since 1945: A Political History* (London: Routledge, 1994).
Glyn, William, and Ramsden, John, *Ruling Britannia: A Political History of Britain, 1688–1988* (London: Longman, 1990).
Jones, Bill, 'British Democracy Today', *Talking Politics*, 6/3 (1994).
Weiner, Martin, *English Culture and the Decline of the Industrial Spirit, 1850–1980* (Harmondsworth: Penguin, 1985).

2 A UNITED KINGDOM?

Reader's guide

The United Kingdom is a four-nation state. Its unitary nature is reflected in the constitutional principle of parliamentary sovereignty. England, Scotland, Wales, and Northern Ireland, have distinctive national identities and political institutions, but powers exercised by subordinate authorities emanate from Parliament. Not only is the UK a multination state, it is also becoming a multiracial society. Despite this varied composition, divisions of nationality, race, and religion have been of less political significance than those of class, which after the industrial revolution emerged to be Britain's primary social cleavage.

This chapter considers the issues of national identity, race, and social class. It examines the extent to which Scotland and Wales have distinct political institutions, it explores the nature and strength of nationalist sentiment within these countries, and it reviews the fluctuating support for devolution. Northern Ireland in respect of questions of nationality, statehood, religion, coercion, and terrorism constitutes a special case that merits extensive treatment. The history of the Northern Irish question, its intractable nature, and the solutions, both proposed and tried, are examined. In the wake of the last great wave of immigration in the 1950s, race has acquired greater political significance in the UK. This chapter examines the way in which governments have responded to this issue by attempting to legislate for racial harmony, passing Immigration Acts to restrict the number of immigrants, and Race Relations Acts to outlaw discrimination. Changes in the occupational structure and rapidly expanding opportunities for higher education have implications for Britain's class structure. This chapter seeks to explain the nature of these changes and their likely impact on a two-party system based on the traditional socio-economic division of middle and working class.

EVERY political system is embedded in a context, which is the product of history, geography, economy, and social structure. Chapter 1 has already noted how Britain's politics has been shaped by its historical development, island position, comparatively homogeneous population, and economic change. This chapter considers the issues of national identity, race, and social class.

The United Kingdom

The total population of the United Kingdom (UK) in 1992 was just over fifty-eight million, and the area over 93,000 square miles. The distribution of the population among the four nations of the Kingdom has consequences for political influence. Scotland, with 5.1 million people, Northern Ireland, with 1.6 million, and Wales, with 2.9 million, can hardly hope to match England, which has over 48.5 million. For many political purposes British society is English, even London-dominated.

The United Kingdom is a *unitary state* with a highly centralized form of political decision-making—a 'top down' model of authority. The essential elements are the supremacy of Parliament in Westminster, ministerial responsibility, and Whitehall control. Westminster has a virtual monopoly of taxation, preferring to provide grants to local authorities and non-English governments rather than increase their taxation powers; it lays down uniform standards for the application of health, welfare, and education services and benefits; and it may reclaim power delegated to subordinate authorities (as it did in Northern Ireland in 1972 and has long done for many local authorities).

The political centralization and integration

BOX 2.1. THE UNITED KINGDOM: A UNITARY STATE

Although the United Kingdom is composed of four nations it is a **unitary state**. All laws are passed by the Westminster Parliament. The state is governed from London. Any power exercised by subordinate authorities is bestowed by, and can likewise be taken away by, Parliament.

For example: an Act of Parliament established a Parliament for Northern Ireland at Stormont Castle in Belfast, but in 1972 Parliament suspended Stormont and imposed direct rule from London.

This **unitary** arrangement in the UK can be contrasted with **federal states** such as the United States of America or Germany, where written constitutions define the legal rights and status of various levels of government. In the UK it is Parliament that is legally and constitutionally supreme.

have been reinforced since the 1920s by the Conservative and Labour parties competing for full political power. This development was not inevitable. Until then, Labour was also a party of Home Rule for the different nations and of municipal socialism, while the Conservatives defended local government and with it the idea of limited national government. However, when Labour was faced with the prospects of political power, it settled for power centralization (for economic planning and redistribution) over decentralization (for participation and local choice). Help for the peripheral parts of the

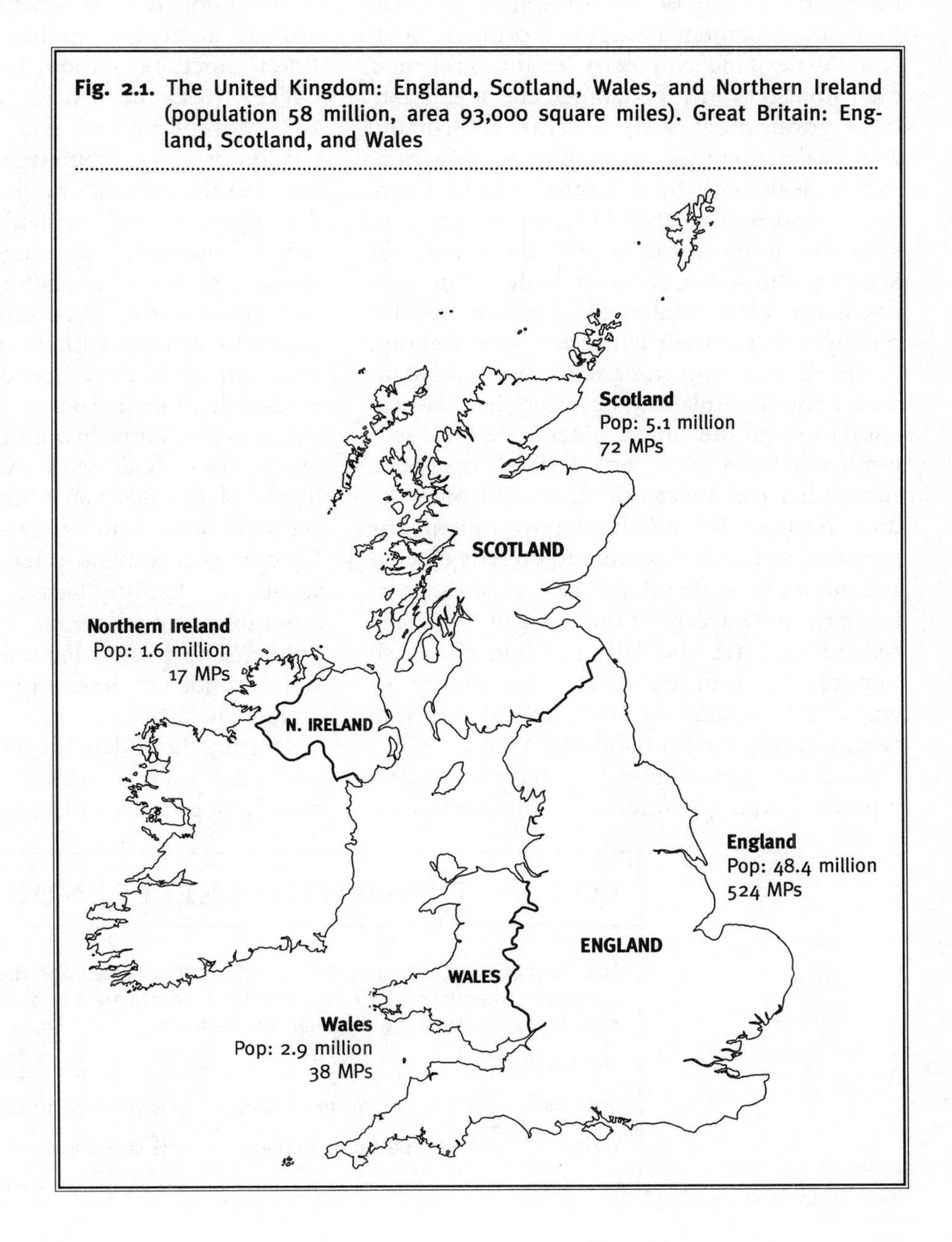

Fig. 2.1. The United Kingdom: England, Scotland, Wales, and Northern Ireland (population 58 million, area 93,000 square miles). Great Britain: England, Scotland, and Wales

UK (and Labour strongholds) would come through central allocation of the government pork-barrel, rather than through constitutional change.

Yet Scotland and Northern Ireland, though subordinate to the Westminster Government, still possess their own distinctive political institutions. Until 1972 Northern Ireland had its own Parliament based in Stormont. Scotland, Wales, and Northern Ireland have their own ministers in the Cabinet and a separate local government system exists in Scotland and Wales. Scottish legislation at Westminster is dealt with by a Scottish Grand Committee and non-Scottish MPs are reluctant to intervene in its work. Neither the Welsh nor Scottish Office has its own budget but they negotiate with Whitehall departments for money to cover their national responsibilities.

The British government has no consistent view about the relations between the constituent nations of the Union. Since 1972 local self-government in Northern Ireland has been suspended and direct rule imposed from London. Yet the 1974–9 Labour government was prepared to devolve powers to directly elected assemblies in Scotland and Wales. British parties have now accepted the right of Northern Ireland to leave the UK and join the Irish Republic at a future date if a majority so wish. Yet separatism and federalism have been rejected for Scotland and Wales.

Scotland and England coexisted as two separate kingdoms under one monarch for a century before the union of *Scotland* with England and Wales in 1707. This involved the creation of a new Parliament of Great Britain within which Scottish MPs were a distinct minority. Scotland, however, was allowed to retain its own legal system, the national Church of Scotland, and many other national institutions which may be said to provide a 'Scottish political system' (Kellas 1989). Since 1885 there has been a Scottish Office, whose head has sat in Cabinet since 1895, except for the war years, and has its headquarters in Edinburgh. It is responsible for health, education, housing, economic development and agriculture, and Home Office matters. Although variations of national policy are permitted by the Westminster government when applied to Scotland, decisions remain with London. The Labour government in 1978 passed bills to create an elected Scottish legislative assembly, to which a Scottish executive would be responsible, and would have dealt with matters then in the hands of the Scottish Office. Nationalists in Scotland called for independence and the Labour government offered *devolution*; this would be a halfway house, creating a Scottish Assembly and allowing it some legislative power. The plan collapsed when an insufficient number of Scots supported it in a referendum in 1979.

During the 1980s, devolution disappeared from the agenda. Mrs Thatcher was vehemently opposed to the idea, the Nationalists

BOX 2.2. DEVOLUTION REFERENDUMS, 1979

The Scotland Act and the Wales Act 1978 provided for the establishment of legislative assemblies in both countries respectively, but subject to approval by 40% of the electorate in national referendums.

The result:

Scotland	33% voted for devolution	31% voted against
Wales	12% voted for devolution	47% voted against

were eclipsed, and Labour, although still in favour of devolution, was in decline. Yet in these years Scotland's sense of distance from England probably increased. Although the Conservatives had only minority support in Scotland they enjoyed a commanding majority in Parliament thanks to their electoral dominance in England. Increasingly, the Conservative governments were seen by Scots as English governments. A particular indignity was the decision to implement the unpopular poll tax in 1989, a year before doing so in England and Wales. According to opinion polls in 1991, one-third of Scots favoured outright independence, 40 per cent devolution, and 20 per cent the *status quo*. The pro-devolution parties, mainly Labour and Liberal Democrats, formed a Scottish Grand Convention to campaign for a Scottish Parliament, with taxing and spending powers, and elected by proportional representation (Marr 1992). In the 1992 general election John Major campaigned strongly in favour of the Union, dismissing devolution as the start of a slippery slope to separatism. Compared with the 1991 survey, votes for independence (for the Nationalists) fell to 21.5 per cent, votes for the status quo (Conservative) rose to 25 per cent, and votes for devolution (Labour and Liberal) remained stable at 52 per cent. There remains the problem, however, that the Conservatives rule Scotland but after the 1992 election had only eleven of its seventy-two MPs.

Wales has been governed virtually as part of England since its absorption in 1536. A Welsh Office was established in 1964 and a Secretary of State for Wales sits in Cabinet. The areas of responsibility are broadly similar to those of his Scottish counterpart, although less significant in total. In the twentieth century, the Welsh political differentiation from England has been expressed in strong support for the Liberal and then Labour parties, rather than the Unionist Conservatives. Support for political nationalism is weak on the evidence of the pro-devolution vote in 1979 and the small vote for Plaid Cymru, the Welsh nationalist party. The appeal of Plaid Cymru is primarily cultural and linguistic, but only a fifth of the population is now Welsh-speaking compared with half in 1901.

The Irish question

Northern Ireland has for long posed a major challenge to many of the assumptions about British politics, not least to that concerning the authority of government. The union of Ireland with the United Kingdom was achieved in 1801, but the present borders date only from 1921. Northern Ireland was created in 1922 under the threat of force. The twenty-six largely Catholic southern counties formed the new Irish state but the six largely Protestant northern counties were allowed to remain part of the United Kingdom. Since the partition, many Irish Catholics and almost all Protestants have continued to protest their exclusive loyalties to the Irish Republic and British Crown respectively, and constitute separate subcultures. In Northern Ireland Protestants outnumber Catholics two to one but in a united Ireland they would be in a clear minority. Many of the Irish felt 'cheated' by the partition and looked to eventual union, while the Protestants in Northern

BOX 2.3. CHRONOLOGY OF KEY EVENTS IN ANGLO–NORTHERN-IRISH RELATIONS

1920	Government of Ireland Act and Anglo-Irish Treaty (1921) partition Ireland
1921	Province of Ulster created with its own Parliament, Stormont
1922	Beginning of Unionist party dominance of Ulster politics, to continue until 1972
1968	The 'troubles' start—Catholics demand equal civil rights. IRA takes up the cause
1969	British government drawn into the conflict; troops deployed as peacekeepers in Belfast and Londonderry
1972	Stormont suspended and direct rule from Westminster begins
1974	Power-sharing Executive established. Brought down by Paisley's DUP and Ulster Loyalist Workers' Council
1985	Anglo-Irish agreement. Recognition of joint role of the Irish Republic and Britain in attempts to find peaceful solutions to Northern Ireland conflict
1993	Downing Street Declaration
1994	Peaceful search for resolution of conflict began

Ireland continued to feel threatened by the claims made by the South.

Northern Ireland's physical separateness from the mainland and its distinctive problems make it seem 'non-British' in many respects. Certainly, many of the statements made about British politics for the last fifty years—on political consensus, the rule of law, tolerance, secularism, class voting—hardly apply to Northern Ireland. In the past two decades, however, the region has mattered increasingly and irritatingly to British politicians, both because of the violence, which has received international publicity, and because of the seemingly impossible task of finding 'a solution' which is acceptable to both Protestants and Catholics. At issue are such fundamental questions as the borders of Ireland and the United Kingdom, national identity, and the legitimacy of British rule (Whyte 1990).

Bitterness has been increased because the national and religious rivalries between the two communities are so long-standing. Catholic Ireland had refused to conform to Henry VIII's Protestant Reformation. For the next two centuries Ireland was a security threat to Britain, as a hostile France or Spain could exploit dissatisfaction there. During the nineteenth century, a party of Irish Home Rule (or greater autonomy) developed and by the end of the century the eighty-plus Nationalist MPs threatened to paralyse Parliament unless their demands for self-government for Ireland were granted. The Liberals were prepared to grant it, but the Conservatives and Protestant Northern Ireland were bitterly opposed. A Home Rule Act was passed and was due to come into effect in 1914, despite threats of insurrection in the North.

After the First World War, the Government of Ireland Act (1921) partitioned Ireland and offered Home Rule to both North and South. The Northern Protestants, ironically, accepted Home Rule, but as a guarantee against separation from the United Kingdom, not a step towards it. In the rest of Ireland, however, Home Rule was rejected as insufficient, and a guerrilla war launched against British rule.

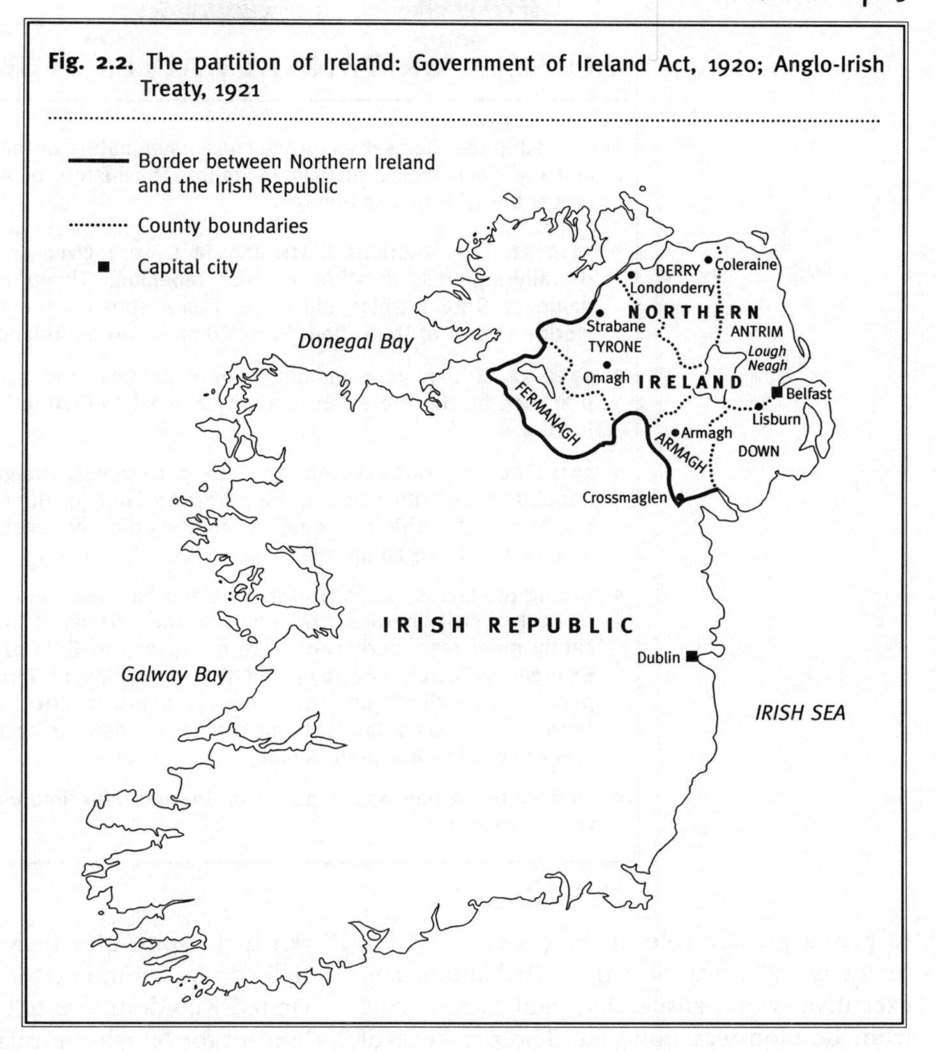

Fig. 2.2. The partition of Ireland: Government of Ireland Act, 1920; Anglo-Irish Treaty, 1921

Eventually a treaty was signed in 1921 by which the Irish Free State was created as a dominion within the Commonwealth, in which it remained until 1949. Gradually, however, Ireland moved towards a republic and left the Commonwealth in 1949. Northern Ireland was given a separate Parliament and control of virtually all administration that did not affect other parts of the UK. The province's public services were to be maintained at the same level as those in the rest of the UK.

Devolution to the Unionist-dominated Stormont Parliament gradually became discredited as the Unionists systematically discriminated against Catholics in housing and public employment. Many Unionists are members of the anti-Catholic Orange Order. Successive British governments were prepared to let the Stormont government go its own way until 1969. In that year the British government intervened to restore law and order, which had collapsed in the face of violent protests by many Catholics at civil and social discrimination and equally violent Protestant reactions. Various reforms were introduced, and the British government gradually came

BOX 2.4. 'BRITANNIA WAIVES THE RULES'

The British pride themselves on the consensual nature of their political culture and a squeaky clean record on human rights. The history of Anglo-Irish relations is incompatible with this self-image.

- **1919–21.** The notorious Black and Tans were given an almost free hand to brutally suppress the IRA and Irish rebellion. 'Things are being done in the name of Britain which make her name stink in the nostrils of the whole world.' Report of the Labour Party Commission on Ireland, 1921.
- **1922–72.** British governments ignored political and economic discrimination practised by the Protestant majority against the Catholic minority in Northern Ireland.
- **1971.** The army was accused of torturing detainees. The government of the Irish Republic took Britain before the European Court of Human Rights. The finding (1978) was that the interrogation methods did not constitute torture but were inhuman and degrading treatment.
- During the last 25 years Northern Ireland has been an exception to the rule in respect of civil liberties. Civil rights of the citizens of the province are significantly more restricted than those of citizens of Britain. The Northern Ireland Emergency Powers Act 1973 and the Prevention of Terrorism Act since 1974 permit trial without jury and allow the police to arrest and detain suspected terrorists for seven days without charge. In 1988 the accused's right to silence was removed in Northern Ireland.
- **1988–1994.** A ban was imposed on broadcasting interviews with terrorists or their spokesmen.

to play a greater role in the province. Eventually, in 1972, the Stormont Parliament and executive were suspended and direct rule from London was imposed. The presence of British troops did not halt the escalation of violence by the sectarian forces of both sides nor prevent the spread of terrorism to the mainland. Since 1969 over 3,000 people have died as a result of violence.

The legacy of history and the conflicts of religion and national identity have made politics in Northern Ireland distinctive in the United Kingdom, and it is an exception to all generalizations about the British political culture. In Britain religion is of little political significance, but it dominates political life in Northern Ireland. Until early 1995 riots, assassinations, and terrorism were everyday occurrences in spite of a heavy military presence. Indeed one commentator has claimed that the United Kingdom was only 'almost a state', or 'Except for Northern Ireland the United Kingdom is a state' (Rose 1982*a*). The claim was reasonable enough, given that the British government lacked support from a substantial minority, its rules were openly defied, it did not have a monopoly of force, and the frontier was not secure. The history of the Irish issue before 1922 and the challenge to British rule in Ulster in the past twenty-five years show that the British have so far been no more successful than other states in dealing with such sectarian conflicts.

Northern Ireland has been a great strain on British governments, making significant demands on the military and exchequer and

BOX. 2.5. GLOSSARY OF IRISH POLITICS

Eire	Gaelic for Ireland, often used to refer to Southern Ireland
Irish Free State	Southern Ireland, 1922–49
Irish Republic	Southern Ireland since 1949
Ulster	Ancient province of Ireland — 9 counties of the north west often used to refer to Northern Ireland
Northern Ireland	6 counties of Ulster politically part of the UK
Dail	Southern Irish Parliament
Stormont	Northern Ireland's Parliament, 1922–72
IRA	Irish Republican Army, a paramilitary terrorist organization formed in 1919. Evolved out of the Fenians
Sinn Fein	(Ourselves alone) — political wing of IRA
Provos	Provisional wing of IRA and Sinn Fein active in Northern Ireland
Unionists	Supporters of union with Britain — usually Protestants living in Northern Ireland referred to as Loyalists, i.e. loyal to the English Crown
Nationalists	Supporters of united Ireland; today the term is used usually in association with Catholics in Northern Ireland
UDF	Ulster Defence Force — Protestant unionist paramilitaries
UVF	Ulster Volunteer Force — as above, but more extreme and violent
OUP	Official Ulster Unionist Party — moderate unionists
DUP	Democratic Unionist Party — populist, extremist
SDLP	Social Democratic Labour Party — moderate Catholic party in Northern Ireland
Alliance	Non-sectarian party of Northern Ireland with very limited support — unsuccessful in breaking the sectarian mould of Northern Ireland politics

BOX 2.6. THE ORANGE ORDER

The Orange Order was founded in response to the Catholic uprising in 1798. Its activities declined during the nineteenth century until revived by the Home Rule crises. Officially the Order describes itself as a religious organization. In practice its objectives blend religion with politics, defending the Protestant religion and supporting union with Britain. For many years the Order has had close links with the Unionist Party and it plays a significant part in uniting unionists. Between 1922 and 1972 some two-thirds of adult male Protestants in Northern Ireland belonged to the Order. It played a key role in maintaining the public displays and ceremonies associated with Protestant Unionism and served as a link between different sections of Protestant society. Since direct rule from Westminster (1972) its influence on government in Northern Ireland has declined.

BOX 2.7. POSSIBLE SOLUTIONS TO THE NORTHERN IRELAND QUESTION

SUGGESTED SOLUTIONS	FAVOURED BY
United Ireland	Catholic Nationalists; Southern Ireland less enthusiastic
Independent Ulster	Some Protestants who suspect Britain will renege on border/nationality pledges
Total integration of Northern Ireland within Britain	Mainstream Unionists (little support in Britain)
Re-partition	Few supporters since recognized as only a partial solution—residual Catholic population
Devolution	Moderates on both sides who are willing to participate in power-sharing
Direct rule	Least worse option; everyone, except Britain, has someone to blame

damaging Britain's reputation as an upholder of democratic standards and the rule of law. Partition was a failure and direct rule was always seen as transitional (although by 1995 it has operated for twenty-five years). But transitional to what? The idea of *an independent Northern Ireland* has so far attracted little support; if violence broke out between Protestant and IRA paramilitary groups, it would probably spill over to Ireland and the rest of Great Britain. *Repartition*, or a redrawing of the boundary to create more homogeneous Catholic and Protestant communities, would require substantial and probably unacceptable shifts of population. The Protestant demand for a restoration of *majority rule* has been rejected by Britain until there are satisfactory safeguards for the minority; suggested safeguards have invariably been rejected by Unionist politicians.

The Westminster system of government has not been regarded by British leaders as suitable for a society divided into two mutually antagonistic religious communities. Majority rule, a basic feature of British politics, is rejected for Northern Ireland because it means permanent Protestant rule. Successive British governments have therefore experimented with forms of *power-sharing*, which effectively allow the Catholic minority a veto. In many West European states (Switzerland and the Netherlands, for example) bitter social divisions have encouraged leaders to adopt power-sharing arrangements (for example, minority rights to veto, legislation, and coalition government). British governments have already recognized Northern Ireland's 'difference' from the mainland by introducing proportional systems for local and Euro-elections.

In November 1985 the British and Irish governments drew up the *Anglo-Irish Agreement*. This pledged both governments to work towards a framework of rule which would recognize the different identities of the two communities. It also accepted that any scheme for a united Ireland would depend on the support of a majority of Northern Irish people. But to the fury of Protestants it also set up an Intergovernmental Conference to allow the two governments to confer on political, security, economic, and legal matters affecting Northern Ireland. Significantly, this was the first formal recognition by the British government of the legitimate interest of Dublin in the Northern Ireland question—hitherto an internal matter exclusive to London. The Dublin and London governments effectively decided that they should do something because the two communities were too entrenched to come up with a solution.

In December 1993, in the so-called *Downing Street Declaration*, the British government offered to hold talks about Northern Ireland's constitution with groups which renounced violence—a clear signal to the paramilitary groups. In 1995 peace broke out and the Dublin and London governments moved closer. British ministers opened negotiations with representatives of Sinn Fein and accepted the right of the people in Northern Ireland to choose to join Ireland. Government by consent—as in the rest of Britain—might require, in the case of Northern Ireland, a redrawing of the boundaries of the United Kingdom. The stance was a contrast to its defence of the union with Scotland, and was an acknowledgement by ministers that Northern Ireland was different (i.e. was less 'British' than Scotland). In early 1996 the IRA resumed bombing in London and the problem of finding a solution appeared as intractable as ever.

Community and nationality

If long-held assumptions about agreement on the boundaries of the United Kingdom have been undermined by events in recent years, the same may be said of the sense of national community and integration in Britain. In the early part of this century the secession of the predominantly Catholic twenty-six counties to form an independent Irish state in 1922 and the failure of Scotland and Wales to develop strong nationalist parties had two important effects. First, it eased potential problems of alternative national loyalties and simplified British politics along lines of social class. It is easy to forget that between 1880 and 1921 territorial and nationalist issues (particularly in Ireland) were at the forefront of British politics—indeed, some historians refer to Britain's 'territorial crisis'. Secondly, issues of religion and nationality, so important in nineteenth-century British politics, waned and social class became more significant (Butler and Stokes 1969). However, since the 1970s questions of national identity have returned as nationalists have demanded either separation or distinctive political institutions for their countries.

Britain is not alone in undergoing such change. The break-up of the Soviet Union has unleashed traditional national and ethnic rivalries, civil war has erupted in the former state of Yugoslavia, and nationalist parties in a number of member states of the European Union (EU) are resisting pressures to merge their countries' sovereignty and identity in a more federal Europe. Paradoxically, the willingnesss of EU policy-makers to negotiate with local and regional governments encourages groups in Scotland, Wales, and Northern Ireland to organize and lobby the Commission in Brussels.

The United Kingdom is a multinational state covering one and one-fifth islands. We

Table 2.1. National and regional identity in England, Scotland, and Wales

Question: Which of these best answers how you regard yourself?			
Base size	England (1,526) %	Scotland (375) %	Wales (240) %
English/Scottish/Welsh, not British	13	30	20
More English/Scottish/Welsh than British	12	34	21
Equally English/Scottish/Welsh and British	45	24	35
More British than English/ Scottish/Welsh	11	3	10
British, not English/ Scottish/Welsh	13	5	12
More English/Scottish/Welsh	25	64	41
More British	24	8	22

Source: MORI 1995.

may talk of a government, but hardly of a society or nationality of the United Kingdom. The four populations are distinctive in terms of their histories, religious affiliations, national identities, and, especially in Northern Ireland, their party systems. Most English people belong to the Church of England, most Scots to the Church of Scotland, and in Ulster most belong to either of the above or to other Protestant religions or to the Roman Catholic Church. Indeed, there is hardly any awareness of being 'British' in the sense of people consciously possessing such an identity (Rose 1982*b*, ch. 1). Most English, Scots, and Welsh people identify with their own respective nationalities, and the Northern Irish population divides its ties between British (Protestants) and Irish (Catholics). These differences are not new, but between the 1920s and the 1970s they lacked political significance because (with the exception of Northern Ireland) they were not expressed in distinctively nationalist political parties and because some five-sixths of the population resided in England (see Table 2.1).

Nationalism

In the 1970s, however, there was an upsurge in support for *nationalist parties*. In 1966 nationalist parties in Scotland and Wales collected only 6 per cent of the two countries' vote and failed to return a single MP. By October 1974 figures had risen to 11 per cent of the Welsh vote and three seats and 30 per cent of the vote and eleven seats in Scotland (the highest share of the vote secured by an ethnic party in Western Europe since the war). The advance was not sustained and in 1992 the figures had fallen back to 21 per cent of the vote in Scotland and 8.8 per cent in Wales, and a total of seven seats. The party systems in the non-English nations differ from that in England. In Scotland and Wales the nationalist parties have attempted to challenge the hold of the traditional (Conservative, Labour, and Liberal) parties. In Northern Ireland the mainland parties have no following at all (although the Conservative party in 1992 began contesting elections in Northern Ireland, and the Liberal Democrats work with the Alliance Party of Northern Ireland). The nationalist parties also vary in their ability to mobilize their core group support. The Scottish National Party gain around a fifth of the votes of the Scottish identifiers, the Plaid Cymru a sixth of the Welsh speakers, and the Social Democratic and Labour Party (SDLP) two-thirds of Catholics in Northern Ireland. Yet in each nation nationalist support at the ballot box has run far behind that of the other parties.

To date, the dominant political response within Scotland and Wales has not been to seek separation but a stronger national voice in Westminster (historically, both Scotland and Wales have had more parliamentary seats than their populations strictly warrant), or to change the British government in the hope of producing a more sympathetic policy. The main British parties and major producer interest groups have not emphasized the territorial aspects of politics: Labour and the trade unions have sought to represent social class and their members across Britain, and the Conservatives have been a party of the Union. Because territorial issues, from Irish Home Rule down to devolution, have been so divisive for the parties and imposed great strains on the political system, most leading British politicians have tried to stifle the expression of the territorial dimension in politics.

The British, or perhaps more accurately the English, have been noted for their conservatism on constitutional matters. Whether or not the reputation is deserved, there has been a flurry of attempted and actual constitutional innovations in Northern Ireland and a failed attempt (1979) to introduce a devolved assembly in Scotland; a future Labour government has promised to promote regional government. But for English politicians, and most of the English electorate, the subject has been, to quote the Labour Prime Minister Harold Wilson, a 'bore'.

The non-English parts of Britain have been less prosperous economically, reflecting their dependence on such traditional and declining industries as coal, steel, and shipbuilding. They have done well economically out of the Union in so far as they are the greatest net beneficiaries from central government expenditure. A division of the United Kingdom into a prosperous South and economically depressed North would include Scotland, Wales, and Northern Ireland with the latter. Labour's strength in the North and weakness in the South adds a political colour to the economic divisions between regions. This geographical division of party political strength is a reflection of the socio-economic composition of the population. It is also part of a long-term trend, as 'The peripheral areas of Britain, with their higher unemployment and the decline in the inner part of conurbations, have become steadily more Labour; while the expanding more prosperous areas have become more Conservative' (Curtice and Steed 1980: 402).

Formal political power in Britain is highly centralized as long as formal sovereignty, or law-making, is concentrated in Parliament. But there is a broader social centralization as well. England has always dominated the United Kingdom, although its position has not been that of a colonial power over subject nations. The dominance of England may be supported by a good democratic argument: it has 80 per cent of the population. London, with a population of over seven million, in turn, dominates England. Only Birmingham among other British cities has a population in excess of one million. London is the centre for so many activities (political, administrative, media, business, commercial, and entertainment) and so many organizations have their national headquarters in London that it is the capital city of the country in a way in which Washington DC, Rome, or Bonn are not. This heavy concentration of decision-makers within the London area makes English society, if not exactly centralized, then compact. A train journey from Liverpool, Manchester, or Birmingham can bring the traveller to London in less than two and a quarter hours. J. Sharpe has commented on the centrality of London as 'a clearly defined status system where the upper levels of all the various occupational and social hierarchies converge and intermingle in one geographical location. Every road leads to the London penumbra' (1982: 136).

BOX 2.8. UK IMMIGRATION LEGISLATION SINCE 1905

1905 Aliens Act	The first modern immigration control, allowing foreigners ('aliens') to be barred as they tried to enter Britain. It mainly affected poor Jews fleeing from repression in Eastern Europe.
1914 Aliens Restriction Act	Aimed at German spies in World War I, it extended powers to refuse entry and order deportations. It came to apply to all aliens.
1948 British Nationality Act	Confirmed status of citizens of Empire as British subjects with right of entry to Britain.
1962 Commonwealth Immigration Act	Ended the 'open door' policy for former British colonial subjects. In future most Commonwealth immigrants would need a work permit (employment voucher) to come to the UK.
1968 Commonwealth Immigration Act	Under this law East African Asians holding British passports lost their automatic right to stay in the UK. The European Commission on Human Rights later ruled that the Act amounted to racial discrimination.
1971 Immigration Act	Introduced the concept of the 'patrial', allowing entry only to those born in the UK or whose parents or grandparents were of British origin. All non-patrials needed permission to enter the UK.
1981 British Nationality Act	Abolished the automatic right to UK citizenship for immigrants' children born in the UK and introduced three new classes of citizenship based on immigration status.

Source: *Education Guardian*, Nov. 1991.

Race

The arrival of a substantial number of *coloured immigrants* between 1950 and 1970 introduced a group of citizens who were visibly different from the rest of the British community. Most were from the new Commonwealth states of India, Pakistan, and the West Indies. In 1992 this group amounted to just under three million or nearly 5 per cent of the total British population. While Irish, French, Huguenot, and Jewish immigrants had been assimilated in earlier waves, the skin colour and traditional religious beliefs of many of the new arrivals made this a more difficult process. Enoch Powell, a prominent opponent of coloured immigration, attracted much support from voters of all parties by articulating

BOX 2.9. LEGISLATING FOR RACIAL HARMONY

1965 Race Relations Act	Illegal to discriminate on grounds of race, ethnic, or national origin, in restaurants, cafes, hotels, cinemas, dance halls, buses. Anyone who felt discriminated against could appeal to the **Race Relations Board,** which could only act on such appeals. It had to seek reconciliation between the parties; only if this failed, would it take the offender to court. The Act also made it an offence to distribute literature or use language likely to stir up racial hatred.
1968 Race Relations Act	Increased the powers of the Race Relations Board. It could initiate investigations without receiving formal complaints, and the scope was extended to include housing and employment. Created a **Community Relations Commission** to encourage harmonious relations.
1976 Race Relations Act (replaced acts of 1965 and 1968)	Intended to encourage racial harmony amongst all groups living in the UK. Established **Commission for Racial Equality** (replacing old Race Relations Board). The main functions are to seek to promote equality of opportunity free from racial discrimination, and to conduct investigations into cases of apparent discrimination.
Points to note	1. The bi-partisan nature of immigration and race relations: neither party wants to grasp this political hot potato. Labour usually condemn Tory immigration policies, but rhetoric does not translate into reality when in office. It was Labour who passed the 1968 Immigration Act to stem the inflow of Kenyan Asians into Britain. 2. Labour has, however, introduced all the race relations legislation, but it has been of limited effect.

Fig. 2.3. Unemployment rates by ethnic group, spring 1994 (%)

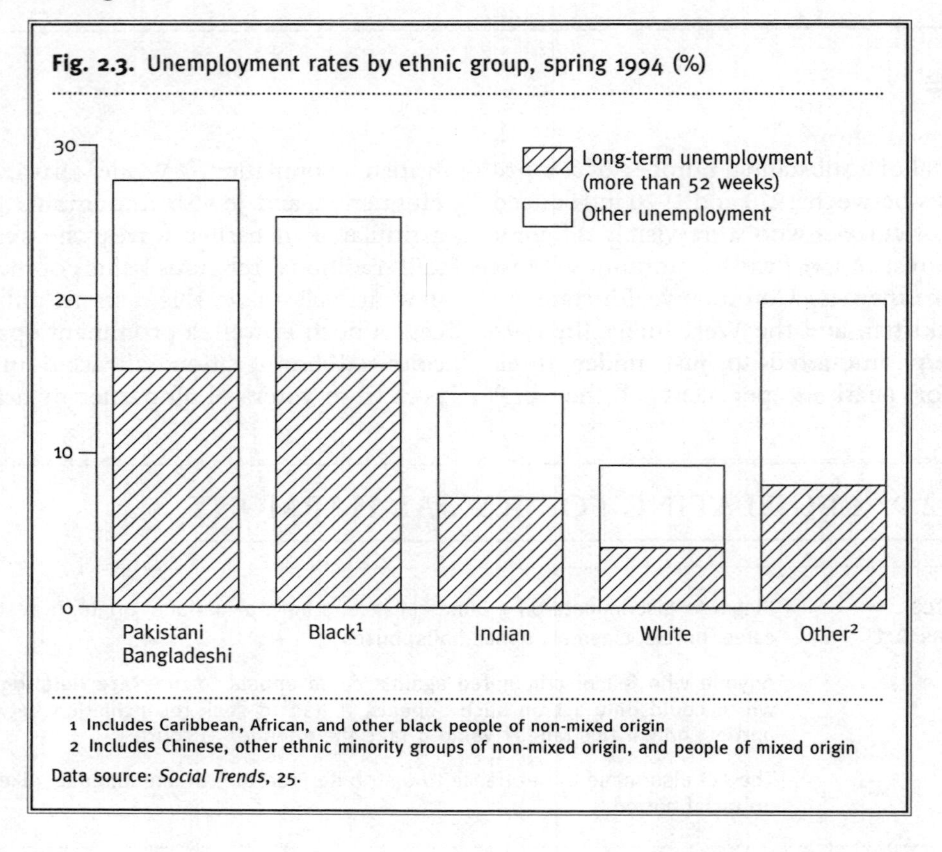

1 Includes Caribbean, African, and other black people of non-mixed origin
2 Includes Chinese, other ethnic minority groups of non-mixed origin, and people of mixed origin
Data source: *Social Trends*, 25.

fears that the immigrants, their British citizenship notwithstanding, were 'alien' in culture and a threat to the 'British way of life'. His outspoken comments led Edward Heath to dismiss him from the Conservative Shadow Cabinet in 1968.

Under the 1948 British Nationality Act a British subject could also be a resident of any of the Commonwealth countries, and citizens of Commonwealth states were free to enter Britain and settle and work in Britain. It was only in 1962, with the passage of the Commonwealth Immigration Act, that the entry rights of overseas British subjects were restricted. Subsequent legislation in 1971 and 1981 has made it more difficult for non-white immigrants to enter. The 1971 Immigration Act allowed free entry to 'patrials', that is, persons who had at least one British grandparent, or who had been naturalized, or who had lived in Britain for five years. The 1981 Nationality Act restricts British citizenship, with rights of residence, to those who are already legally settled in Britain, or who have one British parent and have been registered abroad at birth. Since 1979 immigration from new Commonwealth countries has been reduced to less than 30,000 a year. A source of confusion is that Irish citizens also enjoy full rights of British citizenship, including right of free entry and the rights to vote and stand for Parliament, when resident in the UK. Citizens of European Community states may also enter freely and, under the Maastricht Treaty, vote in local elections in the country in which they are resident.

In contrast to the nationalist parties in Scotland and Wales, whose assertions of identity have a territorial basis, non-whites in Britain have been more concerned to enjoy 'fair' treatment in applications for jobs and housing and respect for their religion and culture. In 1976, the Commission for Racial Equality was set up to receive and investigate complaints of unlawful racial discrimination and to promote harmonious race relations. British governments are committed to policies of equality of opportunity regardless of race. But in spite of legislation, supported by all parties, to outlaw racial discrimination (1965) in housing, jobs, and the possession of goods and services (1968 and 1976), numerous reports document the existence of discrimination and deprivation among non-whites and that the heaviest unemployment is among Pakistani, Bangladeshi, and black youths (see Fig. 2.3). The depressed inner-city areas with heavy concentrations of blacks—London's Brixton, Liverpool's Toxteth, and Manchester's Moss Side—were flashpoints of rioting in 1981.

The number of new immigrants in recent years has been so small that immigration is probably no longer a serious political issue and the majority of black Britons have been born here. In dealing with problems of race relations and racial discrimination, governments may choose to confuse two issues, relying on Acts to limit immigration (1962, 1968, 1971) or Acts to improve race relations (1965, 1968, 1976). Uniform race policies may not be appropriate, for some groups may not want *integration* into 'British culture' but seek to preserve a separate culture (Messina 1989, Layton-Henry and Rich 1986). This was evident in the reaction of many Muslims who were outraged in 1989 over the contents of Salman Rushdie's *Satanic Verses* and supported the death sentence imposed on the author by Ayatollah Khomeini. What is so far striking is that whereas immigration and high levels of unemployment have fostered the growth of far-right and anti-immigrant parties in such states as France, Austria, and Germany, there has been little echo so far in Britain. In the 1992 election, the twenty-seven candidates who stood for the National Front and British National Party gained a total of only 12,000 votes. The explanation may lie in a combination of factors—the absence of proportional representation, strict immigration controls, and anti-discrimination laws.

Class

Social mobility, or the movement of a person to an occupational status different from that of his or her father, has increased in the past generation. The net increase in the numbers who are upwardly mobile reflects changes in the occupational structure, particularly the expansion of middle-class jobs and decline in the number of manual jobs. According to a study of social mobility, nearly half of the professional and managerial respondents had fathers who came from a lower occupational class (Goldthorpe 1987). Though the absolute level of social mobility has grown, there has been no change in the relative mobility rates between the middle and working classes. Education is important in providing the qualifications for entry to high-status occupations since many middle-class professional jobs require higher educational qualifications. As a consequence, the social class of most people has largely been established by the time they start work; there is little social mobility in adulthood.

Social mobility and *changes in occupational structure* have affected the balance between the social classes. Britain was the first state to achieve extensive industrialization and may

Fig. 2.4. Occupational class in Britain, 1951–1981 (%)

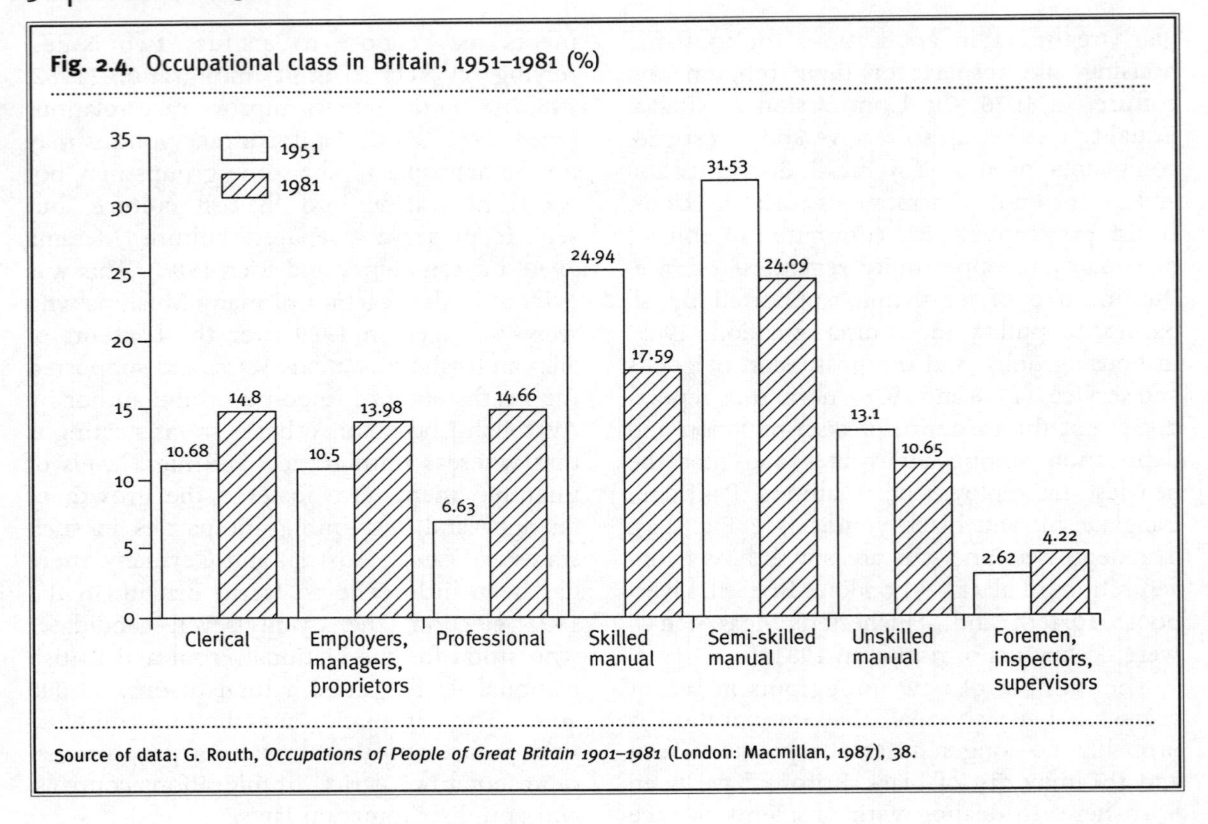

Source of data: G. Routh, *Occupations of People of Great Britain 1901–1981* (London: Macmillan, 1987), 38.

claim to have the oldest working class in the world. The working class, although shrinking, has become more self-recruiting, with most of its members coming from working-class parents. As the middle class has expanded so has it become permeated by 'outsiders', people who are first-generation middle class and come from working-class parents and still retain ties with the working class. About a fifth of working-class boys have moved into the middle class and about a half of the middle class includes men who came from working-class parents. As a result, John Goldthorpe concludes, the middle class has become more heterogeneous and the working class more homogeneous in terms of origins.

The main growth of white-collar jobs in the 1960s and 1970s was in central and local government, particularly the 'service' sector, such as education, health, and public administration. Workers in these sectors are particularly dependent on public expenditure for the provision of the services, their jobs, and their salaries. Other middle-class jobs are in the private sector (dependent on commercial considerations) or in the self-employed category, which has grown since 1979. In the 1980s, with the rundown of manufacturing, the number of working-class jobs fell sharply. Most people are in what might be termed 'mixed' classes, having both middle and working-class attributes. This is particularly so for the 'intermediate' groups of lower middle class and skilled workers who constitute more than half of voters. Over 50 per cent of manual workers now own cars, and 40 per cent of skilled workers are buying their own homes, the same proportion as rent their homes from local councils.

Political significance of social class

There are two important points to consider about the political significance of social class in Britain.

First, *social and economic divisions are likely, other things being equal* (and contrary to Marx), *to place less strain on a political system than divisions involving race, language, religion, or national identity.* The latter are absolute differences and usually prove less amenable to bargaining and compromise. The major challenge to political authority in Britain, after all, has concerned Northern Ireland, an amalgam of national and religious conflicts.

The second feature *concerns the approximate symmetry (until recently) between the middle/working-class division and the two-party system.* The relative weakness of sources of social differentiation, apart from social class, has been important in discouraging the emergence of other significant political parties, based on other social cleavages. Yet the class basis of the party system is subject to important qualifications. Since 1918, when many working-class men and women were effectively enfranchised, the Conservatives have usually been able to attract the support of about a third of that class. The party has managed to come to terms with a largely working-class electorate, and so far in the twentieth century the Conservatives have been in office, alone or in coalition, for sixty-six of the years. With pure class voting, Britain would have had a long spell of Labour rule. Since 1970, that alignment has weakened further as Labour has lost support among the working class and the Conservatives have shed support in the middle class. The loosening relation between social class and party choice is discussed further in Chapter 6.

Since 1970, the fragmentation of the party system, weaker party loyalties, and the growing complexity of social class have made it less

BOX 2.10. BRITISH POLITICS TODAY: HOW LONG CAN THE KINGDOM REMAIN UNITED?

- **Devolution** is back on the agenda: the Labour party are promising devolution to Scotland and Wales. Many Conservatives regard devolution as the slippery slope to separatism. EU membership makes this possibility more viable.
- **Northern Ireland**. The inclusion of the 'Irish dimension' since the 1985 Anglo-Irish Agreement, and the desire of Britain to disengage from the province, seem likely some time in the future to lead to a change in the constitutional status of Northern Ireland.
- **Social cohesion** is threatened by a growing underclass of state dependants living on marginal incomes, long-term unemployment, higher incidence of unemployment among black and Asian youths, and rising crime figures.
- **European Union**. Further integration and monetary union seem likely to erode national identity and increasingly limit the sovereignty of Parliament.

useful to think of a class-based two-party system in Britain. In all general elections since 1974 less than half of voters have supported the party of their class (i.e. Conservatives—middle class, Labour—working class).

The rise of the Labour party after 1918 introduced political issues and rhetoric which were more clearly related to social class differences than applied in the politics of pre-1914. Labour, while condemning class inequalities and working for their elimination, has regarded the accomplishments of this objective as compatible with established parliamentary institutions. Apart from a short spell in the 1930s and in the early 1980s, the Marxist view that social divisions are irreconcilable in a capitalist society has found little support in the party. Until its recent demise, the British Communist party, founded in 1920 and committed to class warfare, had the least effective electoral record of any such party in Western Europe. The economic 'interest' of members of the middle class is divided in terms of public policy between those who work in the public sector and depend on government expenditure, and private-sector workers. The 'interest' of the working class is also divided between the skilled and unskilled workers, the affluent and the poor, those in strong and those in weak trade unions, and those in the public and private sectors.

But class matters in other ways. It is possible to show that class divisions correlate with various social and economic inequalities or life-chances. More middle-class than working-class children, for example, stay at school beyond the school-leaving age, and proceed to higher education. Differentials in income between the classes have narrowed since the war, but there has been little change in wealth, and the richest 1 per cent of the population own over a fifth of the country's wealth. Redistribution of income has largely occurred only within the top 20 per cent, and the top 1 per cent of income earners receive, after tax, the same as the poorest 20 per cent altogether. Sons of middle-class parents have four times the opportunities of remaining in that class as sons of working-class parents have of entering it. Many of the political élite, the higher Civil Service, Cabinet Ministers, MPs, even Labour MPs, are drawn from the professional and business middle class.

Conclusion

The social foundations of modern (say, post-1945) British politics have been changing. The rise of nationalism outside England, shifts in patterns of work and with this the social class structure, the rise of ethnic groups, and developments in Northern Ireland have impacted on the political system, particularly the constitution (see Ch. 3).

Summary

- The UK is a four-nation state in which England, as the centre of the unitary government and having 80 per cent of the population, is the dominant nation.
- Scotland has distinctive political, legal, economic, religious, and cultural institutions. Parliament has granted Edinburgh some power over a limited range of activites, but no power to tax or to legislate. The strength of nationalist feeling in the country fluctuates as do the fortunes of the SNP. Whilst there appears to be considerable support for greater devolution, only about 30 per cent favour total independence.
- Wales has a Welsh Office and cabinet-rank Secretary of State, but nationalist sentiment has focused more on cultural recognition than on independence or even devolution.
- Northern Ireland, though part of the UK, is distinctive. This distinction is borne out in its political system, the divisions over national identity, statehood, and religion that exist in the province, and the fact that throughout its history, but especially in the last 25 years, violence in the form of terrorism and state coercion have been features of everyday life.
- Violence has ensured that Northern Ireland has remained high on the political agenda in Britain in the last 25 years. The search for solutions has been continuous and so far fruitless. The ideological nature of the conflict, coupled with its deep historical roots, makes it less amenable to the British liberal tradition of resolving conflicts by a process of bargaining, negotiation, and compromise. The 1994 Downing Street Declaration has led to a temporary cessation of violence and some hope for the future.
- In Chapter 1 it was noted that persistent economic policy failure places strains on legitimacy. This has also been reflected in a more vociferous pressure for national recognition and devolution within the peripheral nations since the 1970s, pressure that to some extent has been increased by EU membership.
- Throughout its history, Britain has assimilated frequent waves of immigration. Since the 1950s substantial coloured immigration from the 'new' Commonwealth has created a group of citizens visibly different from the indigenous population. Relative economic decline has been reflected in social tension and disharmony. To some extent coloured immigrants and their families have been the greatest sufferers from the economic downturn; at the same time they are blamed for economic problems such as unemployment.
- Race relations and immigration have tended to be bi-partisan issues. Governments have attempted to legislate for racial harmony. Success has been limited: racial discrimination is widespread, if covert. Attempts at integration are not universally welcomed by ethnic minority groups.

- Class has traditionally been the most significant social cleavage in Britain. The relative sizes of the classes are changing. This is forcing a reappraisal of the traditional two-party representation as broadly reflecting class divisions.
- Changes in the class structure and the increasing significance of race and nationality are forces for both integration and fragmentation in Britain. A question for the future is which will be the stronger.

CHRONOLOGY

1536	Wales absorbed into England
1707	Act of Union between Britain and Scotland
1801	Act of Union between Great Britain and Ireland to form the United Kingdom
1905	Aliens Act: the first modern immigration control
1920	Government of Ireland Act: partition, Northern Ireland remaining under British rule
1948	Commonwealth Nationality Act, encouraging immigration from the West Indies
1949	Southern Ireland becomes a republic, but Irish citizens continue to enjoy rights as British citizens
1962	Commonwealth Immigration Act: end of 'open-door' policy
1964	Robbins Report: expansion of higher education creates greater opportunities for children of working-class families
1965	Race Relations Act outlaws discrimination in public places
1968	Race Relations Act establishes Race Relations Board Commonwealth Immigration Act, to reduce flow of Kenyan Asians 'Troubles' begin in Northern Ireland Enoch Powell makes 'rivers of blood' speech in Birmingham
1969	British troops sent to Northern Ireland as peacekeepers
1971	Immigration Act introduces 'patrials' rule
1972	Direct rule in Northern Ireland; Stormont suspended
1973	Referendum in Northern Ireland on border question
1976	Race Relations Act establishes Commission for Racial Equality
1978	Devolution Bills in Scotland and Wales
1979	Referendums in Scotland and Wales on devolution
1981	British Nationality Act tightens immigration controls
1985	Anglo-Irish Agreement recognizes role for Republic of Ireland in Northern Irish question
1993	Downing Street Declaration
1994	Announcement of IRA and Protestant paramilitary ceasefire
1996	Breakdown of ceasefire

ESSAY/DISCUSSION TOPICS

1. Why has the Northern Ireland question proved to be so intractable?
2. 'A disuniting Kingdom'. Is this a fair description of the United Kingdom in the 1990s?
3. Why has devolution to Scotland and Wales returned to the political agenda?
4. Are the changes which are occurring within the class structure in Britain likely to be a force for greater political stability?
5. Are Race Relations Acts and statutory positive discrimination in favour of ethnic minorities an effective way to achieve racial harmony?

RESEARCH EXERCISES

1. Examine the arguments for and against the proposal for the state provision of Muslim schools.
2. Referring to election data since the 1970s, chart and account for the fluctuating fortunes of either the SNP of Plaid Cymru.

FURTHER READING

On different aspects see Madgewick and Rose (1982) and J. McGarry and B. O'Leary, *Politics of Antagonism: Understanding Northern Ireland* (London: Athlone Press, 1993).

The following may also be consulted:

Abercrombie, N., and Wade, A., *Contemporary British Society* (Cambridge: Polity, 1988).

Aughey, Arthur, 'The Downing Street Declaration: A Clarification', *Talking Politics*, 7/1 (1994).

Gamble, A., *Britain in Decline*, 3rd edn. (London: Macmillan, 1990).

McConnell, Allan, 'Scotland after the General Election: The Constitution in Flux', *Talking Politics*, 5/1 (1992).

McCullagh, Michael, and O'Dowd, Liam, 'Northern Ireland: 'The Search for a Solution', *Social Studies Review*, Mar. 1986.

Moran, M., *Politics and Society in Britain* (London: Macmillan, 1989).

Wilding, Paul, 'Poverty and Government in Britain in the 1980s', *Talking Politics*, 5/5 (1993).

3 THE CONSTITUTION

Reader's guide

A constitution is usually assumed to be a document containing a body of laws specifying the powers of, and relationship between, different branches of government and the relationship between the state and its citizens. The British constitution has evolved over a period of a thousand years and is an amalgam of laws and procedures arising from several sources. Many of the most important rules governing the powers of the executive and the relationship between the executive and legislature are governed by conventions — rules of behaviour that are neither written down or enforceable in the courts. The civil liberties of British citizens rest on a tenuous basis of common law. Those parts of the constitution that are written down are mostly in the form of statutes which Parliament can repeal or amend by the same process as passing any less fundamental kind of law. The exact nature of particular rules has frequently been contested, and some dispute that Britain has a constitution at all.

Over the centuries the changing locus of political power has frequently placed costitutional reform on the agenda, and by a gradual process of adjustment a hereditary feudal monarchy has been transformed into a modern, representative liberal democracy. A longish period of one-party dominance, a weakening of the observance of conventions, and the impact of membership of the European Union have, in the last decade, prompted calls for constitutional reform of a more fundamental nature. There has been a growing body of support for those arguing that the time has come for Britain to adopt a written constitution.

This chapter examines the nature and sources of the British constitution, paying particular attention to the key conventions of cabinet government, the role of the monarch, and the fundamental principle of parliamentary sovereignty. It explores the question or whether Britain should have a codified (written) constitution and places contemporary debates about the constitution within their political context.

THERE is a learned debate about whether Britain actually has a constitution. It clearly lacks one in a widely used sense of the term: there is no single written code, or document, which sets out the rules affecting the relations between government institutions and between these institutions and citizens. On the other hand, it does have one in the sense that there are 'laws, customs and conventions which define the composition and powers of organs of the State and regulate the relations of the various state organs to one another and to the private citizen' (Hood Phillips 1978: 6). In other words, there are established procedures affecting the conduct of government and politics, and these are largely adhered to.

This chapter describes how the constitution has evolved, what its main sources are, and the role of the monarchy. It also considers the pressures for reform, including a written constitution, and the obstacles to reform, as well as different interpretations of the constitution.

Britain has for long been widely regarded as a country that illustrates the claim that it is not necessary to have a codified constitution to be a constitutional democracy. Today, however, there are fewer advocates of that position. Some commentators identify the term 'constitutional' with a *system of formal checks and balances on the government* and a separation of powers between government and other bodies. They therefore dispute that Britain is

'The sovereignty of Parliament has increasingly become, in practice, the sovereignty of the Commons, and the sovereignty of the Government which, in addition to its influence in Parliament, controls the party whips, the party machine, and the Civil Service. This means that what has always been an elective dictatorship in theory, but one in which the component parts operated in practice to control one another, has become a machine in which one of these parts has come to exercise a predominant influence over the rest.'

Lord Hailsham, 'Elective Dictatorship', BBC Dimbleby Lecture 1976, cit. *Listener*, 21 Oct., 1976, p. 497

constitutional because of the concentration of formal political power in Parliament, and the principle of that body's absolute and unlimited sovereignty (excluding the European Union, of course).

Growing dissatisfaction with the workings of the political system has, not surprisingly, broadened into a general concern about the

BOX 3.1. CONSTITUTIONS: SOME KEY TERMS AND CONCEPTS

Constitution
A body of fundamental principles and rules according to which a state, or other organization, is governed. A constitution usually specifies the composition and powers of governing institutions, the relationship between them, and the relationship between the state and the citizen.

Flexible Constitutions
No special procedure is required for constitutional change. In Britain there is no difference in the process of passing an act of Parliament requiring that dogs are licensed and one abolishing the House of Lords.

- This can be contrasted with the USA, where constitutional laws are entrenched — a special procedure is required for amending the constitution.

Liberal-Democratic Constitutions
Constitutional arrangements which permit both a high degree of popular participation in politics and the opportunity for institutionalized opposition.

Sovereignty
Concept of the ultimate source of power within the state.

Sovereign
That person or institution which has the final power of decision.

Parliamentary sovereignty

1. Parliament can make or unmake any law.
2. No judicial review is needed to determine the constitutionality of Acts of Parliament.

- Political reality places many constraints on this legal doctrine. Examples: public opinion; pressure groups; EU; NATO; foreign loans.

Unitary state
Subordinate authorities exercise power bestowed by Parliament and Parliament can remove those powers. Example: An Act of Parliament established metropolitan counties in 1972; in 1986 they were abolished by an Act of Parliament.

- This can be contrasted with **federal** states, where powers of subordinate units are set out in a constitution, e.g. Germany.

Civil liberties/ rights
Rights to freedom of speech, freedom of person, etc. In Britain civil liberties are protected neither by a separation of powers nor by a Bill of Rights. Rights in Britain are said to be **residual** — a citizen can do anything not expressly prohibited by law.

- In reality civil liberties are extremely fragile, requiring government willingness to impose restraints.

health of British democracy and fuelled demands for the adoption of a written constitution. In the past, the existence of a competitive two-party system, with its implicit checks and balances, independent groups sharing élite values and traditions, and a broad political consensus may have made constitutional safeguards seem unnecessary. In Britain, party competition, élite culture, and a broad political consensus had provided some insurance against the abuse of power. But critics complain that these safeguards can no longer be taken for granted. In 1976, a Conservative, Lord Hailsham, pointed to the worrying potential for a dictatorial government to emerge on the back of the absolute sovereignty of the Parliament, and more recently a Labour MP has complained that Britain has a 'winner takes all government' (Wright 1994: 26). The party system failed to work in a competitive way in the 1980s. One legacy of the long period of Conservative rule and Mrs Thatcher's radical policy agenda, and such actions as abolishing tiers of local government, imposing broadcast bans, and denying trade union membership to GCHQ workers, has been a quickening of interest in a set of formal checks and balances.

There is a *historical explanation* why the British constitution is not codified in one document. Since the system had evolved over centuries there already existed established ways of conducting politics in Britain before written constitutions became fashionable. It is only in the last two hundred years or so, starting with the United States, that written constitutions have spread. Most constitutions were originally adopted by states when they became independent or suffered a rupture in their evolution through internal collapse or invasion. France has had many different constitutions since 1789 and the present Fifth Republic dates from 1958. The USSR rewrote its constitution a number of times, and on its break-up in 1990 the former member states had to draw up new constitutions. In Britain, neither the system of government nor a formal set or rules has been adopted at one point in time. Instead there is a political system, or set of arrangements, and a style of politics that have evolved over centuries, rather than a constitution.

A liberal democratic constitution

It is a myth that Britain's constitution is the product of a consensus. After all, in the seventeenth century a monarch was executed and another forced to flee, in defence of the idea of Parliamentary against monarchical government. The political nation has often been bitterly divided over such constitutional issues as Catholic emancipation in the 1820s, the 1832 Reform Bill, the powers of the House of Lords, Irish Home Rule, and votes for women. But it is fair to claim that since the flight of James II in 1688, the basic principles of a sovereign Parliament and a limited constitutional monarchy have been securely established. The absence of an upheaval since then and the long period of constitutional stability—always excepting Ireland, of course—provides a sharp contrast to the political evolution of so many other states.

The eighteenth-century British constitution was not democratic; less than 5 per cent of adults had a vote, and ministers were still

In the view of Sir William Blackstone, by the late eighteenth century England had evolved a near-perfect system of mixed government:

'Herein indeed consists the true excellence of the English government, that all the parts of it form a mutual check upon each other. In the legislature, the people are a check upon the nobility and the nobility a check upon the people . . . While the King is a check upon both, which preserves the executive power from encroachment . . . Like three distinct powers in mechanics, they jointly impel the machine of government in a direction different from what either, acting by itself would have done . . . a direction which constitutes the true line of the liberty and happiness of the country.'

Sir William Blackstone, *Commentaries on the Laws of England* (1787)

A close examination of English practice in the eighteenth century, however, reveals considerable divergence from the model extolled by Blackstone.

chosen and dismissed by the monarch. But the executive did not always control Parliament, and neither the monarch nor Parliament could control the courts. Many foreign as well as British observers admired this apparent 'balance of power' between the executive, legislature, and judiciary, and regarded it as the secret of Britain's constitutional government.

The main constitutional and political steps towards *representative democracy* occurred in the nineteenth century. The suffrage was extended and the relative powers of the monarchy, Lords, and Commons were altered in favour of the latter. Successive extensions of the right to vote produced universal adult suffrage by 1928 and made the House of Commons representative of the nation. The principle of Cabinet responsibility to the House of Commons was established by the 1830s. By 1911 the supremacy of the elected Commons over the Lords was formally recognized in the Parliament Act of that year.

Two other reforms—referendums and proportional representation for election to the House of Commons—were often proposed between 1867 and 1918 but not introduced. The first was supported as a means of giving people a say on the issues of the day, the latter as a means of defending minorities against an overbearing party majority in the Commons. The two reforms, as Vernon Bogdanor notes, were envisaged as devices to cope with the rise of organized political parties in the late nineteenth century (1981: 7).

Constitutional change was a live issue before 1914, indeed, until the partition of Ireland. For much of the period after 1918 both Labour and Conservative parties had a vested interest in preserving the winner-take-all electoral system and sovereign Parliament. The Conservatives regarded government as hierarchical and defended its autonomy and authority. Labour also regarded strong government positively, because it was the instrument whereby the working class could overcome the inequalities of a capitalist system (Beer: 1964). The topic, however, has returned to the agenda in the past decade, though largely among critics of the Conservative government and among the politically interested élites.

BOX 3.2. THE BRITISH CONSTITUTION

Characteristics

- uncodified
- flexible
- unitary

Sources

1. **Statutes** Acts of Parliament, which take precedence over other sources

   Examples
   - Parliament Acts 1911, 1949
   - Representation of the Peoples Acts 1932 to 1969
   - European Communities Act 1972

2. **Common law** Rules and customs; judicial decisions; Royal Prerogatives

   Examples
   - Freedom of speech
   - Power to make treaties, declare war, dissolve Parliament

3. **Works of authority** Books and writings which are recognized as sources of guidance on the interpretation of constitutional rules

   Examples
   - A. V. Dicey, *An Introduction to the Study of the Law of the Constitution* (1885)

4. **Conventions** Rules of behaviour considered binding but which lack the force of law

   Examples
   - Monarch assents to bills passed by Parliament
   - Prime Minister is a member of the Commons

5. **European Union law** Since Parliament passed the European Communities Act in 1972 Britain has accepted the superiority of European law. British courts can review statutes in the light of EU legislation

   Examples
   - House of Lords judged the Merchant Shipping Act 1988 to be unlawful because it contravened European Union law

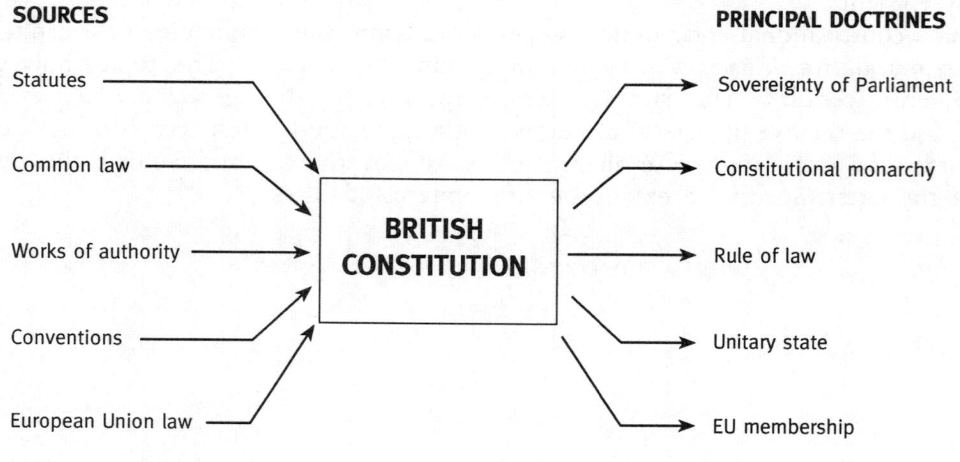

Sources

In a strict sense the British constitution is not unwritten, for large parts are documented. An outstanding feature is that its principles are not codified but dispersed—across statute law, common law, judges' interpretations of these laws, and conventions—though texts and commentaries on the constitution do provide some integration.

There are several sources of the constitution, including:

1. *Statute law*, or law made by Parliament, which overrides common law and provides a substantial part of the constitution. It includes such measures as the Bill of Rights (1689), the Act of Union with Scotland (1707), successive Representation of the People Acts, and the Government of Ireland Act (1920). These laws are made and may be unmade by Act of Parliament, like any other. Even the provision that the House of Commons may not prolong its own life beyond a five-year span without the consent of the House of Lords may be changed by the normal legislative process.

2. *Case law*, or judges' interpretations of statutes. Judges do not rule on the validity of a law, duly passed by Parliament, but they do have the right to decide whether it has been properly applied. By their interpretations the judges have an opportunity to shape the application of the law.

3. *European Community law*, expressed in the European Communities Act to which Britain was a co-signatory in 1972, and subsequently amended by the Single European Act (1987) and the European Communities (Amendment) Act (1993), which gave effect to the Maastricht Treaty; now European Union law. British authorities are required to accept the rules and regulations of the treaties, commitments flowing from them, and future decisions taken by Community institutions. Community laws and regulations are made by the European Commission and the Council of Ministers, and the European Court of Justice declares which laws are self-executing.

4. *Common law*, for example, the traditional rights and liberties of subjects which have been handed down by precedents and upheld by the courts.

5. *Conventions* or rules which, though lacking the force of law, have been adhered to for so long that they are regarded as having a special authority. The conventions differ in their firmness. Firm ones include the expectation that Parliament will be called at least once a year, that the monarch will give her assent to legislation which has duly passed through the appropriate stages in the two Houses of Parliament, and that the Prime Minister and government will resign or dissolve Parliament following defeat on a confidence vote in the Commons. On the other hand, some recent political developments are matters of debate because no conventions have evolved. One can certainly envisage political situations in the future where lack of relevant precedents will create uncertainty about the appropriate course to follow.

Changing conventions

A large part of the constitution is shaped by conventions. Conventions derive from precedent and usage and their force depends on their being observed; continued breaches of a convention weaken its authority. Changes in conventions enable a constitution to adapt and evolve. In the nineteenth century, for example, Prime Ministers sat in either the House of Lords or the House of Commons. Even as late as 1902 the then Prime Minister, Lord Salisbury, sat in the Lords. Today, however, it is accepted that the Prime Minister must be a member of the House of Commons. As the directly elected House, it possesses a democratic legitimacy denied to the Lords, and from the 1920s it was widely understood that the Prime Minister should belong to and be answerable to this body. In 1963, when Lord Home was invited by the Queen to form a government following the retirement of Harold Macmillan, he was required to disclaim his peerage and seek a seat in the Commons on becoming Prime Minister.

At a time of political change or intense political disagreement there may be uncertainty about what is 'conventional'. The Labour Prime Minister of the day suspended the convention of Cabinet collective responsibility in 1975 in favour of a referendum to decide Britain's continued membership of the EC and again in 1977 on the choice of electoral system for direct elections to the European Parliament—in both cases because the party was divided. Mr Callaghan's rather cavalier response to questions about the status of the convention in the latter case was that it still applied 'except in cases I announce that it does not!' (Hansard, 933, 4 C.522, 16 June 1977). His remarks captured the meaning of a convention—its force depends on politicians adhering to it. Taken literally, however, such an approach makes nonsense of the idea of constitutionalism, which entails a set of rules, whether codified or not, independent of and limiting the conduct of government. This is hardly compatible with the idea of the Government deciding for itself what is or is not 'constitutional'. Because a convention has no legal standing, the sanctions against a 'breach' have to be political, such as an adverse vote in Parliament or dismissal of a minister or a government.

Responsible government

The term 'responsible government' has different usages; for example, the Government is responsive to public opinion, is prudent in the exercise of its duties, or is answerable to Parliament (Birch 1964). *This latter is the most important practical constitutional feature.* The idea of responsible government is linked to two important constitutional conventions: (1) *the individual responsibility of ministers* to Parliament for the conduct of their departments, and (2) *the collective responsibility of the Cabinet* for the conduct of policies. If the Cabinet loses the support of the Commons, expressed in a vote on an important measure

BOX 3.3. THEORY AND PRACTICE: RESPONSIBLE GOVERNMENT

MINISTERIAL RESPONSIBILITY

Constitutional theory

Ministers are *answerable* to Parliament for the general conduct of their department and they are *accountable* in the sense that they are expected to resign if serious personal or departmental faults are disclosed.

Examples

- Reginald Maudling, Home Secretary, 1972, regarding his involvement with corrupt architect Poulson
- Lord Carrington, Foreign Secretary, 1982, over Foreign Office misjudgement regarding Argentine intentions in the Falklands

Political practice

In respect of ministerial responsibility for the conduct of departmental civil servants and departmental policy, this convention has weakened in recent years owing to the

- vastly increased workload of departments
- rapid turnover of ministers
- blurred lines of responsibility resulting from the creation of quangos and executive agencies
- closing of ranks of disciplined political parties

In 1996 William Waldegrave and Sir Nicholas Lyell refused to resign, in spite of heavy public and opposition pressure, following the Scott Report's claims that they misled the Commons over the guns for Iraq affair.

Resignation is probably a sign that a minister is unpopular in his party and dispensable to the Prime Minister.

COLLECTIVE RESPONSIBILITY

Constitutional theory

Ministers are required publicly to defend government decisions or resign. Originally collective responsibility was limited to the Cabinet, but now is extended as a tool of party management to include approximately 100 or so members of the governing party.

Example

- Sir Geoffrey Howe, Lord President of the Council and Leader of the House of Commons, resigned in November 1990 because he could not accept the Prime Minister's conduct of policy on Europe.

Political practice

Sometimes Prime Ministers may find it politically expedient to suspend this doctrine. In 1975 Harold Wilson allowed members of his Cabinet publicly to differ during the EC referendum campaign. He justified it as an exception to prove the rule, limited in time and to the specific topic of continued EC membership.

CONFIDENCE VOTE/CENSURE MOTION

Constitutional theory	Collective responsibility implies that the Cabinet resigns if the government is defeated on a major issue in the Commons. *Examples* • MacDonald's government 1924 • Callaghan's government 1979
Political practice	Exceptions: between 1974 and 1979 the Labour government remained in office despite being defeated on a number of votes that would normally have been regarded as confidence issues and entailed resignation. It was, however, finally brought down on a censure motion tabled by the leader of the Opposition, Mrs Thatcher, in May 1979. In November 1994, John Major, with an overall majority of 14, fearing defeat at the hands of Euro-sceptics in his own party on the Commons vote on higher British contributions to the EU budget, turned the issue into a **vote of confidence.** He won the vote by 330 votes to 303. Eight Tory rebels had the whip withdrawn for failing to support the government.

or on a motion of confidence, it is expected either to resign or dissolve the Parliament.

Ministerial responsibility

According to the doctrine of ministerial responsibility each minister is responsible to Parliament for his own personal conduct, the general conduct of his department, and the acts or omissions of his civil servants. The most notable resignation on grounds of policy failure in recent years was that of the Foreign Secretary, Lord Carrington, and two other ministers in his department in the wake of the Argentine invasion of the Falklands in 1982. However, the list of ministers who might have been expected to resign because of policy failures, but did not, is a very long one. It includes:

- James Prior, the Northern Ireland minister, over the breakout of prisoners at the Maze prison in 1986
- Norman Lamont, the Chancellor of the Exchequer, over Britain's exit from the Exchange Rate Mechanism (ERM) in 1992
- The many ministers who were involved in the formulation and implementation of the poll tax 1990–1

The resignation in the 1992 Parliament of ministers like David Mellor, Tim Yeo, Michael Mates, and Neil Hamilton were all over private conduct, as were the earlier resignations from Cabinet of Cecil Parkinson (1983) and Nicholas Ridley (1990).

Of more political significance is the idea that a minister is responsible to the House of Commons for the conduct of departmental policy. At one level, responsibility simply means answering questions on policy and departmental matters from MPs. A stronger interpretation is that Parliament can force the resignation of a minister in a case of demonstrated negligence. Demonstrating negligence, however, is a difficult task. Departments contain so many civil servants, some of whom exercise a necessary degree of initiative and discretion, that it hardly seems practical politics to visit the personal responsibility for their mistakes on the minister. A hundred years ago, a conscientious minister might have been acquainted with most of the work done by his department. Today, however, so much is delegated that it is impossible for the minister to keep abreast of all the actions taken by civil servants. Moreover, the turnover of ministers in office is so rapid (a little over two years on average) that by the time negligence or a mistaken policy has been uncovered, the minister will probably have moved on. Yet the convention remains.

The meaning of the convention has altered as the political climate, particularly following the rise of disciplined political parties, has changed. Only exceptionally does a minister resign or suffer dismissal because of a policy mistake; as S. E. Finer (1956) noted, if a minister goes, it will probably be because he is unpopular and/or because the Prime Minister regards him as dispensable (Marshall 1984). Here we have to take account of two important political barriers to the ability of the Commons to force the dismissal of a minister. One is *party loyalty*, as the majority of party backbenchers rally behind 'their' hard-pressed minister. The other is *collective responsibility*. Ultimately, the Cabinet is also responsible for policy, and a minister's resignation, however personally honourable, or even appropriate, often reflects badly on the government as a whole and is a source of encouragement to the opposition. Other Cabinet ministers will tend to support a minister under attack, though he may be moved or sacked. In 1986 Leon Brittan resigned as Trade and Industry Minister after vehement backbench criticism of his role in the Westland affair (Oliver and Austin 1987). Alternatively, a Prime Minister may reshuffle a minister who has been associated with policies which have failed or which are unpopular. The distinction between a minister's responsibility for policy and a chief executive's responsibility for operation matters is in practice not clear-cut (see below, p. 74).

The doctrine, and the practice, still have some important consequences, however. It gives MPs and the opposition an opportunity to ask questions of ministers. Questions keep the civil servants in the department on their toes: their jobs may be virtually safe but their reputations and their career prospects are not. If an official has made a mistake—for example, if he has provided a poor briefing, failed to anticipate reasonable questions or criticisms from the opposition, has not carried out instructions, or has given an incorrect version of the minister's viewpoint—it is soon known in a department and circulated on the Whitehall grapevine. Finally, the convention helps to personalize politics in that the public tends to praise ministers for apparent successes and blame them for apparent failures.

What remains unclear as yet is the extent to which the creation of agencies by departments will weaken the principle of the minister's accountability to Parliament.

Collective responsibility

The convention of collective Cabinet responsibility has two features. The first is that the Cabinet resigns if the government is defeated on a major measure in the House of Commons; this was clearly established in 1830 when the government resigned after losing an important vote and faced almost certain defeat on the issue of parliamentary reform. After the Reform Act of 1832 the principle was established that a Cabinet depended on the support of the Commons. The second is that all ministers must, unless the principle is relaxed, accept and support in public Cabinet decisions or keep their dissent private; if they choose to express disagreement they should resign or expect dismissal.

The House of Commons may force a government's resignation by carrying a censure motion against it or by defeating it on a vote on an issue of confidence. An *issue of confidence* is one which the Prime Minister declares to be so, the implication being that rejection of a government-sponsored measure entails the resignation of the Government. Defeat on a vote on the Budget, on the second or third readings of government bills, or on the address in reply to the Queen's Speech have usually been regarded as votes of no confidence, because they are central to a government's programme. Such a defeat in the House of Commons has twice overturned governments in the twentieth century, in 1924 and 1979. In addition, in 1940, although Mr Chamberlain won a vote of censure on his handling of the Norwegian campaign, his normal party majority was so reduced that he felt he had to resign.

The convention, however, has become an instrument of Cabinet dominance of Parliament and is now caught up in the politics of party management and stable providing government. For much of the period since 1945 it has been usual for one party to have a reliable majority in the Commons and not lose important votes in Parliament. Even if a government faces a rebellion in its own ranks, it often pressures the potential rebels into line by making the issue one of confidence. The extension of the convention to include even parliamentary private secretaries (PPSs), or personal assistants to ministers, means that nearly a third of the governing party's MPs are on the 'payroll' (although PPSs are unpaid) and expected to support the government. Collective responsibility implies cohesion, and obviously fits more easily with a Cabinet whose members are drawn from one party, have been elected on a programme, and share a common philosophy. It also implies secrecy, so that the differences which precede or follow a Cabinet decision are not made public. *Presenting a united front to the opposition in Parliament and to the public springs at least as much from considerations of political management as from constitutional propriety.*

Yet there are signs that the convention, as applied to Cabinet unity, is declining. In this century there have been three notable agreements to differ; each occurred in circumstances when insistence on a common line would have split the Cabinet of the day. They concerned (1) the adoption of tariffs by the coalition National Government in 1932, when dissenters, notably the Liberal ministers, were allowed to support free trade; (2) the referendum on membership of the EC in 1975; and (3) a vote on the European Assembly Elections Bill in 1977. Between 1976 and 1979 a minority Labour Government lost votes on a number of key issues and so did the Conservative government on Maastricht in 1993. A constitutional consequence of the Cabinet waiving collective responsibility on such questions is that it is difficult for the House of Commons to hold it accountable for the result of a parliamentary vote. The myth of Cabinet unity has also been weakened as accounts of Cabinet proceedings and disagreements have been 'leaked' to the media more extensively in recent years.

The monarchy

Apart from a brief period (1649–60), England has had a monarchy since the tenth century. The monarchy has proved to be a durable institution, largely because it has acquiesced

BOX 3.4. THE MONARCHY

- Britain is a hereditary but constitutional monarchy.
- The Queen is head of state, but her role is largely ceremonial: she reigns but does not rule.
- The monarch is expected to remain politically impartial.
- The monarch can do no wrong as long as she acts on the advice of a responsible government with a majority in the Commons.

Functions

1. a symbol of continuity of the state and the constituition
2. a symbol of national unity
3. the formal head of state

Royal Prerogative

Traditional powers of the monarch which, by convention, are now exercised by ministers of the Crown (usually the Prime Minister), and which do not require authorization by Parliament.

Examples

- the power to dissolve and summon Parliament
- the power to declare war
- the power to make treaties

in pressures to change its role. Only a handful of monarchies remain in the world today. In the twentieth century they have usually collapsed as a result of (1) events associated with military defeat (e.g. Germany, 1918), (2) the break-up of multinational empires, again usually after defeats in war (e.g. Austria-Hungary and the Ottoman Empire after 1914–18), or (3) failure to come to terms with the growth of democracy. *The key to the monarchy's survival in Britain had been its willingness over the last three centuries to concede power in good time to head off demands for its abolition.*

When we say that *Britain is a constitutional monarchy we mean that the monarch plays a largely ceremonial role*—for example, representing Britain on overseas tours, attending public functions, and so on—*and on political matters acts on the advice of ministers.* When Parliament is sitting, the monarch receives daily reports on parliamentary proceedings, government papers, and reports of Cabinet meetings, and meets the Prime Minister weekly for a confidential discussion of government policy. *The Royal Prerogative* was described in the nineteenth century by the constitutional lawyer Dicey as 'the residue of a discretionary or arbitrary authority legally left in the hands of the crown'. This means that some powers may be exercised by the Crown without parliamentary authority; they include summoning, dissolving, and proroguing Parliament, making treaties, declaring war, command of the armed forces, appointment of judges, conferment of honours, creation of peers, appointment of ministers, initiation of criminal proceedings, granting of pardons, and so forth. Most of these acts, performed in the name of the Crown, are independent of the courts or Parliament.

Today, however, these decisions have been assumed by the Prime Minister and Cabinet. The right to *appoint peers* was of more significance before 1911, when the House of Lords was able to veto legislation, and was last exercised independently by a monarch in 1712. The monarch's threat to create extra peers to overcome the opposition of the Lords to the Reform Bill in 1832 and the Parliament Bill in 1911 was not carried out, because on each occasion enough members of the Lords eventually gave way to the governments of the day. The monarch's *choice of a Prime Minister* is normally uncontroversial, being the elected leader of the party able to command a majority in the House of Commons, and *appointment of Cabinet ministers* is made on the recommendation of the Prime Minister. Until 1963, whenever a Conservative Prime Minister resigned his office, the monarch's choice of successor was based on advice from senior party figures. The fact that the major parties now elect their own leaders has virtually removed any element of the choice of Prime Minister from the monarch.

There are two areas which are more problematical, however. The first concerns *the dissolution of Parliament*. It is widely agreed that the monarch must generally agree to a request from the Prime Minister to dissolve Parliament. With one party or a group of parties having a clear parliamentary majority in all but one of the general elections since 1931, this has so far presented little difficulty. But what happens if a Prime Minister, who lacks a majority in the Commons, seeks a dissolution, although another politician might be able to command a majority? Should the monarch agree to the request or invite somebody else? A second area of potential controversy may arise if no one party has a clear majority of seats or if the parties to a coalition do not agree on a leader; the monarch might then become involved, however unwillingly, in delicate negotiations *to choose a Prime Minister*. Such circumstances may draw the monarchy into controversy, by expanding its political role and that of its advisers. Conduct which does not follow clear precedents is liable to offend somebody and perhaps prompt allegations of political bias. The monarch of the day was criticized in 1931 over the formation of the National Government and in 1957 and 1963 over the appointments of Mr Macmillan and Lord Home respectively as Prime Minister. One way of avoiding controversy would be to transfer the decision to grant a dissolution or make appointments to the Speaker of the House of Commons or some other impartial figure.

The monarch's role has hardly changed from that laid down by Bagehot in 1865: the right to be informed, to encourage, and to warn. The constitutional dictum that the monarch reigns but does not rule is linked to two principles. The first is that of *political impartiality*: the monarch must act, and be seen to act, in a non-partisan manner and be willing to co-operate with any properly elected government. To play the role of constitutional referee requires the monarch to be even-handed between the parties, to avoid politically controversial statements or behaviour, and to adhere to accepted rules. Problems are likely to arise, therefore, when the 'rules' are not clear or are in dispute. The second and more crucial principle is that *the monarch can do no wrong as long as he or she acts on the advice of a responsible government* with a majority in the Commons. In this way the government assumes political responsibilities, and the monarch is kept 'above' party politics and protected from criticism.

So much for the theory. In recent years, however, the institution of the monarchy has attracted criticism. Much of this has been stimulated by negative media reporting of the behaviour of the younger royals and their failure to set a moral example to the nation, the size of the Civil List, or revelations that the Queen did not pay income tax until recently. Other critics complain that the premodern and undemocratic values associated with hereditary monarchy are ill suited to the twenty-first century.

Parliamentary sovereignty

The principle of *parliamentary sovereignty* is perhaps the outstanding feature of the British constitution, a feature which renders the political system 'unconstitutional' in the eyes of some observers. Sovereignty, or the power to make law, is exercised by the Queen, Lords, and Commons assembled. An Act of Parliament is not constrained by any higher laws and no other authority may rule on its constitutionality. This means that, apart from the case of European law, there is no judicial review: the courts cannot set aside, but only interpret, statute law. Parliament may not bind its successors and legislation may be overridden by subsequent laws. This was not always so. Before the mid-eighteenth century, courts were prepared to void a statute if it was deemed to clash with common law. It was only in the nineteenth century that the idea of the absolute supremacy of Parliament developed. According to Dicey (1952: 39):

The principle of Parliamentary sovereignty means neither more nor less than this, namely, that Parliament thus defined has, under the English Constitution, the right to make or unmake any law whatever: and further, that no person or body is recognised by the law of England as having a right to override or set aside the legislation of Parliament.

Other countries with written constitutions (notably the United States) usually have a Supreme Court or equivalent body which is authorized to decide whether the actions of a government or legislature are in line with the Constitution.

The *Factortame* legal case in 1990 concerned a conflict between Community law and the British Parliament. As a result of a ruling from the European Court of Justice the House of Lords suspended the operation of a British law affecting the registration of fishing vessels for EC fishing quota purposes until a definitive ruling from the ECJ. The ruling starkly illustrated the subordination of British to European law. It is also clear that, as long as Britain is a member of the European Union, the action of one Parliament can bind its successors, and therefore Parliament is not sovereign.

The fact (except for Europe) that there are no formal checks on the sovereignty of Parliament does not mean that in practice there are no limits. There are *political* checks on the government in the form of its sense of self-restraint, its need to bargain with powerful pressure groups and appease its own backbenchers, its duty to respect the right of Opposition parties in the Commons, and, of course, fear that the voters may turn it out at the next general election. There have also been cases of groups successfully defying Parliament, such as the trade unions' obstruction of the 1971 Industrial Relations Act, the government pay policy in 1974, or the strike in 1974 in Ulster that brought about the collapse of the power-sharing executive there. But these are not *constitutional* checks and balances.

Pressures for change

In recent years there have been several signs of a loss of support for our constitutional arrangements (for a defence of the *status quo* see Norton 1991). One set of pressures has been clearly *political*. These include opposition complaints about the 'illiberal' actions of the Thatcher governments, the rise of the large centre party vote in the 1980s, which added weight to traditional Liberal demands for proportional representation, and the rise of nationalism in Scotland. The trend to multi-partyism and the possibility that one party will not have a clear parliamentary majority has created additional uncertainties. As David Butler observes (1986), much of our practice regarding the formation and conduct of government assumes that government will be in the hands of a single party that commands a majority in Parliament.

Other pressures are more narrowly *constitutional* (F. Ridley 1988, Mount 1992, Wright 1994). They relate, particularly, to the weakening of conventions, such as those of collective and ministerial responsibility, the lack of settled rules regarding the conduct of referendums, the frequent rulings of the European Court of Human Rights against Britain on civil liberties cases, and the effects of membership

BOX 3.5. PRESSURES FOR CHANGE: THE CONSTITUTION

- Constitutions are not above politics.
- At any given moment they are more or less advantageous to some groups and disadvantageous to others.
- Current demands for change include:

Reform	*Advocated by*
proportional representation	third parties
devolution	nationalist
Bill of Rights/written constitution	Labour party
freedom of information	third parties, civil rights groups
halt to further EC integration	Euro-sceptics
reform of House of Lords	Labour and Liberal parties
reform of monarchy	republicans
written constitution	Charter 88, some Labour, some Liberals

Likelihood of change

Piecemeal change frequently occurs, but fundamental change in the shape of a written constitution, introduction of PR, etc. is unlikely:

- only those out of power want change, only those in power can bring it about
- the nature of change is a matter of dispute among advocates

of the European Union. In ruling on a dispute between domestic law and European legislation, British courts are bound, under the Treaty of Accession, to give precedence to the latter. The constitutional effects of membership are many and are likely to continue.

The agenda of reform encompasses various proposals. One is for an *entrenched Bill of Rights* to protect citizens' rights. Britain is one of the few democratic states that does not provide a constitutional guarantee of individual rights to its citizens; the Bill of Rights of 1689 is largely concerned with asserting the rights of Parliament against the monarch—for example, it protects MPs from being sued in the courts for anything they say in Parliament and forbids the maintenance of a standing army in Britain in peacetime without the consent of Parliament. *Entrenchment* of a law means that Parliament either cannot override it or can do so only in exceptional circumstances.

Traditionally, the common law provided some protection for civil liberties. However, Lord Scarman (1974), among others, has suggested that more interventionist government, combined with the courts' passive interpretation of their role, has made the common law a frail defence of individual liberty. In fact, there are signs that the courts are changing. Because of the obstacles to entrenchment some reformers would settle for the formal incorporation of the provisions of the European Convention on Human Rights, which Britain ratified in 1953 but has not yet enacted into law. At present, many aggrieved citizens or groups resort to the Court in Strasburg because Britain does not have its own procedures. An objection, as to all proposals for entrenchment, is that it would limit the sovereignty of Parliament. Tony Blair, like his predecessor John Smith, has declared himself in favour of incorporation.

The interest of Westminster politicians in *devolution for Scotland and Wales* was prompted primarily by the rise of electoral support for the Scottish Nationalist party in 1974. Labour, as the major party in Scotland, had the most to lose and hurriedly prepared devolution proposals, outlined in the Scotland and Wales Acts. The proposals were opposed not only by most Conservatives but also by some Labour backbenchers; opponents succeeded in requiring 40 per cent of the electorate to vote for the Act in referendums in Scotland and Wales before it could become operative. The proposals eventually lapsed because they did not attract sufficient support in referendums held in the two countries in March 1979. Labour still promises to establish a Scottish Parliament, elected by proportional representation, with tax-raising and legislative powers, as well as an elected assembly for Wales. In a political system such as Britain's, without federalism or a written constitution, interest centres on how cases of conflict between Westminster and the Scottish Parliament would be resolved. Critics of the proposals for Scotland point to the paradox of Scottish MPs in the Commons having the right to vote on non-Scottish matters, while non-Scottish MPs are precluded from voting in Scottish matters. The weakness of Conservatism in Scotland in the 1980s and 1990s has inevitably fuelled claims about rules by an alien England and kept demands for devolution and even separatism on the agenda (Marr 1992).

The position of Northern Ireland is discussed elsewhere (see Ch. 2). What is clear is that no major British party regards the *status quo*, that is, direct rule from London, as a long-term policy, but as a transition to something else. The British government has declared that any change in the constitutional status of Northern Ireland, including incorporation in the Irish Republic, is acceptable as long as it has the consent of a majority of the local population.

The absence of a regional tier of government and marked reduction in the powers of local government since 1979 make Britain

BOX 3.6. THEORY AND PRACTICE: REFERENDUMS

Theory

The electorate is asked to vote yes or no to a specific question.

Example • June 1975, 'Do you think Britain should stay in the European Community?'

The referendum is regarded as an alien device in British politics since it is problematic in relation to the constitutional doctrine of **sovereignty of Parliament.**

On the rare occasions it has been used, potential conflict has been avoided by making the referendums 'advisory'.

To date referendums have been used on only three occasions (except in local examples such as pub-opening times in some parts of Wales).

Each time the question was a **constitutional issue**:

- 1973 in Northern Ireland on the border question
- 1975 throughout the UK on continued European Community membership
- 1979 in Scotland and Wales on devolution

In theory referendums provide an additional opportunity for popular participation, the government consulting the governed on important constitutional questions.

Practice

In practice the referendums had more to do with political expediency than constitutional principle or democracy.

1973 The Conservatives were attempting to reassure the Protestant community that Northern Ireland would remain part of the UK unless the majority wished for change. The result: **57.4% in favour of *status quo*** was a foregone conclusion. The referendum was boycotted by Catholic parties and Catholics abstained.

1975 The Labour Cabinet and party was split over the issue of continued EC membership and Harold Wilson resorted to a referendum and 'agreement to differ' to enable the Labour government to survive in office. The result: **64.5% voted in favour of remaining in Europe.**

1979 James Callaghan's minority Labour government depended for its survival on the Lib–Lab pact. Devolution for Scotland and Wales was introduced in return for continued Liberal and Nationalist support, but was unpopular with many Labour MPs. Devolution was, therefore, made dependent upon **40% of the electorate** in each country voting in favour. The result: **in Scotland 33% voted for devolution, 31% against, whilst in Wales 12% voted for and 47% against**. Devolution was abandoned and the government lost a censure motion in May 1979.

During the autumn of 1994 and spring 1995 use of a referendum has again returned to the agenda. Many Euro-sceptics argue that further integration and monetary union with Europe is such an important constitutional issue that the British electorate should be allowed to express its opinion in a referendum. In that all the main parties have pro-European policies there is some validity in the argument that such a bi-partisan approach leaves the electorate with no choice.

distinctive in the European Union (EU). It remains probably the most politically centralized state in Western Europe (Crouch and Marquand 1989). Again, it is the non-Conservative parties which favour a more balanced relationship between central and local/regional tiers of government.

There is also uncertainty over the form of Britain's *second chamber*. At present, the unelected House of Lords possesses only a twelve-month power to delay legislation passed by the Commons, and it still consists largely of hereditary peers. It is the voting rights of the latter that offend reformers. A popularly elected body would have a mandate to challenge the Commons, while one that was appointed by the government of the day would lack independence. Yet the Lords remains the sole formal check on the power of a majority in the Commons. Labour has switched from proposing to abolish the Lords (in manifestos for the 1983 and 1987 elections), to one of electing it by proportional representation and removing the voting rights of hereditary peers. Dissatisfaction with the Lords is not new, but the chamber has endured because reformers have not agreed on an alternative.

Finally, *referendums*, long considered 'un-British' and incompatible with a sovereign Parliament, have also been introduced. So far the most powerful demands for the referendum have been pressed in connection with constitutional issues, for example, reform of the House of Lords in 1910–11, membership of the EC in 1975, and devolution for Wales and Scotland in 1979, and carried through for the last two. Party leaders have usually turned to it as a political device when their own party was divided, as Labour was over the EC and Scottish devolution. In 1970 Mr Wilson, as Prime Minister, had opposed proposals for holding a referendum on entry to the EC, adding 'I shall not change my attitude on that'. Pressure within his party gradually brought him round to the idea, although he claimed in 1975 that it was a 'unique event'. A divided Labour party, faced by a backbench revolt, again turned to a referendum in 1977 over the devolution proposals for Scotland and Wales. In the 1990s Conservatives, particularly the Euro-sceptics, have called for a referendum to decide on measures which involve a further integration of Britain into the European Union or the creation of a single currency. The device is now probably established as part of the constitution: the important question is how to develop procedures for its use. For example, what rules should be attached to the holding of a referendum, on what sorts of issues should one be held, should a minimum turnout figure be specified, and what proportion of votes should count as a majority?

The above does not exhaust the list of demands made by reformers. They also call for legislation to deliver a freedom of information act, fixed-term Parliaments, and proportional reform. Liberal Democrats support all these measures; Labour supports the first two and officially has an open mind on the third.

A written constitution?

Political debate in the 1980s and 1990s has returned to constitutional issues and at times divided the two major parties. The reform agenda is now well established: proportional representation (PR), a bill of rights, devolution, an elected second chamber, freedom of information, and reform of the monarchy. Many of the issues have arisen piecemeal, in response to changing political circumstances and/or calculations about party gain. Reformers often want to promote certain policy objectives and/or strengthen their own parties. But, apart from these pressures, there has also been a growing body of support for a comprehensive restatement of the rules and even a written constitution. *Charter 88*, an all-party movement but dominated by Labour and Liberal Democrat supporters, has called for a written constitution, including devolution of power to elected regional authorities, a Bill of Rights, a Freedom of Information Act, fixed Parliaments, and proportional representation. The centre-left think-tank Institute for Public Policy Research (IPPR) drafted a written constitution in 1991, which included most of the 'reform' agenda. The programme closely resembled the manifestos of the two parties in the 1992 general election. But from another part of the political spectrum Ferdinand Mount (1992) and Frank Vibert (1991) have also expressed interest.

There are several arguments for such a far-reaching step. First, some claim that the reforms would give Britain *a 'better' political system*, that is, it would produce a more representative House of Commons, protect citizens against arbitrary government, or bring the United Kingdom into line with other Western democracies.

Second, some have regarded constitutional change as a reform for *promoting a more efficient policy-making process*. For example, more open government might encourage more informed debate about policy, or proportional representation might weaken a class-based two-party system, loosen the producer-group links with the Labour and Conservative parties, and promote more continuity and consensus in policy. Others regard a constitution as a device for placing certain issues 'above party politics', as it were, and limiting the impact of partisanship.

A third claim is that *limited government is good per se*, and should be safeguarded. This is more likely to appeal to free-market Conservatives.

Fourth, some simply want to bring Britain into line with other member states of the European Union by, for example, introducing a Bill of Rights, greater decentralization, or proportional representation. Britain remains the only member state not to elect its MPs to the European Parliament by a form of PR.

Finally, those who regret the waning of agreement and certainty about the constitution want an *authoritative statement* which will preclude the rules of the game from being at the mercy of a temporary parliamentary majority.

One should not lose sight of the interplay of party advantage in the constitutional debates. Governments, enjoying the full powers of a sovereign Parliament, usually have a different vantage point from the opposition parties. In the past Labour's left has regarded proposals for limits on the executive as part of a campaign against a socialist government which would want to use its parliamentary majority to make radical social and economic reforms. Mrs Thatcher and her followers, who wanted to reverse the collectivist drift of much post-war policy, also resisted constitutional reform. Both the Labour left and Thatcherites have regarded parliamentary sovereignty, exercised by a majority party, as the means through which they could achieve radical change.

The demand for *checks and balances* is often made by Opposition MPs who may simply not like what the Government of the day is doing or who may regard it as an additional weapon with which to curb ministers. Questioning the legitimacy or constitutional propriety of an action by the executive is a useful device for the Opposition. It is remarkable how politicians in opposition, notably Lord

Hailsham (1978) and other Conservatives between 1974 and 1979, Liberals (and Social Democrats in the early 1980s), and Labour since 1987 have advocated reforms which aim to check the power of government. When their own party has been in office, however, their complaints have been more muted. The more distant a party's prospects of office the more advantageous to it are changes which limit the powers of government. There is nothing exceptional in this. Politicians calculate the advantages and disadvantages about changes in the rules of the political game, as they do about other issues.

One means—not persuasive to a strict constitutionalist—of checking government is the existence of a competitive two-party system in which the main opposition party has a realistic prospect of displacing the government. In other words, political competition, with the threat of losing an election, is a discipline on the government of the day. Party politics substitutes, as it were, for constitutional checks. In the 1980s, however, the Conservative government enjoyed handsome majorities in the Commons and Labour was not a credible alternative party of government. This imbalance in Parliament, combined with measures of the Thatcher governments, which critics have regarded as attacking freedoms (for example, abolition of the Greater London Council and denial of trade union rights to Government Communications Headquarters (GCHQ) employees), and the lack of checks on parliamentary sovereignty have encouraged the opposition to shift from political to constitutional reform. Constitutional reform now unites the Labour and Liberal Democrat parties in Britain more than any other issue at present and clearly divides them from the Conservative party. Not surprisingly, the leadership of the two parties also strike a more pro-European stance than does the Conservative leadership.

The Conservative party has been unsympathetic to demands for change. In 1989 Mrs Thatcher replied to *Charter 88*'s appeal for support: 'The government considers that our present constitutional arrangements continue to serve us well' (Wright 1994: 29). John Major has been as resolute, warning, for example, of the extra layers of bureaucracy and tax which the creation a Scottish Parliament would entail, the threat which devolution might pose to the unity of the United Kingdom (the 'slippery slope' argument), and the unstable government that would be encouraged by PR which led to coalition or minority governments and a foothold for extremist parties. In 1992 John Major's government helped the cause of open government by publishing details of the security services (including identities of the heads of the services) and the Cabinet committee system, and points to its *Citizen's Charters* as providing greater rights for users of public services.

Obstacles

Drawing up a new constitution—let alone implementing it—is no easy matter. In Britain, the main practical difficulty involved in having a written constitution or an authoritative formal statement of the rules of the game is presented by the doctrine of *the sovereignty of Parliament*. This is the argument often used to oppose a Bill of Rights, devolution, the right of the courts to challenge the constitutionality of legislation, or a 'pooling' of sovereignty in the EU. Wright (1994: 3) complains that 'It is the doctrine to be summoned up to explain why things cannot be done . . . It is the no-doctrine of British politics.'

As noted, the British have no experience of a system of government or set of rules being constituted at one point in time. In the United States, by contrast, one thinks back for a starting-point to 1787, in France to 1958, and in West Germany to 1949. Similarly, the British have no concept of *fundamental, or entrenched (i.e. difficult to change) rules* but have taken pride in the flexibility provided by their unwritten constitution. Strict adherence to the doctrine of parliamentary sovereignty is essentially incompatible with genuine constitutional principles. A written constitution is likely to increase the political role of the courts and the possibility of their coming into conflict with Parliament, as they would have to decide whether statutes infringed the constitution.

The main difficulty, however, in finding a set of rules which is acceptable to different viewpoints is likely to be *political*. A large measure of agreement is essential if a constitution is to be legitimate, that is, voluntarily accepted and widely regarded as 'above party politics'. A first step would presumably be to establish a body, say a Constitutional Commission, which in turn would formulate and then submit proposals to Parliament. These, when approved, might be subject to a referendum and if approved then entrenched. Yet many of the possible topics for inclusion (such as the right to belong or not to belong to a trade union, parental choice in education, private health care, referendums, the voting system, or the existence and role of a second chamber) have until recently been highly political issues. An additional difficulty is that governments are unlikely to be willing to sacrifice the necessary parliamentary time for a new constitutional settlement, when there are so many other pressing economic and social issues competing for attention. Legislation affecting membership of the EC and devolution for Scotland and Wales, both of which were also subjects of referendums, dominated debate and the legislative timetable in Parliament in the 1970s, and the Major government spent the best part of two years in battling to pass the Maastricht bill through Parliament. Governments also have little incentive to introduce any reforms which might constrain their own powers or undermine their authority in office. *Political reality, therefore, is the major barrier to constitutional reform.*

There are other reasons for caution. Constitutions, like laws, do not always have the effects which were originally intended or anticipated. Some countries provide constitutional guarantees for civil rights yet have abysmal records in human rights. As noted, the purposes of reformers vary, and a 'constitutional' remedy is not always appropriate. Some reformers are simply unsympathetic to the clash of ideas and interests that is inseparable from democratic politics.

The adoption of a written constitution might well challenge many traditional assumptions about the conduct of British politics. The political élite has managed to operate without a formal constitution, in large part because the informal 'rules' were widely understood, and governments practised self-restraint and respected the rights of other groups, including the Opposition in Parliament. In other words, there was a political style which was 'constitutional', without being based on law. In many West European states which have a Roman law background and authoritative written constitutions, the political culture is more legalistic and based on formal rules. In Britain the tradition has been more one of laws following behaviour than vice versa. Ultimately, however, constitutions need to rest on popular support and be acceptable to the major sources of political and economic power in society.

A merit often claimed for Britain's unwritten constitution is its flexibility. Statutes can be changed or conventions altered and brought into line with political practice. Yet this flexibility is also what concerns critics of the present system: the British constitution is largely what the government of the day says it is. It is also worth noting that the possibility of change is also usually provided for in written constitutions; most written constitutions may be amended through a formal process, for example, by referendum or a vote by a fixed majority in the legislature. In the United States, there have so far been twenty-six amendments to the constitution since it was adopted in 1787. Some of the legislative changes affecting constitutional issues in Britain have been controversial and politically divisive—extensions of the suffrage in the nineteenth century, reform of the Lords in 1911, partition of Ireland in 1921, entry to the EC, and proposals for Scottish devolution. Constitutional change has often come about less by general agreement than by power politics, by one party using its parliamentary majority to push through a change which was then accepted by opponents. This is true of reforms of the Lords, the expansion of the suffrage, and Irish Home Rule. Changes in conventions tend to be more gradual and recognized as binding only after the passage of time, like the convention that the Prime Minister is a member of the Commons or that the monarch acts on the Prime Minister's advice on political matters.

Interpreting the constitution

A Frenchman or an American writing about his country's constitution will inevitably concentrate on the formal constitutional document and legal decisions relating to it. Accounts of the British constitution, on the other hand, usually describe the way the system works, a situation that has allowed different writers to use evidence selectively in support of differing interpretations of the constitution. In 1865 Bagehot's *The English Constitution* distinguished between the 'real' and 'dignified' parts of the constitution, between the 'paper description' and 'living reality'. The 'outward show', provided by the monarchy, did not, as many people believed, betoken effective power. Although the monarchy could 'excite and preserve the reverence of the population', it was the sober-suited Cabinet ministers who ruled. The distinction *between* the 'real' or 'secret' and 'formal' constitution has also been invoked by critics of unelected power-holders, like civil servants, business, and the European Commission (Sedgemore 1980, Harden and Lewis 1988).

A. H. Birch, in his *Representative and Responsible Government* (1964), has drawn on historical evidence to organize ideas and practice of government around two traditions. A *liberal* or *Westminster-based* view of the constitution is drawn from nineteenth-century experience and emphasizes the importance of Parliament, particularly its ability to hold the government accountable. It takes literally the fact that the Cabinet emanates from the House of Commons and depends on a majority there, the idea of ministerial responsibility, and the claim that the elected Commons represents the electorate. Most observers would accept that this view may accurately describe the mid-nineteenth-century situation but is inadequate today because it ignores the role of organized mass parties, pressure groups, a large civil service, and interventionist government. But it is still important in colouring the views of those who mourn 'the decline of Parliament'.

A contrasting '*Whitehall*' view emphasizes the independent status of the Crown and its consequences for the relatively autonomous conduct of civil servants and ministers who lead Parliament and public opinion. According to the 'Whitehall' model the executive is the initiator in the policy process, and Parliament's main function is largely reactive and supportive.

Reviewing an array of different models through time is useful as long as one bears in mind the different context in which each arose. It gives a sense of historical perspective to current political debates and proposals for constitutional reform. A politician's perceptions of the rules of the game may depend not only on whether or not his preferred party is in government or opposition, but also to which wing of the Conservative or Labour party he belongs (Tivey 1988).

Conclusion

Whether or not Britain adopts a written constitution depends largely on the willingness of a government to bring it about. Interest in constitutional reform has traditionally been 'the poetry of the politically impotent' (Mount 1992: 2). What is new in the 1990s is the attitude of Labour. It seems to have turned its back on the idea of a House of Commons party majority as a battering ram of change. It set up a party commission in 1991 which recommended electoral reform and its 1992 manifesto proposed a wide range of constitutional reforms. Labour's view has been moulded in part by its long exclusion from government and in part by closer relations with socialist parties in Western Europe, which take a written constitution and PR for granted. Some changes to the working of the constitution seems likely if there is a Labour government, although the prospects of a written constitution are still remote. Sceptics may wonder if Labour will still be as supportive of reform once it is in power. But there is more uncertainty and dissatisfaction with the working of the constitution than has existed for many years.

Summary

- The British constitution is partly written but uncodified.
- It has evolved in response to economic, social, and political change.
- It has five main sources: statute law, European Community law, common law, works of authority, conventions.
- A constitutional monarch has a symbolic role.
- Britain is a parliamentary democracy: the executive is drawn from and accountable to Parliament.
- Parliamentary sovereignty is the fundamental principle of the constitution. There is no judicial review. Britain is a unitary state; subordinate authorities are empowered by Parliament and can likewise be disempowered. This principle also partly explains some of the constitutional problems raised by Britain's membership of the EU.
- Failure to halt relative economic decline and its associated social problems; a longish period of one-party dominance; a weakening of the observance of key conventions; and membership of the European Union have raised questions about Britain's governing institutions and placed constitutional reform on the agenda. There has been much debate about whether the time has come for Britain to adopt a written constitution.
- The flexible nature of the constitution has facilitated piecemeal adjustments but constitutions are not above politics, and this partly explains the failure to introduce fundamental reform. Those out of power want change but can't deliver it. Those in power can introduce change but don't need it.

CHRONOLOGY OF IMPORTANT DATES IN THE HISTORY OF CONSTITUTIONAL EVOLUTION

1215	Magna Carta: documentary recognition of the need for rule by consent; the king is not above the law
1265	De Montfort's Parliament
1640	Civil War
1649	Charles I executed; England governed as a republic under Cromwell
1660	Restoration of the monarchy
1688	Glorious Revolution
1689	Bill of Rights limits monarch's power
1707	Act of Union between England and Scotland
1800	Act of Union between Great Britain and Ireland
1830	Collective responsibility of the Cabinet established
1832	Reform Act enfranchises the middle class
1867	Urban skilled workers get the vote
1872	Secret Ballot Act
1884	Rural and urban labourers get the vote
1911	Parliament Act: House of Lords' veto is replaced by delaying powers
1918	Women 30 and over and men 21 and over get the vote
1920	Government of Ireland Act: Ireland partitioned
1928	Women 21 and over get the vote
1949	Parliament Act: reduces Lords' delaying power Abolition of plural voting (businesses and universities)
1958	Life Peerage Act: non-hereditary peerages created; first women in Lords
1963	Peerage Act: hereditary peers can resign title for their own lifetime
1965	Parliamentary Commissioner for Administration (ombudsman) created
1969	Voting age lowered to 18 years
1972	European Communities Act
1973	Referendum on border issue in Northern Ireland
1975	Referendum on continuation of EC membership
1979	Referendums on Scottish and Welsh devolution

ESSAY/DISCUSSION TOPICS

1. Discuss the view that 'Procedure is all the constitution the poor Briton has' (Pickhorn).
2. Why is constitutional reform often mooted but rarely introduced?
3. Why can't the British constitution be 'exported'?
4. The time has come for Britain to adopt a written constitution. Discuss.
5. Can the British government be described as constitutional?
6. Discuss the view that 'Our so-called constitution allows a resulted politician who doesn't want to play according to the rules, who isn't an "officer and a gentleman", almost unlimited power' (F. F. Ridley).

RESEARCH EXERCISES

1. Briefly outline the circumstances in which each of the following ministers resigned, paying particular attention to any constitutional principle involved in each case.

   Cecil Parkinson 1983
   Michael Heseltine 1986
   Leon Brittan 1986
   Edwina Currie 1988
   Nigel Lawson 1989
   David Mellor 1992

2. Use examples from the twentieth century to show the different ways in which the British constitution can be amended.

FURTHER READING

For an introduction see P. Norton, *The Constitution in Flux* (Oxford: Martin Robertson, 1982). More and different views are given in F. Vibert, *Constitutional Reform in the United Kingdom: An Incremental Agenda* (London: Institute of Economic Affairs, 1990); Institute of Public Policy Research (IPPR), *The Constitution of the United Kingdom* (London: IPPR, 1991); and F. Mount, *The British Constitution Now: Recovery or Decline?* (London: Heinemann, 1992).

The following may also be consulted:

Holme, R., and Elliot, M., *1688–1988: Time for a New Constitution* (London: Macmillan, 1988).
Norton, Philip, 'Should Britain Have a Written Constitution?', *Talking Politics*, 1/1 (1988).
——'Europe and the Constitution', *Talking Politics*, 6/3 (1994).
Ridley, F. F., 'There is no British Constitution: A Dangerous Case of the Emperor's Clothes', *Parliamentary Affairs*, 41/3 (1988).
——'What Happened to the Constitution under Mrs Thatcher?', in B. Jones and L. Robins (eds.), *Two Decades in British Politics* (Manchester: University of Manchester Press, 1992).

4 BRITAIN AND EUROPE

Reader's guide

The 'Europe' debate — what is the appropriate relationship between Britain and the EU? — is one of the most important questions on the political agenda. It divided the Labour party in the 1970s and 1980s, it was a contributory factor in the formation of the SDP, and, therefore, one explanation for Labour's long period in Opposition. Now in the 1990s, it poses the most serious threat to Conservative party unity since the free trade issue at the turn of the last century. It is not a question that lends itself to easy resolution or final settlement. The changes in the name, from European Economic Community (EEC) to European Community (EC) to the present European Union (EU), reflect the dynamic nature of the European project. This presents member states with a continuous need to reappraise their relationship with Europe and to adjust to its increasing impact on their domestic and foreign policies. What is the appropriate relationship between sovereign nation states and the aspiring supranational institutions of the EU? Will political union be an inevitable outcome of monetary union? Such questions about sovereignty, statehood, and national identity are of fundamental importance to states which are members of the EU but which do not necessarily share a common vision as to its future development.

This chapter explores the history of Britain's relationship with the EU. It outlines the key institutions of the Union and explains some of the reasons that underlie Britain's rather turbulent relationship with Europe. The chapter concludes by examining the impact of Europe on Britain's constitution, its government, and its political institutions.

THE questions of whether Britain should join the European Community (EC) and then what its role as a member should be have plagued political leaders for much of the post-war period. By 1996 Britain's membership of the European Community had lasted for twenty-three years, although the first (unsuccessful) application for membership was made thirty-five years earlier, in 1961. The relationship has come to dominate British debates affecting domestic and foreign issues and has destabilized both Labour and Conservative parties. As Nugent (1993: 40) has commented: 'Increasingly, EC politics go to the heart of British politics.'

This chapter reviews the reasons which led Britain to seek entry, and the factors that have created tension with other member states.

Britain's international position

At first, Europe was seen by most British politicians largely as a foreign policy issue. Over time, however, it has raised important constitutional questions and influenced debates on many key domestic and economic policies. Neither the Treaty of Accession, which Britain signed in 1972, nor the referendum or membership in 1975, nor the ratification of the Maastricht Treaty in 1994 have finally *decided* the question of Britain's role. The European Union (EU), as it is now known, is dynamic: its institutions acquire new powers and the pressures for integration or centralization are growing. This agenda alarms the so-called Euro-sceptics in Britain. They call for resistance to further integration and entertain the prospect of Britain's withdrawal.

The reactions in Britain to the end of the 1939–45 war differed from those in most of Western Europe. On the Continent there was a greater awareness of the limitations of the sovereign state and nationalism, and of the need for greater cooperation between states. Many political leaders resolved that the Continent, having been a theatre for two grand conflicts in the first half of the century, should take steps to avoid a repetition. Political and economic cooperation and, eventually perhaps, a political union, was seen as an answer to traditional Franco-German rivalry and the devastating consequences this had caused.

In Britain, however, the war and its outcome pointed to different lessons. British leaders claimed that it was the country's independence and separation from the European mainland by the English Channel that had saved it in 1940, and eventually Europe as well. The rest of Western Europe, except for Spain and Portugal, had been defeated or occupied during the war. *If nationalism and the nation state were to some degree discredited on the Continent, they were vindicated in Britain.*

Britain therefore stood aside from the early stages of building the European Community. The European founding fathers began by seeking cooperation in steel and coal. They hoped that success in these limited areas would lead to cooperation elsewhere, and eventually to political union. France and Germany were joined by Belgium, the Netherlands, Luxemburg, and Italy in a customs union in which the countries agreed not to impose tariffs on goods imported from member states. Founding treaties were signed for steel and coal in 1951, atomic energy and the EC itself (the Rome Treaty) in 1957.

Entry to Europe

The Community thus began without Britain, although the original states envisaged and hoped that Britain would in time join. Contrary to the expectations of some British politicians and officials that the venture was bound to fail, it turned out to be a great success. In the first ten years of its existence the economies of the member states grew at a rate twice as fast as Britain's. When the Conservative government under Harold Macmillan first applied in 1961 it was as a supplicant. Britain was still a major power with many global interests but was finding it increasingly difficult to sustain that role. Neither the Commonwealth nor the so-called special relationship with the United States provided a suitable alternative international role. If Britain was not a member of the EC, a danger was that it might be bypassed in a world of emerging power blocs. The Conservative government hoped that Britain would be able to exercise political leadership in the Community and improve relations between the EC and the United States. Membership also appealed to British leaders because it would enable her to become part of a larger market and share in the EC's economic dynamism. Except for the small Liberal party, the political parties were both divided over the application, though the Conservatives were on balance a pro-European party and Labour more anti-European. The issue temporarily disappeared from the agenda when de Gaulle, the French President, vetoed the application in 1963. He suspected that Britain was still too closely tied to the United States, and that Britain would also be a rival for France.

A further application was made by the Labour government in 1967—again rejected by de Gaulle—and another in 1970. This last application was picked up by a new Conservative government under Ted Heath. Heath has to date been Britain's most unambiguously pro-European Prime Minister. His government signed the Treaty of Accession in 1972, and entry took effect on 1 January 1973. The two main parties remained divided and entry was only achieved because sixty-nine Labour MPs defied the party whips on a key House of Commons vote to support the Treaty. Labour, now in Opposition, rejected the terms and promised that when in government it would allow the people to decide on membership in a referendum. When it returned to government in 1974, it conducted renegotiations and held a referendum in June 1975, in which membership was upheld by a 2 to 1 majority. At the time, advocates of membership argued that a 'pooling' of British sovereignty in the EC would not involve a reduction in the sovereignty of Parliament. Yet under the Treaty, Community institutions were empowered to administer and even enforce some laws directly in member states.

After more than twenty years of membership, Britain has proved to be a *reluctant partner* (George 1990). Party leaders have had to take account of significant pockets of scepticism or even hostility in their parties. The Labour Prime Ministers, Wilson and Callaghan, had to cope with opposition, mostly from the political left and some trade unions, who dismissed the EC as a capitalist club, hostile to policies of public ownership and controls on capital which a left-wing Labour government might want to impose. Indeed in the 1983 general election a left-wing Labour manifesto promised to withdraw from the EC—this time without bothering with a referendum.

Mrs Thatcher also took British sovereignty seriously. If Labour was impatient with checks from Brussels on its socialist plans, she regarded Brussels as a barrier to the realization of her free-market vision. She had little time for the proposals for subsidies, promoting harmonization of social rights (which she dismissed as socialism by the back door), and what she regarded as the excessive bureaucratization of the European Commission. She jealously guarded her image as a prudent housekeeper and resented the fact that Britain was a substantial net contributor to the European budget. Unlike Heath she was more Atlanticist than European and felt more at ease with Presidents Reagan and Bush in the White House than she did with European leaders. In contrast, the pro-European wing of the Conservative party envisaged a European Union of interdependent states, welcomed monetary union, and did not exclude the eventual possibility of political union. They considered that economic interdependence made talk of national sovereignty outdated and that a 'pooling' of sovereignty was sensible.

There were few European initiatives in the second half of the 1970s and in 1978 Britain remained outside the newly created Exchange Rate Mechanism (ERM). It was unfortunate that Britain's entry coincided with an economic recession and a sharp slowdown of economic growth in Western states. During the 1980s, however, the EC expanded and more steps were taken to achieve greater integration. In 1986 the number of member states increased to twelve and then in 1995 to fifteen, as Austria, Sweden, and Finland joined, leaving Switzerland and Norway as the only West European states outside. The Single European Act (SEA) came into effect in 1987, to create a single market of goods,

BOX 4.1. LABOUR AND THE EC, 1983

The Labour party manifesto of 1983, 'The New Hope for Britain', proposed:

'The next Labour government, committed to radical, socialist policies for reviving the British economy, is bound to find continued membership a most serious obstacle to the fulfilment of those policies. In particular, the rules of the Treaty of Rome are bound to conflict with our strategy for economic growth and full employment, our proposals on industrial policy and for increasing trade, and our need to restore exchange controls and to regulate direct overseas investment.

For all these reasons, British withdrawal from the Community is the right policy for Britain—to be completed well within the lifetime of the parliament.

Withdrawal would be achieved by

(a) amending the 1972 European Communities Act, ending the Community's powers in the UK; and then

(b) repealing the 1972 Act, and abrogating the Treaty of Accession.'

BOX 4.2. TOWARDS AN EVER CLOSER UNION

1972 European Communities Act

Under the provisions of this act Britain accepted the terms of EC membership. EC law became applicable in the UK, and no assent of Parliament was required. EC law took precedence over British law.

1986 Single European Act (SEA)

- extended EC competence into new policy areas: environment, research and technology, regional policy
- goal of completion of internal free market in goods, labour, and capital by 1992
- extended areas covered by qualified majority voting (QMV) in the Council of Ministers (see below)
- increase in the role and potential influence of the European Parliament

1992 Treaty of European Union (Maastricht)

- proposed monetary union, a single currency, and a central bank
- new areas of EU competence, including education, cultural policy, health, immigration, criminal justice, and consumer protection
- Social Chapter: minimum level of working conditions
- extended areas covered by QMV in Council of Ministers
- some increased power for European Parliament
- citizenship: citizens of member states are also citizens of EU

The term **federal** is avoided, but the Treaty represents a substantial commitment that the EC would evolve into the EU.

Voting in the Council of Ministers

Votes are allocated as follows:

Britain, Germany, France, and Italy	10 votes each
Spain	8 votes
Belgium, Greece, Netherlands, and Portugal	5 votes each
Austria and Sweden	4 votes each
Finland, Denmark, and Ireland	3 votes each
Luxemburg	1 vote

Decisions can be made in three different ways:

- simple majority vote: used mainly for procedural matters; each member has one vote
- qualified majority vote: 61 out of the possible 86 votes are required for such issues as completion of the internal market
- unanimous vote: required to enlarge the Community, increase the powers of the EU, change the treaties, harmonize the tax system

services, capital, and labour on the basis of free and fair competition between member states by 31 December 1992. More states joined the ERM of the Economic Monetary Union.

The Conservative leadership, as well as the financial and business communities, favoured British participation in the ERM. Membership would fix the value of the pound to a basket of European currencies, so providing more price stability for the prices of goods traded. But Mrs Thatcher, true to free-market principles, opposed entry on the grounds that it was futile to try and 'fix' the exchange rate. She was not prepared for Britain to lose control over its monetary policy and interest rates to decision-makers outside the country, particularly the German Bundesbank. She was sustained in her opposition by her economic adviser Sir Alan Walters. Her reply to pressure from Cabinet colleagues was the formula that Britain would join the ERM 'when the time was right'—which was a roundabout way of saying never.

This was not good enough for her Chancellor of the Exchequer, Nigel Lawson, who began in 1985 to shadow the German mark as a substitute for and a prelude to entering the ERM. Mrs Thatcher subsequently claimed that she did not know of this policy and relations between the two deteriorated (Lawson 1992, Thatcher 1993: 690–726). On the eve of a conference of European leaders in Madrid in 1989 she was forced to accede to a demand from Lawson and her Foreign Secretary, Sir Geoffrey Howe, to specify more clearly the conditions under which Britain would join. Soon afterwards Howe was removed from his post. An unintended consequence of the tension was the resignation of Nigel Lawson in October 1989, following the publication of criticisms of the ERM by Sir Alan Walters. A year later, when the new ministers at the Foreign Office (Douglas Hurd) and the Treasury (John Major) approached her in September 1990 to press the case for Britain's membership, her position was weaker and she agreed. In the 1992 general election Britain's ERM membership was supported by all the major political parties as well as most economic commentators, the Bank of England, and the City. Indeed a key part of the Conservative campaign message was to warn that Labour's economic policies would imperil Britain's membership of the ERM and that withdrawal would damage the economy. In the end Britain was driven out in September 1992, and this proved to be the springboard for an economic recovery.

Europe was a running sore throughout the latter stages of Mrs Thatcher's premiership, a cause of Cabinet instability and ruptures with senior colleagues, including:

- 1986 January *Westland*. The Cabinet was faced with a choice between supporting a European-backed or American-backed rescue of the ailing Westland helicopter company. Clashes led to the resignation of two senior Cabinet ministers and weakened Mrs Thatcher's position.
- 1989 June. *Madrid summit*. Mrs Thatcher was presented with a demand—accompanied by threats of resignation from Lawson and Howe—for a more positive British attitude to ERM membership.
- 1989 October. Nigel Lawson's resignation following Mrs Thatcher's refusal to dismiss her economic adviser, Sir Alan Walters, who openly opposed ERM membership.
- 1990 July. Nicholas Ridley's resignation, for expressing anti-German and anti-Europe comments in an interview with the *Spectator* magazine. He was a close ally of Mrs Thatcher.
- 1990 October. The Government entered the ERM.
- 1990 November. Geoffrey Howe's resignation. In his resignation speech Howe blamed Mrs Thatcher's opposition to European integration for the party's divisions and for weakening Britain's influence.

Many of the clashes arose from the conflict between Mrs Thatcher's negative attitude to the EC and those colleagues who believed that Britain's future was to be at the heart of the EC. Mrs Thatcher's view was that decisions should be made on an intergovernmental basis rather than by EC institutions and should involve cooperation between independent states where this was mutually beneficial. Her vision of 'nationalist conservatism and economic liberalism' (Wincott 1992: 16) was boldly expressed in a speech at Bruges in September 1988. She warned that granting powers to the supranational institutions would lead to a 'European super-state'. Her views made her an enthusiastic signatory to the Single European Act (SEA) in December 1985. Critics have noted that the Act also allowed for majority voting and actually strengthened EC institutions as opposed to member states. But it excluded a common foreign policy, exchange rate and monetary policy, and defence, essential features of a sovereign state.

It was during this period that the Labour party moved from its anti-European stance. The Commission under President Jacques Delors was developing a social dimension to accompany the single free market. Labour, with little influence in Westminster and its local government base deprived of much power, was now turning to Europe to compensate. The failure by 1983 of the French government under President Mitterrand to implement socialist measures showed again that the interdependence of national economies ruled out socialism in one country as a practicable project. Europe was Labour's only means of promoting its social and industrial policies. But Mrs Thatcher, having vanquished socialism at home, had no intention of allowing it to enter via Brussels in the form of a European super-state based on more regulations and subsidies.

The *Social Charter* was a divide between the parties. This was an agreement to give common workplace and bargaining rights to employees and was signed in 1989 by all member states, with the exception of Britain. Conservative leaders complained that the Charter would nullify some of the economic gains of recent industrial relations reforms and add to labour costs. This stance was confirmed two years later at Maastricht when John Major negotiated an opt-out for Britain from the Social Charter.

The trend in the 1990s has been for the Commission and Council of Ministers to take more initiatives, together with a strengthening of the role of the European Parliament to overcome the democratic deficit, and more qualified majority voting to speed decision-making. The thrust of these broad changes has been to weaken the national veto, for national parliaments to cede more power, and for the sovereignty of individual states to be diluted.

Maastricht

The heads of the member states' governments decided in 1989 to convene an intergovernment conference on monetary and political union at Maastricht in the Netherlands in December 1991. At Maastricht the leaders signed a treaty in which they agreed to work together towards an 'ever closer union' among the peoples of Europe. The Treaty accepts the continuance of the authority of the Treaty of Rome but extends the European Union's competence into new fields of (1) foreign and defence policy, and (2) immigration, asylum, and policing. These are decided on an intergovernmental basis, so that in these areas the European Court has no jurisdiction and the Commission's role is limited. It was also agreed that qualified majority voting in the Council of Ministers would be extended to these new fields.

In addition, the participants (except for Denmark) agreed a timetable for achieving economic monetary union with a single currency and central bank. This was to be achieved between 1997 and 1999 for those countries meeting the convergence criteria (covering inflation, interest rates, budget deficits, and currency stability). Britain negotiated an opt-out which left it for a future Parliament to decide on the adoption of a single currency. The Social Charter, governing worker–employer relations, was excluded from the draft treaty. Eleven member states agreed under a protocol to accept the Chapter, which Britain did not sign. Britain regards the so-called 'social dimension' a key test of *subsidiarity*, or allowing states to solve problems at the national level where this was appropriate.

John Major and Europe

If Europe was troublesome in the last years of Mrs Thatcher's premiership, it has dogged John Major throughout his. At first he struck a different note from his predecessor by declaring that Britain should be at the heart of Europe. The negotiations at Maastricht of the opt-outs from the Social Charter and monetary union managed to hold the party together before the 1992 general election. But as the EC has developed, questions of national sovereignty have emerged and the Tory party has become more divided.

After the Treaty was signed by the heads of state it was then referred to the individual states for ratification. This proved to be a painful process for the British government, not least because membership of the ERM until September 1992 was increasingly damaging Britain's 'real' economy. The regime of high interest rates, required to keep Britain in the agreed range of exchange rates, damaged exports and added to bankruptcies, unemployment, and mortgage failures. When Britain was finally forced out of the ERM, it was the prelude to sustained economic recovery—of falling inflation, unemployment, and interest rates, and increased growth—all contrary to the disasters that the government had foretold would occur if Britain left the ERM.

The failure of Britain's membership of the ERM only emboldened critics of the whole integrationist thrust in the European Union and weakened the hand of supporters. Among Conservative critics, Lady Thatcher and Lord Tebbit both declared that they would not have signed the Treaty. Tebbit indeed said that support for Maastricht was tantamount to treason. Within Parliament, the process of ratifying the treaty caused spectacular divisions within the Conservative party as well as defeats for the government. The best part of two years was spent in laborious debates and votes as the government eventually achieved ratification of the Maastricht Treaty in the European Communities (Amendment) Bill in June 1993, but only after making it a vote of confidence. The government lost key votes, had to accept hostile amendments and make concessions to party rebels, eventually removing the whip from eight persistent dissenters (effectively expelling them from the parliamentary party), and thereby imperilled its overall majority (Baker *et al.*, 1993, 1994). Ministers spoke with different voices about whether Britain would rejoin the ERM or join a single currency in the foreseeable future. Pro-Europeans like Michael Heseltine and Kenneth Clarke struck a positive note. The Euro-sceptics opposed monetary union on the grounds that it was intolerable for Britain to cede control of economic policy, and two Cabinet ministers, Jonathan Aitken and Michael Portillo, declared that they could not envisage it happening at any time. John Major ruled it out for the foreseeable future and said that it was not an immediate issue.

Maastricht was not the end of John Major's problems. The government then had to agree to a rise in the numbers for qualified majority voting in the expanded EC. Again, this was a bitter blow for the sceptics.

Key institutions of the European Union

Policies and legislation for the European Union are decided by four main institutions, each of which has its separate powers and responsibilities. They are:

- **European Council.** Consists of all member states' heads of government and foreign secretaries as well as the President and Vice-President of the European Commission. It meets two or three times a year to direct strategy and settle key issues. The agreements have no legislative force, but must be translated into legislation on the basis of a proposal from the Commission.
- **The European Parliament.** Elected every five years. In the most recent elections in June 1994 Labour gained 62 seats, Conservatives 18, and other parties 7. The body has increased its powers as a result of Maastricht. It can veto the membership of the Commission and laws covering, for example, the single market, health, education, and environmental protection. Before the Council of Ministers can put new proposals to the Commission these have to be considered by Parliament.
- **Council of Ministers.** Consists of the ministers from equivalent departments in each state. Ministers of Transport, for example, attend when transport issues are discussed. The Council is the supreme legislative authority in the EU, and decides matters by unanimity, or simple majority or qualified majority. As a result of the Maastricht Treaty, the veto has gone. The last takes account of the fact that each member state's vote is 'weighted' according to its population. The four largest states, Germany, France, Italy, and the UK, have 10 votes each, and the total for the fifteen states is 87. A qualified majority is 62, and a 'blocking majority' is 26.
- **European Commission.** Consists of twenty members drawn from member states. Each member state has at least one

BOX 4.3. THE EUROPEAN PARLIAMENT

Membership

567 members

- directly elected since 1979
- 5-year term
- last elections June 1994
- MEPs sit in ideological rather than national groups: British Labour MEPs sit with the Socialist Group, British Conservatives with the European Democratic Group

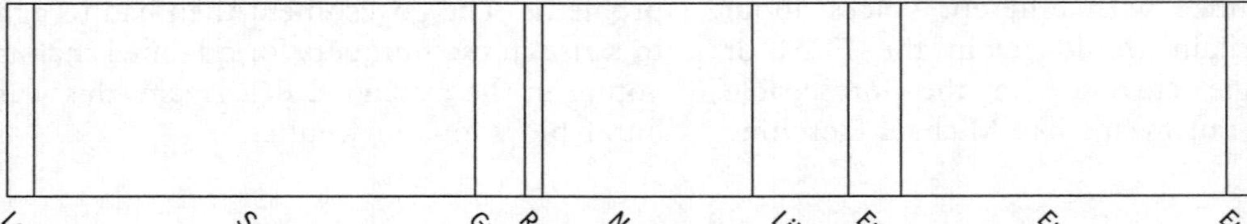

Powers

- few, since it is not the legislature of the EC and has no formal powers over the Council of Ministers
- can suggest ideas for legislation to the Commission
- has the right to be consulted on legislation but can be ignored in all but a few instances
- involved in the budget process and can reject the draft budget
- can dismiss the Commission *en bloc* but has no power to dismiss individual commissioners

Britain elects 87 MEPs

June 1994 results:

Labour	62 seats
Conservatives	18
Liberal Democrats	2
Scottish Nationalist	2
Northern Irish parties	3
Turnout	36.5%

Britain is the only member state not to use some form of PR (except in Northern Ireland). Turnout in Britain for Euro-elections is low compared with that in other member states and compared with turnouts at British general elections (usually around 75%).

BOX 4.4. A SIMPLIFIED MODEL OF KEY STAGES IN THE LEGISLATIVE PROCESS OF THE EUROPEAN UNION

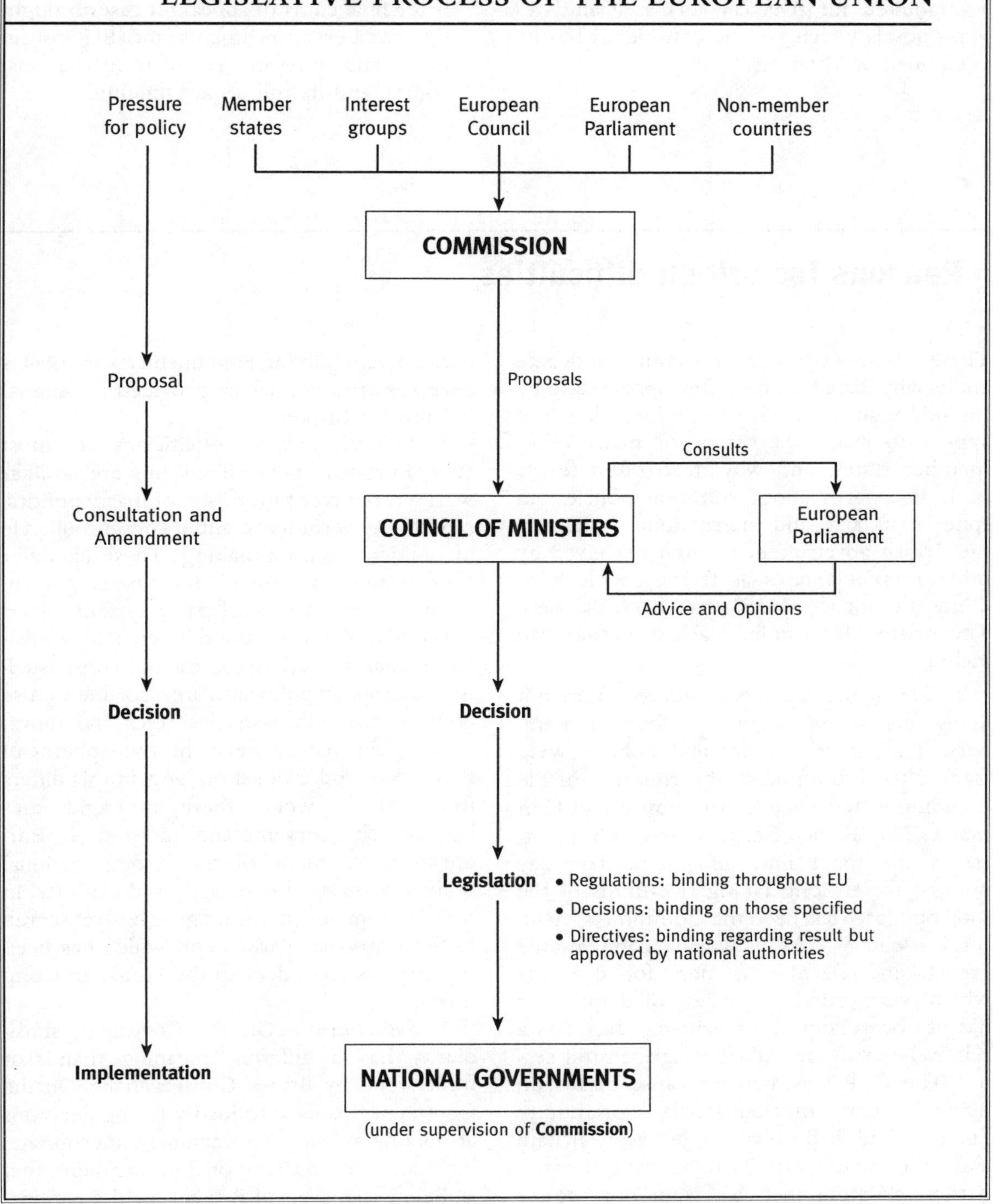

Commissioner, with the five biggest states having two Commissioners each, who are appointed for five-year terms. It initiates proposals which will be considered by the Council of Ministers.

- **European Court of Justice.** Consists of one judge per member state, plus one other. It is the final court of appeal for cases brought by member governments and EU institutions and relevant cases from national courts, and its rulings are binding.

Reasons for British difficulties

There is both a scholarly and a political debate about why Britain has so often appeared to be the odd man out in Europe, why it does not appear to share the vision of many other member states, and why it so often resists or is lukewarm about common policies. In spite of its size and international influence, the British government is often perceived by other member states as trying to block or dilute the impact of initiatives from Brussels. The reasons for Britain's lack of enthusiasm include:

1. *Timing*. Britain has suffered from not being one of the original six founder members. The original treaty and policies were framed to accommodate the concerns of the founding states, particularly France and Germany. The Franco-German axis still determines the main lines of policy. The key policy—for so long taking two-thirds of the total budget—has been the Common Agricultural Policy (CAP), which compensated French and German farmers for products which were grown regardless of demand. It has not been helpful for Britain, which has a relatively small and efficient agricultural sector. The CAP was a major cause of Britain being a net contributor to Community finances and bad relations between Britain and other states. Mrs Thatcher waged many battles, claiming that 'we want our money back'. Eventually at Fontainebleau in 1984 a deal was arranged which provided for annual refunds for Britain.

2. *Political Culture*. Politicians in most West European political systems are familiar with written constitutions, multiparty politics, coalition governments, and a consensual style of political decision-making. These all differ from Britain's system of single-party government and an adversarial party system. Hearl (1994: 520) describes the difference as 'British adversarialism versus continental consensualism'. European politicians are also able to fuse political and administrative roles and movement is allowed between the two spheres of Parliament and civil service. Again this differs from Britain, where there are rigid lines between civil servants and ministers. In European states many of the leading national politicians have also served as Euro-MPs. In Britain no major political figure has yet served in the European Parliament, which has been regarded as secondary to the House of Commons.

3. *Subsidiarity*. On the Continent, subsidiarity has a different meaning than that understood by British Conservatives. On the Continent it is used to justify taking decisions at the lowest level of government appropriate, including local and regional government, that is, below the levels of Brussels or the national

government. Some member states have more experience than Britain of power-sharing and federalism. In the United Kingdom, however, Conservative ministers have used the subsidiarity argument to justify Westminster rule as against Brussels—in other words, to assert the sovereignty of the nation state and decisions by the national government.

Effects of membership on British politics

Some of the effects have been clear cut but are still working their way through in other areas. They include:

1. Constitutional, across many areas:

(i) *A written constitution*. By signing the Treaty of Accession in 1972, Britain incorporated a lengthy written element in her constitution. Like any written constitution the treaties and subsequent laws take precedence over 'ordinary' statute law.

(ii) *An enhanced role for the courts*. The courts of each nation are required under the treaties to give precedence to European law. For example, the government has had to comply with rulings on granting equal pension rights for men and woman and social security entitlements for part-time and full-time workers. The House of Lords, in considering appeals, is also expected to turn to the European Court of Justice for a definitive ruling—which will give precedence to European law. Particularly significant in this respect has been the *Factorame* case (1990), in which the European Court ruled that British courts could suspend the provisions of an act of Parliament if it appeared to conflict with EC law. Not only do British statutes have to be amended to remove any conflict, but plaintiffs are also able to claim damages against national governments which do not give effect to EC laws.

(iii) *Sovereignty of Parliament*. The supremacy of European law is a major limitation on Parliament's sovereignty. Strictly speaking, however, sovereignty could be said to remain intact because Parliament may at any time repudiate membership of the European Union and the obligations consequent on that membership and leave the EU.

(iv) *Referendums*. A nationwide referendum was held for the first time in 1975 on Britain's continued membership of the European Community. It had been called by the Labour government, largely as an answer to its internal divisions. Although the electorate handsomely endorsed the favourable verdict of the House of Commons, one unanswered question is what would have happened if Parliament had voted one way and the electorate another.

(v) *Collective responsibility* of Cabinet. The Labour government suspended the convention for the duration of the 1975 referendum campaign and again in 1979, allowing a free vote over the system for electing the European Parliament. John Major's government in 1994 and 1995 has found great difficulty in keeping to an agreed line about monetary integration and future developments in the EU.

2. *Parliament*. Both Houses of Parliament have created select committees to consider the increasing body of draft European legislation. The House of Commons in 1990 added two standing committees on European legislation to cope with the work. As more decisions are made in the Council of Minis-

ters by majority voting, so British ministers can hardly be 'responsible' to Parliament for Council decisions which they are not able to veto.

3. *Local Government.* A number of local authorities have established direct links with Brussels, for local and regional authorities are the main beneficiaries of the EU's Regional and Social Funds.

4. *Whitehall.* Whitehall is becoming more European-minded: more civil servants are having to travel to Brussels to take account of the European dimension of policies, and an increasing number are being seconded to Brussels as part of their career development. Departments which have a strong European dimension in their work, for example, the Treasury or DTI, have European sections. Ministers are also spending more time negotiating with their ministerial counterparts in the other member states. Although the Cabinet is collectively responsible for policy, much is actually left to the Prime Minister and Foreign Secretary. The European dimension of a department's work also has to be coordinated so that a national response can be made, and this gives some scope for Foreign Office intervention. The views of departments are coordinated through the European Secretariat in the Cabinet Office.

5. *Political Parties.* It is difficult to think of any other issue in post-war politics which has caused more *divisions within as much as between the parties.* Both major parties have always had considerable dissenting wings to whatever line was propounded by the leadership. Yet the parties' leaderships have often been united, for example, in applying in 1970 and in remaining in the ERM in 1992. Only in 1983, when Labour favoured withdrawal, has there been a clear choice over membership between the parties at a general election. Within the Conservative party the pro-Europeans are organized by the Tory Reform Group and Conservative Groups for

BOX 4.5. PRESSURES FOR CHANGE: BRITISH OPTIONS FOR FUTURE DIRECTIONS

The next Intergovernmental Conference is scheduled for 1996. Continuation of the movement towards 'ever closer union' — monetary and political union — seems likely. The options are:

1. **Withdraw** — the least likely option now that the British economic and political system is oriented towards Europe.
2. **Resist further integration** and risk being on an 'outer' track of membership, with reduced influence over future development of the Union but probably being equally affected by any developments.
3. **Accept further integration and monetary union,** recognizing the implications for sovereignty and risking damaging party splits at home, whichever party is in power.
4. **Persuade EU partners of the virtues of widening** (additional member countries) as opposed to deepening (increasing the areas of EU policy competence). But widening would also bring budgetary (many would-be EU members in Eastern and Southern Europe have poorer economies), institutional, and organizational problems, as well as additional language problems.

Europe, and opponents by Fresh Start and Conservative Way Forward. The Liberals and Liberal Democrats have been consistently pro-European.

The parties also elect members to the European Parliament. After the 1994 Euro-elections Britain had 87 of the 587 MEPs, overwhelmingly Labour.

6. *Pressure Groups.* As an increasing number of policies are shaped in Brussels so groups have to lobby there. The Commission may take action if it is able to prove that a policy concern falls within the harmonization provisions of the Single European Act. Interest groups regularly lobby both the European Council and Parliament on proposals. They also have to remain in touch with the 'sponsoring' department in the national government (e.g. British farmers with the Minister of Agriculture) as well as forming cross-national alliances with similar groups in other member states.

Conclusion

In the 1990s Europe has come to dominate high politics in Britain, almost to the extent that Ireland did a century ago. One says 'high' politics because the European issue has troubled the leaderships of the two main political parties more than it has the general public. It has been the single most continuing divisive issue within the Conservative and Labour parties within the past four decades. It nearly caused Parliamentary defeats for the Heath government in 1972 and the Major government in 1993, divided the Labour party during the 1970s and encouraged the SDP breakaway in 1981, and contributed to Mrs Thatcher's loss of Cabinet support and to her eventual downfall in 1990. The failure to achieve membership in 1963 weakened the Macmillan government, the apparent opportunistic change of heart on membership in the early 1970s lost Labour some support, and since 1992 the failure of the ERM and party divisions on Europe have also sapped the authority of John Major. The issue has proved more persistently troublesome in British politics than it has in the domestic politics of any other member state.

For most of the past three decades there has been something of a consensus among élites—media, business, finance, and Whitehall—about the benefits for Britain of membership and of greater integration of the EU. The public, however, has shown little enthusiasm for such goals as a single currency or a united Europe. Turnout in the European elections is about 35 per cent, by far the lowest in the EU, and the British are usually among the least enthusiastic nations in support for Europe-wide initiatives. Surveys also show that European issues are not a salient issue for voters.

For the twenty-three years of her membership Britain has usually resisted or acquiesced in the initiatives from the Commission or the Council of Ministers. This has been true of the creation of monetary and economic union, the Social Chapter, and efforts to forge a common defence and foreign policy. Britain was, however, in the vanguard of the Single European Act.

The future seems to promise more difficulties for Britain's political leaders, particularly in the Conservative party. The Conservative government has been unhappy about the extension of qualified majority voting and moves to even more common policies. It is also divided over whether Britain should sign up for a single European currency, targeted for completion by 1999. The Commission wishes to extend integration even further and disapproves of Britain's existing exemptions or opt-outs from common policies. Britain has welcomed suggestions for the enlargement of the Union, including membership of the former Soviet bloc states in Eastern Europe. Their membership, however, may lead to the EU's agenda being dominated by the concerns of small and medium-sized states and be a force for extending the number of policy areas which will be subject to qualified majority voting.

Summary

- There have been several phases in the history of Britain's relationship with the EU. In the early years (1950s) of the formation of the EEC, Britain remained aloof. By the 1960s Britain was eager to participate in the European economic success story. The 1970s brought entry to the Community, but after Britain's own early coolness, followed by two rebuffs and an eventual 'late' entry during the onset of a recession in the world economy, membership got off to a bad start. In the 1980s the British government, under Mrs Thatcher, was eager to pursue the benefits of a free market, embodied in the SEA, but flatly rejected Europe's proposed intervention in social questions. In the 1990s Britain still has not decided what is to be the nature of its relationship with the EU and continues to be a reluctant partner, showing little enthusiasm for further integration or monetary union.

- In 1994 Parliament ratified the Treaty of European Union (Maastricht) but only after a long and difficult process which exposed clear divisions within the government, stretching to its limits the convention of collective responsibility.

- At present the decision-making powers of the key institutions — the European Council, Parliament, Council of Ministers, Commission, and Court of Justice — reflect the intragovernmental nature of Europe. Controversy about a 'democratic deficit' and the weakening of national vetoes, resulting from an extension of the number of policy areas covered by qualified majority voting, is an indication of a gradually evolving supranationalism.

- Some of the difficulties that have characterized Britain's rather turbulent relationship with Europe can be explained by late entry into the EC, a different political culture, and different understanding of concepts such as subsidiarity and federalism.

- Membership of the EU has made a notable impact on Britain's constitution, particularly in respect of the fundamental doctrine of the Sovereignty of Parliament. British government departments now have to take account of European law and relations with Europe when formulating domestic policy, and the Foreign Office and Cabinet Secretariat have acquired new supervisory powers to coordinate relations with the EU.

Europe has also had an impact on the political system in the shape of elections for the European Parliament; many interest groups now focus their efforts on Brussels. The European perspective also effects the shape of persistent divisions within, as well as between, the parties.

- 'Europe' has presented party leaders with a more difficult task of party management. It has foreshortened the careers of leading politicians and at present poses a serious threat to the unity of the Conservative party.
- Europe poses questions of fundamental importance for Britain's future; it is little wonder, therefore, that it exercises the hearts and minds of politicians. What is somewhat surprising is its apparent failure to engage the interest of the British electorate.

CHRONOLOGY

1952 Six states — France, West Germany, Italy, Luxemburg, Netherlands, and Belgium — form the European Coal and Steel Community (ECSC)

1957 Treaty of Rome establishes European Economic Community (EEC 'Common Market') between the six

1961 UK applies to join EEC, Macmillan — Conservatives

1963 President de Gaulle of France vetoes UK membership

1967 UK reapplies, Wilson — Labour. Again vetoed by de Gaulle

1969 Pompidou replaces de Gaulle

1970 UK reapplies successfully

1972 The Six become Nine — UK, Ireland, and Denmark join. Parliament passes European Communities Act 1972

1973 UK membership takes effect

1974 Labour government replaces Conservatives and commences renegotiation of 'adverse' entry terms

1975 Referendum on continuation of membership. 67.2% yes vote

1979 First direct elections to European Parliament

1984 Margaret Thatcher demands Britain's money back, gets a budget rebate, and negotiates a lower budget contribution. European Parliament election

1986 Nine become Twelve — Greece, Spain, and Portugal join EC. Single European Act (SEA) agreed

1988 Bruges speech: Mrs Thatcher says 'Non! Non! Non!' to European superstate and intervention in social policy

1989 'Delors Report' outlines scheme for EMU. European Parliament elections

1990 UK joins ERM (Oct.)

1991 Maastricht European Council agrees to Treaty of European Union (TEU)

1992 TEU formally signed. European Parliament elections. UK leaves ERM (Sept.)

1994 Parliament passes the TEU Act. In November Euro-rebels vote against increased contribution to the EU budget and have the whip withdrawn

1995 The Twelve become Fifteen — Austria, Finland, and Sweden join EU

ESSAY/DISCUSSION TOPICS

1. To what extent is it fair to describe Britain as an 'awkward partner' in the European Community?
2. How important has the European issue been in the conduct of British politics?
3. What basis is there for arguing that the European Union is becoming a super-state by stealth?
4. Why is Europe a question of great importance to Britain's policy-makers but of great indifference to Britain's electorate?
5. Europe is said to suffer from a 'democratic deficit'; why might attempts to address the problem meet with resistance from the British government?
6. To what extent does the history of the European Union provide support for the view that greater political integration is an inevitable outcome of economic integration?
7. In what ways does British membership of the European Union undermine the sovereignty of Parliament?

RESEARCH EXERCISES

1. What was the *Factorame* case (1990)? Explain its significance for the British constitution.
2. Examine the arguments for and against adopting a single European currency.

FURTHER READING

For an introduction see S. George, *An Awkward Partner: Britain in the European Community* (Oxford: Oxford University Press, 1990) and id. (ed.), *Britain and the European Community: The Politics of Semi-Detachment* (Oxford: Oxford University Press, 1992). On the EC itself, see Neill Nugent, *The Government and Politics of the European Union*, 3rd edn. (London: Macmillan, 1994).

The following may also be consulted:

Carr, Fergus, and Cope, Stephen, 'Implementing Maastricht: The Limits of the European Union', *Talking Politics*, 6/3 (1994).

Jones, A., '1994 Euro-Election Results', *Talking Politics*, 7/1 (1994).

Norton, Philip, 'Europe and the Constitution', *Talking Politics*, 6/3 (1994).

5 POLITICAL RECRUITMENT AND POLITICAL LEADERSHIP

Reader's guide

The sudden deposition of Mrs Thatcher, a serving Prime Minister, from the leadership of the Conservative party in 1990, the dramatic throwing down of the gauntlet by John Major in 1995, the growing presidentialization of British politics reflected in the almost continuous media focus on the party leaders, and the frequently published polls of the party leaders' popularity have all served to increase interest in questions about leadership. How are leaders chosen? From what pool of talent are they drawn? What is their social and educational background? What kind of apprenticeship do they serve? What qualities enable them to attract and retain followers? What makes a good leader? What do leaders need to 'arrive' and survive?

The almost continuous tinkering with candidate selection processes and with methods of electing leaders in both the main parties over the last thirty years testify to the fact that these questions are of more than academic interest. Choosing electable party candidates and selecting leaders who are electoral assets are as crucial to party survival and success as is a programme of popular policies.

This chapter examines political participation and political recruitment in Britain. It analyses the kind of social and educational background from which political leaders are drawn and notes a gradual change towards a strain of meritocratic, professional politicians. It also examines the methods by which the parties chose their leaders and the characteristics of the leaders and prime ministers that emerge from this system. The chapter concludes with some reflections on the unrepresentative nature of political recruits in Britain and on the increased opportunities for patronage offered by the developing 'quangocracy'.

POLITICAL participation includes all those activities which citizens engage in to influence or implement public policies. These range from voting by ordinary citizens to the actions of political leaders. The chapter discusses low-level participation of ordinary citizens as well as recruitment into more politically active roles such as an MP, minister, and party leader.

Political activity

Forms of political participation range from the relatively 'easy', such as voting every few years in a general election or signing a petition, to those which are more demanding, such as taking direct action or standing for election to Parliament. Some behaviour, which is not overtly or consciously political, may also have political consequences. Tax evasion, strikes, demand for higher education, failures to fill teaching posts, and changes in patterns of expenditure all convey signals to policy-makers.

For most people political activity is confined to voting; only a fifth of the public are politically active apart from voting (see Table 5.1). What often distinguishes the political activists from the general public is their greater interest in politics and sense of political efficacy, or belief that they can influence political outcomes.

Small proportions claim to be involved in the actions of a group which tries to influence national or local government. Clearly, the more 'demanding' the activity in terms of motivation, time, and complexity the fewer

Table 5.1. Levels of political participation in the UK

	Electorate (%)
Voting (1992 general election)	77
Contacted local councillor	20
Party member (individuals paying subscriptions)	3
Contacted MP	10
Campaign activist (canvassing at elections)	8
Organized petition	8
Political-group activist	7
Participated in protest march	5
Stood for public office	1

Adapted from *State of the Nation 1995*, Joseph Rowntree Reform Trust, and Parry, Moyser, and Day (1992).

the participants. The most frequent activity, voting, requires less personal motivation and initiative than canvassing, writing letters to newspapers, and making speeches. It is also worth noting that the British have fewer opportunities to vote than in many other Western states. There is no popular election for a head of state (cf. Presidential elections elsewhere), second chamber, or regional government, few referendums, and no opportunity to decide who a party nominates as candidate, as happens in the American party primary.

We may also distinguish different forms of political participation, according to each act's required amount of initiative, extent of influence, degree of conflict raised, and scope of outcome. Voting, for example, scores low on the first three criteria, a political speech scores highly on the first and third, and standing for office scores highly on all four. Most people, though prepared to vote, simply lack the interest to become more politicallly active. Yet it is from those who are interested in politics (less than 20 per cent) and active in local party politics (about 4 per cent) that the political recruits come.

BOX 5.1. EXTENSION OF THE FRANCHISE

1832	Reform Act: middle classes can vote, but there is a property qualification
1887	Reform Act: vote extended to urban skilled workers
1872	Secret Ballot Act: eradication of corrupt practices
1884	Reform Act: vote extended to rural and urban labourers
1918	Representation of the People Act: all men 21 years and over and women 30 years and over can vote
1928	Representation of the People Act: voting age for women reduced to 21 years
1949	Representation of the People Act: abolition of plural voting (businesses, universities)
1969	Representation of the People Act: everyone 18 and over can vote
1985	Representation of the People Act: those living and/or holidaying abroad can apply for a postal vote

Electoral participation

In Britain, the various extensions of the suffrage have now made *the electorate almost identical with the adult population*. The expansion of the electorate has been gradual, extending from 1832 to 1969, when voting rights were granted to 18-year-olds. In 1948, the principle of one person, one vote was finally accepted and the extra vote for business proprietors and university graduates was abolished. The electorate now embraces all British citizens (as well as citizens of all Commonwealth states and of the Republic of Ireland) who meet the residence requirements for voting in a constituency and are on the *electoral register*, which is compiled in October of each year.

Over time the register becomes less accurate of the electorate due to moves, deaths, comings of age of voters, or people simply not being registered, and the final turnout needs to be adjusted upwards if one wants to estimate the proportion of effectively eligible voters who voted. Turnout in 1992 was 77.7 per cent and has ranged between 72 per cent and 84 per cent since 1945. The mean turnout at the twelve post-war general elections has been 77 per cent. (If one adjusts for the inaccuracy of the register, the effective figure is nearer 81 per cent.) This is not high by the standards of other Western democracies, and in no recent general election has it approached the figures for the general elections of 1950 (84 per cent) and 1951 (82.5 per cent).

Compared with non-voters, voters score higher on levels of political information, sense of political effectiveness, commitment to a political party, and concern about issues. The non-voters do not appear to be habitual abstainers. A panel survey found that only 1 per cent claimed not to have voted in all four general elections between 1966 and 1974. Electoral participation is related to political attitudes: the non-voters are less interested in and informed about politics, less attached to a party, and less concerned about the election, compared with those who vote.

Political recruitment

Complaints that politicians are *socially unrepresentative* of the general population have to take account of the characteristics of those who present themselves in the first place. The great majority effectively exclude themselves from becoming MPs simply because they are not very interested in politics. Those who are interested, ambitious, self-confident and prepared to pursue a political career tend to be disproportionately male, middle-aged, and highly educated, as well as middle class. They also tend to be drawn from occupations which are easily combined with a parliamentary career—law, journalism, party and pressure-group activity, teaching, and some business roles. A background in these occupations, university education (particularly Oxford and Cambridge), service in local government, or

BOX 5.2. WOMEN IN BRITISH POLITICS

1900	Suffragettes resort to riots, arson, and hunger strikes to gain women the vote
1918	Women 30 and over get the vote
1919	Sex Disqualification Removal Act: women can stand for Parliament Lady Nancy Astor elected MP for Sutton, a division of Plymouth
1924	Margaret Bondfield becomes Parliamentary Secretary for Labour
1928	Women 21 and over get the vote
1929	Margaret Bondfield becomes Minister for Labour — the first woman Cabinet minister
1958	Life Peerage Act: women allowed to sit in House of Lords
1964	Barbara Castle becomes Minister for Overseas Development — the first post-war woman Cabinet minister
1975	Margaret Thatcher becomes first female party leader
1979	Margaret Thatcher becomes first woman Prime Minister
1989	Neil Kinnock introduces quota system for Shadow Cabinet posts for women: three extra positions are created; Shadow Cabinet is increased from 15 to 18
1990	Labour announces ten-year plan to increase number of Labour women MPs
1992	A record 341 (18%) women candidates stand for the three main parties, of whom a record sixty women (11%) are elected to Parliament Betty Boothroyd becomes the first woman Speaker of the House of Commons
1993	Labour promises a Ministry for Women in a future Labour government. Conference adopts women-only shortlists for some safe seats
1995	Signs that the Labour party intends to abandon all-female shortlists after next election

having parents interested in politics may create or cultivate the expectations, skills, contacts, and commitments which help launch a person into politics. In other words, the social class bias starts early in the recruitment process. In all, the number of such 'eligibles' for a political career may amount to at most 300,000 persons, less than 1 per cent of the electorate.

What is striking in Britain, as in many other countries, is the number of politicians who come from political families, that is, whose parents or relations sat in Parliament or were active in local politics. The Churchills, Chamberlains, Macmillans, and Salisburys spring readily to mind; of the nineteen twentieth-century Prime Ministers six had fathers who sat in Parliament and seven have had children who entered Parliament. In the Parliament elected in 1992, twenty-four MPs had fathers who had been MPs and three had grandfathers who had been MPs.

A more blatant form of social 'misrepresentation' among politicians is that of gender. The proportion of women who are informed about and interested in politics is not far short of men. But few women present themselves as prospective parliamentary

Fig. 5.1. Women as a proportion of parliamentary candidates and MPs

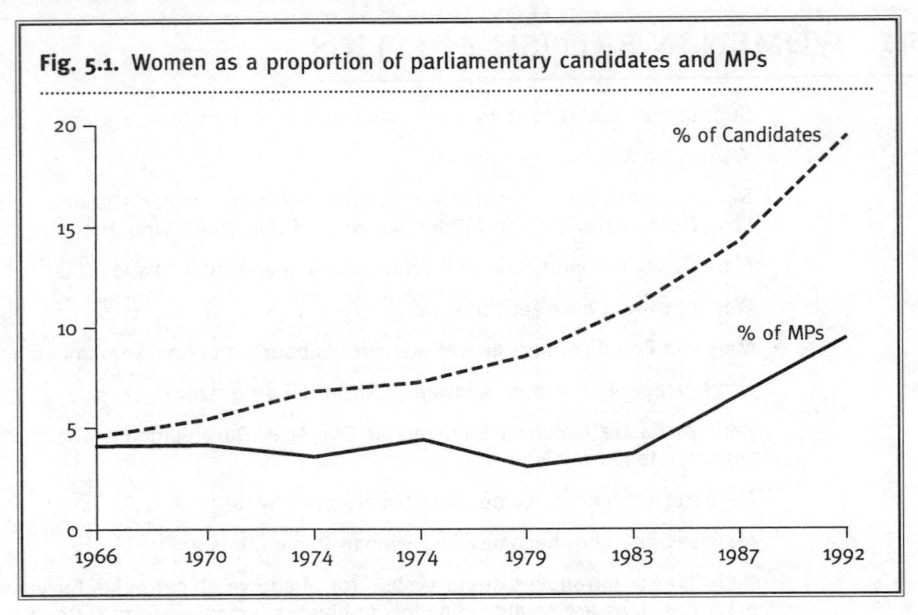

candidates, and when they do, they are usually offered the more hopeless constituencies. The strong pressures in all political parties to recruit more women MPs are now having some effect. In 1992 a record 341 (18 per cent) of the three main parties' candidates were women and in the new House there were a record sixty, or 11 per cent of the total (see Fig. 5.1). The Labour and Liberal Democratic parties now require women to be shortlisted when parliamentary candidates are being selected. Labour has pledged to achieve 50 per cent female membership of the PLP (Parliamentary Labour Party) by the end of this century and, to make progress, has agreed on all-women shortlists in a number of safe seats. In spite of encouragement from Central Office, Conservative associations, jealous of their autonomy, have been slower to adopt women in winnable seats. Women still face handicaps in trying to combine marriage and family responsibilities with a career, as well as prejudice from selectors (though not from voters) (Norris and Lovenduski 1994).

Candidate selection

To become an MP it is virtually necessary to stand on behalf of a major national party. After all, 96 per cent of all MPs since 1945 have sat for the Labour or Conservative parties. Some MPs break with their parties during the lifetime of a Parliament but since 1945 only four have been able to win re-election as an independent at subsequent general elections—three dissident Labour and one dissident Conservative.

There are few formal disqualifications to taking a seat in the House of Commons. Clergymen of the Church of England, Church of Scotland, and Roman Catholic Church may not sit, but Nonconformists may. Lunatics, some classes of criminals, and holders of a Scottish or English peerage are also excluded. The main categories of disqualification, however, are covered in the exclusion of holders of all offices of profit under the Crown, and this covers holders of full-time judicial posts, as well as regular members of the armed forces, police, and permanent Civil Service.

In the Conservative party the Standing Committee on Candidates maintains a list of 'approved' candidates from which constituencies may select. Like the other main British parties it has a panel of senior party officials and MPs who screen the applicants before allowing them on to the approved list of candidates. The panel holds weekend selection conferences in which aspiring candidates take part in interviews, debates, and discussions and are graded for their suitability. When a vacancy is advertised, the local party executive draws up a shortlist of aspirants, who make speeches and answer questions before a general meeting of members, which then chooses the candidate.

In the Labour party, the party headquarters also maintains lists of 'approved' candidates. A Labour candidate has to be nominated by a group affiliated to the local party (e.g. a trade union branch or constituency ward), and the local party's executive selects a shortlist. The candidates appear before the local party's general management committee, which then makes the final choice. Just over half of Labour MPs in the newly elected 1992 Parliament were sponsored by a trade union (143 of 271), which is allowed to contribute up to 80 per cent of a candidate's election expenses and a sum to the constituency party. The financial help that accompanies sponsorship makes such a candidate attractive to hard-pressed local parties. In return for sponsorship, the member is supposed to represent the union's interest in Parliament.

For the minor parties, including most Liberal Democrat seats, the traditional shortage of suitable candidates has meant that the reality has been less one of making a selection than of pressing somebody to stand. The local party's ability to choose depends to a large extent on the 'winnability' of the seat.

The selection of candidates is a jealously guarded prerogative of the constituencies. Although each party headquarters has the 'passive' control of maintaining a list of candidates and approving or vetoing the constituency's choice, endorsement is rarely withheld from a nominated candidate. Conservative Central Office has long exhorted local parties to recruit MPs from a more diverse social background, but with little success. The power of the constituencies to make nominations severely limits any portrayal of the British parties as being centrally controlled. But under Neil Kinnock Labour's ruling body, the National Executive Committee or NEC, intervened to check the influence of far-left groups, and two MPs were expelled from the party in 1991. The NEC has also taken powers to impose candidates on local parties in by-elections.

In 1980 the Labour party adopted a system of *mandatory reselection*, under which MPs had to undergo a formal process of reselection in mid-Parliament (within three years of a previous general election), with possible challenges from other candidates. Critics warned that the change might make MPs more beholden to the pressures of local (and perhaps unrepresentative) activists than to the national party leaders or to Labour voters. In the 1979 and 1983 Parliaments fourteen Labour MPs were denied renomination, and replacements were usually to their Left; fear of dismissal was certainly a factor in the defection of perhaps another half a dozen Labour MPs to the SDP in 1981.

As part of his plan to modernize the Labour party Neil Kinnock set out to reverse the impact of the change. In 1989 selection was removed from the activist-based management committees and given to the local membership as a whole and reselection was henceforth to be carried out only if favoured by a majority of members. Most enforced retirements of MPs are on grounds of old age, personality clashes with local officers, drink problems, or neglect of constituency duties, and only a small minority of MPs are denied renomination by their local parties. Alterations to constituency boundaries, as a result of the recommendations of the Boundary Commissions, are a greater threat. In the majority of seats which are politically 'safe' for a party, the member has virtually been guaranteed tenure for as long as he chooses to remain. In the typical post-war Parliament some two-thirds of members have been returned at the subsequent general election and on average only forty-seven incumbents have been defeated.

It is possible to exaggerate the degree to which local Labour party activists veer to the ideological left of most of their MPs and Conservative activists to the right of most Conservative MPs, and chose candidates for ideological reasons. A study of the motives of selectors shows that party members select candidates for their perceived vote-winning abilities more than any other reason (Norris and Lovenduski 1994: 139). Support for Tony Blair's campaign in 1995 to reform the party's Clause Four was stronger among constituency party members than activists, and among unions which polled members than among union activists.

Members of Parliament

In the nineteenth century positions of political leadership were often closely associated with social and economic leadership. Many MPs were substantial landowners or employers in their constituency. Before 1911 they did not draw a parliamentary salary and were expected to be financially independent. The rise of the Labour party, before 1914, introduced men of working-class backgrounds to the House of Commons. The party now plays a diminishing role in providing the working class with entry to the political élite, and the Conservative party has recruited and advanced MPs from modest social backgrounds.

Before the Second World War Conservative MPs were drawn largely from the upper middle and middle classes and had usually been educated at a public school, often followed by attendance at Oxford or Cambridge. Most Labour MPs, by contrast, came from the ranks of manual workers or trade union officials and, before 1922, few had attended a university. *The parliamentary composition of the parties in the inter-war years institutionalized class differences.*

In recent years, as the Labour party has shed its manual workers and the Conservative party its aristocrats, the social distance between Tory and Labour MPs has narrowed. Both parties now draw their MPs mainly from the ranks of the professions and the graduate middle class. There has been a steady decline in the number of Conservative MPs from such élite public schools as Eton and Harrow, and from Oxbridge.

Labour has a (shrinking) minority from the working class and trade-union sponsorship (originally a device for recruiting people from a working-class background) now supports middle-class university graduates (Criddle 1992). At least a fifth of union-sponsored Labour MPs in the 1980s were university graduates, and a number had no prior experience in the unions.

Educational and occupational backgrounds

The main 'springboard' for entry to politics has been attendance at a university, usually by way of public school and then Oxbridge for a Conservative MP, or by grammar school and then a red-brick university for a Labour MP (though Oxbridge for Labour ministers). Emphasis on academic achievement for political recruitment and promotion has particularly affected the new type of Labour MP. The shift to recruitment by educational merit is shown in Table 5.2, which presents figures on the educational backgrounds of MPs returned in the general elections of 1945, 1959, and 1987. It distinguishes between MPs whose education has been (1) *purely prestigious* (attendance at public school only), (2) *purely meritocratic* (attendance at non-public school and university), (3) *prestigious and meritocratic* (attendance at both public school and university), or (4) *without prestige and merit*. Since the elections of 1945 and 1959 there has been a decline in the numbers of Labour MPs whose education lack both merit and prestige and in the number of Conservatives who have a purely prestigious education. A mark of the rising meritocracy is that between 1945 and 1992 the proportion of graduates on the Conservative side has grown from 59 to 73 per cent, and on the Labour side from 33 to 61 per cent.

Table 5.2. Prestige and merit in the education of British politicians

General election of:	1945	1959	1987
Conservative MPs			
number returned	213	365	376
educational background (%)			
pure prestige	29	24	16
pure merit	3	12	18
prestige and merit	56	48	51
neither	12	16	14
Labour MPs			
number returned	400	258	229
educational background (%)			
pure prestige	4	2	0
pure merit	18	23	43
prestige and merit	15	15	14
neither	61	58	43

Source: Data for 1945 extracted from Mellors (1978). For subsequent elections they are drawn from Nuffield election studies.

Most MPs have turned 40 by the time they first enter Parliament, and politics is a second career for them. There are two distinct types of middle-class members on the Labour and Conservative benches. Labour MPs are often *first-generation middle class*—they come from working- or lower middle-class homes and have made their way via grammar schools and university into the public service sector, as teachers, lecturers, or welfare and social workers. Conservative middle-class MPs, on the other hand, usually come from *established middle-class families*, have been to public schools, and are drawn from the commercial and private sector, being lawyers, accountants, and business executives.

Over the post-war period there have been subtle changes in the occupational backgrounds of MPs in the post-war period. The decline of manual workers on the Labour side has been matched by a fall among Conservative MPs of farmers and those who had been in the armed services. The proportions of Conservative lawyers and company directors and executives (amounting to a half of Tory MPs in 1992) has been pretty steady. A crucial distinction is between *established* professions (law, medicine, accountancy) and *communicating* professions (lecturing, teaching, journalism, and political and interest groups). The proportion of Labour MPs from the established professions has remained steady at between a fifth and a sixth in post-war Parliaments, while the communicators rose from 26 per cent in 1951 to 42.5 per cent in 1992. But the proportion of communicators has also increased on the Tory side, from 6 per cent in 1945 to 22 per cent in 1992 (Norris and Lovenduski 1994: 99).

An increase in the number of professional politicians, who have no paid careers outside of Parliament, has long been favoured by some Parliamentary reformers. Such a step, they claim, would provide MPs with more scope for scrutinizing the executive and helping constituents. An unanticipated consequence of professionalism, however, may be that MPs become more ideological, more remote from the electorate, and perhaps more interested in self-enrichment. As early as 1981 Anthony King warned that the trends to professionalism had 'led to a certain loss of experience, moderation, detachment, balance, ballast even, in the British political system' (King 1981: 205). Indeed the pressures to make an early start in a political career encourage aspiring politicians to choose jobs which will help their political careers (for example, in public relations, political parties, interest groups, or local government) and not to acquire commercial or industrial experience (Kavanagh 1992). In other words, *the new professionalism and pressures to make an early start to a political career have produced a narrowing of pre-parliamentary experience among MPs*. Of the new intake of MPs after the 1992 election, 29 per cent had work experience confined to politics, compared with 14 per cent of all MPs (Riddell 1993). In the 1992 Parliament two-thirds of Labour MPs and nearly half of Conservatives had been involved in local politics (Rush 1994: 573).

These changes in the background of the two main parties' MPs have coincided with a change in their behaviour. MPs appear to be more full-time and ambitious. The old 'knights from shires' on the Conservative benches, or trade union officials on the Labour side, were 'loyalists', there to support the party rather than make speeches or hold office. They are fewer now, and the newer, young MPs have proved to be more rebellious over the past two decades (Norton 1980).

Ministers

An MP's opportunities of becoming a minister depend partly on factors outside his or her control, particularly the frequency with which the party forms the Government and the

Prime Minister's appreciation of his or her talents. The long periods which Labour spent out of government between 1951 and 1964 and since 1979 effectively blighted the ministerial chances of many of the party's experienced MPs. A number gradually left Parliament, while others, having spent some of their best years in Opposition, were too old by 1964 (and will be in the late 1990s) to anticipate a long spell in office in a Labour government. The emergence of a formal Shadow Cabinet and ministers goes some way to helping members to anticipate and prepare for office. In March 1974 and May 1979 respectively, all members of the Labour and Conservative Shadow Cabinets were given Cabinet posts, usually those which they had been 'shadowing'. To a large extent, therefore, a new Prime Minister's 'choice' of Cabinet minsters has been shaped by the pattern of previous appointments made in opposition. Malcolm Punnett's fair judgement is that the Shadow Cabinet is less 'a Shadow of the Past' than 'a Cabinet of the Future' (1973).

A member's short-term ministerial prospects *crucially depend on his standing with the Prime Minister.* Yet in deciding whom to appoint, and to which office, a Prime Minister does not have a completely free hand. Some colleagues may have such a standing in the party or country that it may be inconceivable not to offer them office. A Prime Minister will not only consider the likely competence of a person in a particular ministry but will also want to achieve a politically balanced Cabinet. He or she may consult the Foreign Secretary about who should head Defence, and the Chancellor of the Exchequer about the economic posts. In each case the Foreign Secretary and Chancellor have to work particularly closely with these other departments. A Prime Minister is usually careful to reward some colleagues for their past loyalty and some in the expectation of their future loyalty. The Scottish and Welsh departments obviously require ministers with Scottish and Welsh connections respectively although Conservatives struggle with Welsh. The Law officers (Solicitor-General, Lord Chancellor, Attorney-General) have to be parliamentarians who have a legal background. A Prime Minister may be inclined to placate rivals by including them in the Cabinet; the conventions of secrecy and collective Cabinet responsibility have often restrained critics from mounting a public opposition to a Prime Minister. On the other hand, a politician who is likely to be too outspokenly independent may strain the unity of the Cabinet or the goodwill of the party leader. Mrs Thatcher excluded Mr Heath from her government in 1979 and did not seek to retain Mr Heseltine in 1986. Both were powerful figures but they had dissented from, or were not closely associated with, the thinking of the leader on key issues.

A Prime Minister, hedged in with other political restraints, may feel it necessary to make concessions over ministerial appointments. In forming his administration in 1963, Sir Alec Douglas-Home regarded the participation of his defeated rival, R. A. Butler, as crucial and acceded to the latter's request for the Foreign Office. Mrs Thatcher may have sacked many dissenting Cabinet ministers but for most of her premiership she had a Cabinet dominated by non-Thatcherites. Her supporters on the backbenches were not thought to be of sufficient calibre to merit Cabinet appointment. John Major once spoke unguardedly to a political journalist about his Cabinet dissenters as 'bastards', ministers who supported Mrs Thatcher's brand of conservatism more than his own, and were more sceptical of Europe than the rest of the Cabinet. But Major retained them, partly in the interests of political balance and partly out of the calculation that it was better to have them in the Cabinet than on the backbenches, perhaps causing trouble. According to a study of the subject: 'Bargaining rather than "command" tends to be the characteristic feature of the relationship between Prime Ministers and candidates for office far more frequently than is allowed' (Alderman 1976: 101). At the end of the day, once a Prime Minister has taken account of those MPs deliberately excluded—on grounds of personal choice, inexperience, old age, ill health, incompetence, or personal and political objections—he may be able to offer a post to about half of the 'eligible' MPs.

Ministerial ladder

The ladder of an MP's career advancement from parliamentary private secretary to Minister of State, minister without Cabinet rank, and then membership of the Cabinet is now firmly established. *The track for political advancement in Britain is very narrow.* Since 1945 only a handful of ministers have entered the Cabinet having served less than five years in the House of Commons. If non-parliamentarians are to be given office they must either find a seat in the House of Commons or be given a peerage. In 1940 Ernest Bevin, the General Secretary of the powerful Transport and General Workers' Union, and in 1951 Lord Woolton, a businessman, were given Cabinet posts and are generally considered to have been successful. Two other 'outsiders' appointed to Cabinet posts without previous parliamentary experience were the trade union leader Frank Cousins (1964), and the business leader John Davies (1970). Neither was a success, particularly in parliamentary debates—a skill which might not be regarded as vital as others for managing a large organization like a government department.

In the twentieth century, MPs have, on average, been in the Commons for fourteen years before first entering Cabinet. In Britain, the restriction of recruitment to experienced MPs places a *premium on the skills of party and parliamentary management.* In drawing attention to the largely parliamentary background of ministers, one might also note what this implies about their lack of other sorts of experience. Ministers in many other Western countries are precluded from being MPs and this encourages them to come from a wide variety of backgrounds—the civil service, banking, business management, military life, and local and regional politics (Norris and Lovenduski 1994: 197). In other countries, the ministerial role is more frequently weighted to subject expertise, commitment to policy, and management skills. The comparatively short tenure of British Cabinet ministers in a department (just over two years on average) and the movement of ministers between departments may also reflect a modest assessment of the importance of subject expertise among ministers.

The importance of an early entry to Parliament followed by an apprenticeship through the ministerial ladder is hardly an incentive for distinguished 'outsiders' to enter politics, or for MPs to develop skills which are not prized outside the parliamentary arena. As in many other professions an early start and the accumulation of experience also help. Growing specialization and career pressures have probably sharpened the barrier between a successful career in Parliament and one in many other professions. More than a third of MPs are drawn from a handful of occupations—like law, education, and communications—which permit relatively easy movement into and out of politics. Whether the skills associated with these occupations need to be so heavily represented in Parliament is questionable, in view of the relative absence of others.

A feature of political recruitment in Britain, as in other societies, is that the higher one ascends the political hierarchy, the more socially and educationally exclusive it becomes. Cabinet ministers have usually been of higher social and educational status than those ministers outside the Cabinet and they, in turn, usually stand above backbenchers. In the post-war period, prominent Conservatives like R. A. Butler, Eden, Whitelaw, Soames, Carrington, Home, and Macmillan came from privileged backgrounds. This may be changing with Heath, Thatcher, and Major, all ex-grammar school products, self-made, and professional politicians. Many of Major's Cabinet colleagues, for example, Kenneth Clarke, Tony Newton, David Hunt, Michael Howard, and Gillian Shephard, come from less privileged backgrounds than earlier Conservative ministers; the change in social background has been from the upper to the middle class (Kavanagh 1980, 1992).

In terms of social background, most Labour ministers have fallen into one of three broad groups. The party has always found a place for the *patricians* (MPs from established middle-class professional families, and who attended the public schools and Oxbridge and entered a profession). Such men included Attlee, Dalton, Cripps, Crossman, Gaitskell, Crosland, and, more recently, Blair. This group supplemented the *proletarians* (MPs from working-class families who left school at an early age and then became manual workers, trade union organizers, or lowly clerical workers). Many of the pre-1945 leaders, like Morrison, Bevan, and Bevin, and, today, John Prescott, came from this second background.

The third group, now numerically and politically the most significant, consists of the *meritocrats*. These come from working- or lower middle-class backgrounds and have attended state schools (usually winning scholarships to grammar schools) and gone on to university. The group has been represented by such people as Wilson, Healey, Barbara Castle, Hattersley, Shore, Kaufman, Kinnock, John Smith, Jack Straw, Gordon Brown, and Robin Cook. These are scholarship boys and girls whose parents are from the working class or the lower ranks of the professions and who have become academics, journalists, and lawyers. In contrast to the proletarians their social mobility has been achieved prior to a political career, by dint of going to university. These are the 'new' types of men and women who have identified their career advancement with the Labour party.

Political leadership

Political leaders in Britain emerge in the House of Commons and in a political party. This is an alternative way of stating that Prime Ministers sit in the Commons and usually lead the largest party. Leaders, therefore, have to conform to the expectations of followers. For the most part, Prime Ministers have been political *reconcilers*, representing and responding to

BOX 5.3. POLITICAL MOBILIZERS: TONY BENN AND ENOCH POWELL

TONY BENN, on the left of the Labour party, and ENOCH POWELL, on the right of the Conservative party (until 1974), seem strange bedfellows but they are both anti-marketeers (EU), though for different reasons, and, in 1969, they joined forces, again for different reasons, to defeat proposals to reform the House of Lords. Both have been would-be leaders but both have also been *enfants terribles* of their respective parties; in 1974 Enoch Powell even advised his constituents to vote for Labour candidates. They have added colour to the British political scene and have helped to invigorate debate over a wide range of constitutional and political issues.

Tony Benn

Anthony Wedgewood-Benn, born 1925, son of a Labour hereditary peer, Viscount Stansgate. Educated at Westminster School and Oxford. 1950 elected MP for a Bristol constituency, but in 1960 he inherited his father's title and was disqualified from sitting in the Commons. He pressed for the passing of the 1963 Peerage Act which permitted peers to renounce their title and was elected to Parliament again in 1964 and served in the Labour government as Postmaster General 1964–6 and then Minister for Technology 1966–70. In Opposition, 1970–4, he became spokesman for trade and industry, becoming Secretary of State for Trade and Industry in 1974 and then Energy 1975–9. He lost his seat in the 1983 election but was elected MP for Chesterfield in a by-election in March 1984.

He was elected to the NEC in 1959 and served as party chairman, 1971–2. He made an unsuccessful bid for the Labour party leadership in 1976 after the resignation of Harold Wilson, and again in 1988 when he challenged Neil Kinnock. In 1981 he unsuccessfully contested the deputy leadership.

During the 1970s Tony Benn became a spokesman for the left of the Labour party, calling for Labour to adopt a wholeheartedly socialist programme. He was a prominent anti-marketeer and was instrumental in persuading Harold Wilson to call the 1975 referendum. In the 1980s he helped to form the Campaign Group which succeeded in pushing through measures to democratize the Labour party, including mandatory reselection of MPs and an electoral college for choosing the leader.

After his unsuccessful bid for the deputy leadership and the leadership in the 1980s, he has become marginalized in the drive to create 'new' Labour under Kinnock, Smith, and Blair. In 1994, after thirty-five years, he lost his place on the NEC.

Tony Benn has written several books on British politics, including fascinating and detailed diaries of his years in politics. He has also accumulated an extensive archive which he intends to leave to the nation. Harold Wilson said of Tony Benn 'He immatures with age.'

Enoch Powell

Born 1912, educated at King Edward School, Birmingham and Trinity College, Cambridge, where he became a fellow. He served in the army during World War II, rising from the rank of private in 1940 to brigadier by 1944. After the war he joined the Conservative party, was elected MP for Wolverhampton 1950 and in 1955–7 was junior Minister of Housing. In 1957 he was appointed Financial Secretary to the Treasury, but as a free-market Conservative he disagreed with the government's policy of economic intervention, and resigned. In 1960 he returned to government as

Minister of Health under Macmillan, but resigned again in 1963 in protest at Home's 'emergence' as party leader. He made a bid for the leadership himself under the new electoral system adopted by the Conservatives in 1965, but only received 15 votes (4.9%) and therefore withdrew after the first ballot.

In 1968 his provocative 'rivers of blood' speech warning of racial disharmony if immigration were not curtailed, resulted in Heath sacking him from the Shadow Cabinet. Unrepentant, in 1971 he urged the government to adopt a policy of voluntary repatriation of immigrants.

In February 1974 he resigned his Birmingham seat in protest against Britain's EC membership and advised his constituents to vote Labour. In the October election that year he returned to Parliament as an Ulster Unionist member. He retired after losing his seat in 1983.

Enoch Powell is not only a politician but also a scholar. He is the author of numerous academic and political works, including *The History of Herodotus* (1939); *Common Market: The Case Against* (1971); *Joseph Chamberlain* (1977); *Enoch Powell on 1992* (1989).

the different interests in the party, and being prepared to sacrifice policy goals in the interest of party unity. Leaders are expected to provide not only a sense of direction, but also one which reflects 'the sense' of the party or Cabinet. Robert Peel (Conservative) and Ramsay MacDonald (Labour), whose actions split their parties in 1846 and 1931 respectively, are negative models. Lloyd George, or would-be party leaders like Joseph Chamberlain, Enoch Powell, or Tony Benn, have been *mobilizers*, primarily concerned to achieve policy goals and, usually, radical change. Chamberlain successively split the Liberal and Conservative parties, first with his rejection of Irish Home Rule in 1886 and then with his advocacy of tariff reform in 1903; Lloyd George was usually advocating some special cause—social reform (pre-1914), conscription and armaments (1916), and the attack on unemployment (post-1928)—and regarded party as an instrument to promote it.

The Conservative Enoch Powell made immigration a key issue, and his speech on the subject led to Ted Heath dismissing him from the Conservative Shadow Cabinet in 1968. In the 1970 Parliament he regularly voted against his party and appealed to voters in 1974 to vote Labour, as the best way of repealing Britain's membership of the EC. Mr Benn's attempts to shift the Labour party to the Left on policies, and since 1979 to engineer radical changes in the power structure of the party, divided the Parliamentary Labour Party. Earlier, he was successful in campaigning to allow peers to disclaim their titles (and thus become eligible for membership of the Commons) and for a referendum on membership of the EC. All four men were widely distrusted among parliamentary colleagues, all took their campaigns outside Parliament, and Chamberlain, Lloyd George, and even Powell were not respecters of party lines.

Political style and skill

Routine politics has nourished the reconciling style of leadership, and reduced the scope for the mobilizing style. The other side of this relationship is that only in a crisis, such as war, have the British been responsive to energetic or mobilizing political leaders. But a period of dissatisfaction with the *status quo*, or a sense of national crisis, opens the way for the *mobilizer*. The political and national security crises associated with the two great wars of this century provided opportunities for Lloyd George in 1916 and Winston Churchill in 1940. Their alleged dynamism, ambition, and party 'unreliability' made both men widely distrusted. Lloyd George after 1922 and Winston Churchill before 1939 spent long periods in the political wilderness. The types are not mutually exclusive. A leader may start out as a mobilizer and end up trying to be a conciliator: the contrast in Harold Wilson between 1963–4 and 1974–6, or in Edward Heath between 1970–2 and October 1974, illustrates the development.

There is little doubt that Mrs Thatcher belonged throughout to the camp of the mobilizers. She was impatient with the *status quo* (see her remarks on consensus politics, p. 149) and the traditional style of decision-making by reliance on departmental 'wisdom' and bargaining with the major interests. She divided public opinion and, by her repudiation of policies associated with previous Conservative leaders, her party also. She was critical of many established institutions, including the senior Civil Service, Foreign Office, Bank of England, universities, Church of England, and local government. Her views on capital punishment, immigration, and the trade unions resemble those of the right-wing tabloid press. John Major has shown himself to be a classic conciliator—how successfully remains to be seen (Kavanagh and Seldon 1994).

Political skill in Britain is shown within the parliamentary arena. The approval of fellow politicians has been more important than a popular following in reaching the top of British politics. *Fourteen of the nineteen Prime Ministers of the twentieth century first assumed office without the sanction of a general election.* A Prime Minister or Conservative or Labour party leader will have spent, on average, over twenty years as an MP before appointment. This lengthy apprenticeship provides ample opportunity for learning the skills appropriate to managing Parliament and party, and usually ensures that leaders have those skills. Acceptability to colleagues at the levels of party, Parliament, and Cabinet has attracted and rewarded a personality and style in which qualities of reliability, self-restraint, and trustworthiness have figured prominently. Managing the party and Cabinet places a premium on the avoidance of conflict.

The Conservative party has always vested great authority in its leader. Only as late as 1965 were formal arrangements made for the leader to be elected by MPs, and in 1975 to be re-elected (or possibly dismissed). The culture or character of the Labour party has been rather different. Its values of egalitarianism and collective decision-making, allied to the rejection of both deference and élitism have downgraded the influence of the party leader and the idea of leadership. In Opposition, members of the Shadow Cabinet are elected by the MPs, and these are expected to form the Cabinet if Labour forms a government. Until 1980 the leader and his deputy were elected by the Labour MPs, but since then they are chosen by an electoral college

Table 5.3. Mean public approval of Prime Ministers, 1945–1995 (Gallup Poll)

Period	Prime Minister	% approval
1945–51	Attlee	47
1951–5	Churchill	52
1955–7	Eden	57
1957–63	Macmillan	52
1963–4	Douglas-Home	45
1964–70	Wilson	45
1970–4	Heath	37
1974–6	Wilson	46
1976–9	Callaghan	46
1979–90	Thatcher	39
1990–5[a]	Major	34

[a] to March 1995 (cf. Rose 1995)

Note: All figures are the average (mean) monthly rating.

weighted in favour of extra-parliamentary groups, the trade unions, and local parties.

The Labour party's structure and its tendency to break out into factional fighting has, perhaps paradoxically, often led to the election of leaders whose main political skill is that of reconciliation, acceptable to both the left and the right wings, and concerned to promote and preserve party unity. In Philip Williams's judgement (1982), Labour leaders have more often been 'Stabilizers', who play this unifying role, than 'Pathfinders', who try to move the party's policies in a new direction. Attlee, Wilson, Callaghan, and Foot fit the first type; Hugh Gaitskell, a leader from the right, Kinnock from the left, and Blair fit the second.

A party's choice of leader is electorally important in so far as the person influences its electoral image. Surveys indicate that most voters' comments on the leaders are concerned more with their personal qualities—such as likeability, experience, strength, and intelligence—than with their policies. Mass media coverage of the elections has become more 'Presidential' in the concentration on the party leaders over the secondary leaders. But popular leaders cannot win elections on their own. In 1992, although John Major was widely preferred over Neil Kinnock as Prime Minister, a detailed study concluded that the *net* impact on votes of his popularity was tiny (Crewe and King 1994). *British party leaders are not popular figures with the public*; party loyalties limit their national appeal. According to Gallup polls, since 1957 only Macmillan and Wilson have maintained the support of more than half of the electorate for two years or more. Mr Heath, Mrs Thatcher, Mr Foot, Mr Kinnock, and Mr Major have attracted very low levels of support among the electorate. Governments and parties have become unpopular in recent years, and the standing of Conservative and Labour leaders has not escaped from the disenchantment (see Table 5.3).

Eligibility

The fact that party leaders are drawn from the ranks of a party's MPs obviously restricts the list of eligible candidates. Politicians have also usually had long periods in the House of Commons—on average, twenty years—before becoming party leader (see Table 5.4) and have usually held Cabinet office for some time, factors which further reduce the number of 'eligibles'. *The British system tends to produce leaders who are well known to collea-*

Table 5.4. Leaders of the Conservative and Labour parties, 1935–1995

	Length of service as MP before becoming leader (yrs)	Age at election to leadership	Length of service as leader (yrs)
Conservative			
Stanley Baldwin (1923–37)	15	55	14
Neville Chamberlain (1937–40)	19	68	3
Winston Churchill (1940–55)	38	66	15
Sir Anthony Eden (1955–7)	31	58	2
Harold Macmillan (1957–63)	30	62	7
Sir Alec Douglas-Home (1963–5)	15	60	2
Edward Heath (1965–75)	15	49	9½
Margaret Thatcher (1975–90)	15	49	15
John Major (1990–)	11	47	5[a]
Average	21	57	8
Labour			
Clement Attlee (1935–55)	13	48	20
Hugh Gaitskell (1955–63)	10	49	7
Harold Wilson (1963–76)	17	46	13
James Callaghan (1976–80)	31	64	4
Michael Foot (1980–3)	30	67	3
Neil Kinnock (1983–92)	13	41	9
John Smith (1992–4)	22	53	2
Tony Blair (1994–)	11	41	1[a]
Average	19	51	7

[a] As of October 1995

gues, experienced parliamentarians, and skilled party managers. There is a process of sifting and screening among MPs and frontbenchers which provides an informal shortlist of possible leaders. It has also meant, until Labour changed its rules for electing leaders in 1981, that approval among a few hundred MPs rather than support among the party activists was decisive for gaining the leadership.

These comments apply also to the party leaders who become Prime Ministers. Premiers have usually had *experience of one or more of the three leading departments* (Foreign Office, Treasury, and Home Office). Mr Callaghan had experience of all three, Churchill, Macmillan, and Major two, and of the ten Prime Ministers who have held office since 1935 only Mr Attlee, Mr Wilson, Mr Heath, and Mrs Thatcher did not have previous experience in at least one of these departments. *Luck or good fortune* is also extremely important, particularly in the form of rivals' deaths, retirements, or unavailability. It is very doubtful that Winston Churchill would have come to the top but for the outbreak of war in 1939. It was also important in 1940 that he was acceptable to the Labour opposition as leader of a coalition government. Lord Home, a member of the House of Lords in 1963, would not have been able to renounce his peerage and thus become Prime Minister, had not Anthony Wedgwood Benn recently succeeded in his campaign to allow such renunciation.

Leaders usually emerge after a lengthy and varied apprenticeship. They ascend a ministerial hierarchy and over time prove their political and administrative ability to parliamentary colleagues. The elections as party leader of Mrs Thatcher, John Major, Neil Kinnock, and Tony Blair are therefore rather surprising in light of this historical review.

Mrs Thatcher first entered Parliament in 1959, aged 34, and was a junior minister between 1961 and 1964. She held the Cabinet post of Education in the 1970–4 Cabinet of Ted Heath. When she decided to stand against Mr Heath for the party leadership in 1975 she had been an MP for 'only' sixteen years, her ministerial experience had been confined to a relatively minor department, and she had not been a senior minister in Mr Heath's Cabinet. She had little support among former ministers. She defeated him on the first ballot and on the second she gained enough votes to win an outright victory. She was elected primarily because she was the only substantial politician prepared to stand against Mr Heath on the first ballot in 1975; the many MPs who wanted a change of leader had to vote for her as the instrument of removing Mr Heath.

John Major entered Parliament in 1979 and, following spells in departments as junior minister, joined the Cabinet in 1987 as Chief Secretary to the Treasury. Gaining Cabinet rank after only eight years was an achievement in itself. He held this post for two years and then attained two of the top posts, in quick succession. He became Foreign Secretary when Sir Geoffrey Howe was demoted in 1989, and held the post for four months. He then became Chancellor of the Exchequer, after the sudden resignation of Nigel Lawson in 1989, and held this post for thirteen months. By this time Major was being talked of as a likely successor to Mrs Thatcher if and when she stepped down. The opportunity came when Mrs Thatcher was challenged by Heseltine and failed by four votes to be re-elected as leader in November 1990. John Major and Douglas Hurd entered a second ballot, and ended up with the most votes. Major presented himself as a unifying candidate and was the only one of the three contenders who had the support of much of the right. His apprenticeship, of eleven years in Parliament and three years in Cabinet, was even shorter than Mrs Thatcher's.

Neil Kinnock entered the House of Commons in 1970. He was briefly a parliamentary private secretary in 1974–5, and served as Labour's spokesman on education between 1979 and 1983 when the party was in Opposition. When Michael Foot resigned the leadership in 1983, Mr Kinnock, aged 41, was easily elected. He was helped by the party's desire for a younger leader (so weakening the chances of established figures in their 50s and 60s) and by the defection of other contenders to the SDP. But it was remarkable that Mr Kinnock became leader without ever having held a Cabinet post.

Tony Blair was elected an MP in 1983. He entered the Shadow Cabinet in 1988 and was quickly identified as one of Labour's coming young men. His prospects improved even further when Neil Kinnock resigned after the 1992 election and other senior figures like Gerald Kaufman and Roy Hattersley declined to stand or serve on the front bench. The new leader, John Smith, gave Blair a key post as Shadow Home Secretary. The sudden death of John Smith in 1994 opened the field for a younger generation. Blair and Gordon Brown were leading figures from the centre-right and once Brown stood down it was no surprise that Blair won a landslide victory.

The Thatcher, Kinnock, Major, and Blair records show that younger and less experienced politicians have been emerging in the top posts. Yet accident and good luck continue to be important factors: the decision of senior figures not to stand against Ted Heath and the changes in the rules to allow a leadership challenge (Thatcher); the decision again of senior figures not to stand for the leadership

BOX 5.4. ELECTING A LEADER

<table>
<tr><th></th><th>CONSERVATIVE PARTY</th><th>LABOUR PARTY</th></tr>
<tr><td>The Challenge</td><td>Within three months of the opening of a new parliament or fourteen days of the start of a new session
or the leader can resign, e.g. John Major
Challenger needs backing of 10% of Conservative MPs.</td><td>Contest at annual conference
or under special circumstances as arranged by NEC, e.g. after John Smith's death.
Candidates need backing of 12.5% of PLP or 20% of PLP if challenging an incumbent.</td></tr>
<tr><td>The Constituency</td><td>Only Conservative party MPs vote for the leader but they are expected to seek the views of constituency parties, and party leaders are expected to canvass Conservative peers.</td><td>The leader is elected by an electoral college made up in equal thirds of:
• MPs (PLPs and MEPs)
• constituency parties
• trade unions
Postal ballot.</td></tr>
<tr><td>The Process</td><td>1st Ballot: The winner needs an overall majority of those eligible to vote plus a 15% lead over the nearest rival.
If none achieves this, a 2nd ballot is taken.
2nd Ballot: New candidates can stand. The original candidates stay in or withdraw.
The winner needs an overall majority.
If not achieved a 3rd ballot is taken.
3rd Ballot: Only top two names from 2nd ballot; if a tie, 4th ballot, etc.</td><td>A preferential voting system is used.
To win outright the candidate needs over 50% as an average percentage of votes across three sections of the electoral college. If this is not achieved on the first count, then the bottom candidate is eliminated and the second choices on those papers which had him/her first are distributed to other candidates.</td></tr>
<tr><td>The Example</td><td>In June 1995 John Major challenged his critics to 'put up or shut up' and resigned the leadership.
1st Ballot: Of 329 Conservative MPs 218 voted for Major, 89 voted for John Redwood, and 20 abstained.</td><td><table><tr><th>Candidates in 1994</th><th>MPs %</th><th>Unions %</th><th>Members %</th><th>Overall %</th></tr><tr><td>Tony Blair</td><td>60.5</td><td>52.3</td><td>58.2</td><td>57.0</td></tr><tr><td>John Prescott</td><td>19.9</td><td>28.4</td><td>24.4</td><td>24.1</td></tr><tr><td>Margaret Beckett</td><td>19.6</td><td>19.3</td><td>17.4</td><td>18.9</td></tr></table></td></tr>
</table>

in 1983, the exit of right wingers to the SDP, and the new system of election (Kinnock); the patronage given by Mrs Thatcher to John Major which promoted him to two key posts and the combination of factors which forced her resignation (John Major); and the sudden death of John Smith (Tony Blair).

Selection rules

By tradition the Conservative leader *'emerged'* after consultation among senior party figures. Because the party was so often in office when a leadership vacancy arose, the monarch's choice of Conservative Prime Minister usually became the party leader. The consultation was a kind of consensus system which tended to favour the person who encountered the least opposition, particularly among more influential colleagues. (It is doubtful that Mrs Thatcher would have 'emerged' by this route.) But the system was discredited by the acrimony surrounding the private soundings which led to the selection, over a number of other declared candidates, of Lord Home to succeed Macmillan in 1963.

Under the new rules adopted in 1965, each MP was allowed to vote for a candidate and a winner emerged on the first ballot only if he or she gained a majority plus a margin of 15 per cent of the total vote over the runner-up; on a second ballot, in which new candidates were allowed, the winner required an overall majority. Ted Heath was elected leader. In 1975 provision was made for an annual election of the leader, as in the Labour party. To be elected on the first ballot the winner had to have an overall majority and a lead of 15 per cent of all MPs *eligible* to vote (compared to those who voted under the old system), and a simple majority in a second or subsequent ballot. In the first election held under the new rules Mr Heath was defeated by Mrs Thatcher, who went on to win a second ballot and the leadership. She in turn was unseated in 1990 when she failed to defeat a challenge by Michael Heseltine by a sufficient margin to secure re-election on the first ballot. The rules were amended again in 1991, and now require the request of 10 per cent of Tory MPs to force a contest.

The new rules have clearly strengthened the role of the backbench MPs in the election of the leader. They have also made the leader vulnerable to attacks and encouraged speculation about the leadership. The 15 per cent hurdle can make it difficult for an incumbent to win on the first ballot, if there is a serious challenge. In June 1995 John Major felt that mutterings about a leadership challenge to him were contributing to the instability of his government, and called an election five months before one was due. He was, in effect, calling for a vote of confidence and won it, by defeating John Redwood on the first ballot. Interestingly, the *present Conservative system is quite similar to the old Labour one.*

Before 1981 the Labour leader was elected exclusively by the party's MPs. The party's rule provided for an exhaustive ballot until one candidate emerged with an overall majority. In 1975 Mr Callaghan won after three ballots, in 1980 Mr Foot after two. Demands by the extra-parliamentary groups of the party that they should have a voice grew in the 1970s, as the left increased its influence.

Most MPs resisted such a change on the grounds that the leader in Parliament should command their confidence and that they were the best judges of the contenders' abilities. Under the new rules, adopted in 1981, the names of candidates, who must be Labour MPs and supported by at least 5 per cent of the PLP, were submitted to a special delegate conference. Voting was divided between MPs (30 per cent), trade unions (40 per cent), and constituency parties (30 per cent) until 1993, when the proportions were divided equally into thirds. The groups constitute an electoral college. Neil Kinnock in 1983 was the first party leader to be chosen in a contested election under the new rules. Following Tony Benn's unsuccessful challenge to Kinnock in 1988, the threshhold for making a challenge was raised, by increasing the number of Labour MPs required to support a candidate to 20 per cent, subsequently reduced to 12.5 per cent. In government, however a challenger to a Labour Prime Minister needs the 12.5 per cent support among MPs, as well as two-thirds of Conference through a card vote. *While the Conservatives have made it easier to challenge the party leader, even when in government, Labour has moved in the opposite direction.* Although the new rules were introduced to constrain the leader there has to date been only one challenge to the leader, in 1988.

Deposition of leaders

The popular image of the Conservative leader is of an authoritarian figure who is backed by loyal and even docile MPs. In fact, since 1937 Conservatives have proved much more willing to change their leaders. In every case, including the voluntary retirements of Baldwin (1937), Chamberlain (1942), Churchill (1955), Eden (1957), Macmillan (1963), and Douglas-Home (1965), the party leader was under some pressure to stand aside before he actually did so or was defeated in a leadership election. In spite of the difficulties which Labour leaders often have in managing their party, all who have retired since 1935 did so in their own good time. There were attempts to remove Mr Attlee in 1947 and Mr Wilson in 1969, and Mr Gaitskell was formally challenged by Mr Wilson in a leadership election in 1960 and Mr Kinnock by Tony Benn in 1988. The cases of Attlee and Wilson involved attempts to unseat a Prime Minister at times when the Labour government was unpopular. Both failed, largely because of lack of agreement on a successor among the plotters.

An important influence on the emergence and style of Conservative party leaders is that the party has enjoyed several prolonged periods in office: most of the years from 1886 to 1905, most of the inter-war years, and from 1951 to 1964 and post-1979. For much of the party's modern history the leader has also been the Prime Minister. Since 1922 only Mr Heath and Mrs Thatcher have assumed the leadership of the party without already being Prime Minister. The ability to win or retain office has been perhaps the major influence on the Conservative leader's security, and the system of annual elections makes the leader vulnerable. The poor electoral records of Mr Balfour, under whom the party lost three general elections (1906 and two in 1910), and Mr Heath, who was on the losing side in three out of four elections (1966 and two in 1974), certainly weakened their positions. In 1990 the fear of a likely defeat in a general election undermined Mrs Thatcher's support among MPs when she was challenged by Mr Heseltine.

Internal disagreement over policy may weaken a leader's security or desire to continue. It is worth repeating that a major task of a party leader, particularly when in opposition, is not so much to develop an alternative and workable line of policy to that of the government, as to produce one that is acceptable to the main factions and tendencies in the party. In the Labour party, bitter divisions over defence, membership of the EC, and economic policy have sometimes made it almost impossible for the leader to make an authoritative pronouncement on policy, particularly in opposition. Because Conservatives have had a special view of Britain's place in the world, many of the disputes within the party have been over foreign and imperial questions (for example, Home Rule for Ireland, Dominion status for India, appeasement of Nazi Germany, entry to the EC, and steps to greater integration in that body).

We may infer what leadership qualities MPs have sought by examining the leaders who have emerged. It is not necessary to be particularly extrovert, sociable, or likeable to become leader. Neither Mr Heath nor Mr Attlee had much small talk, and Mrs Thatcher was abrasive and opinionated. Hugh Berrington's fascinating study, 'The Fiery Chariot: British Prime Ministers and the Search for Love' (1974) observes that Prime Ministers have, contrary to stereotype, frequently been introspective, sensitive, and withdrawn. A large number suffered bereavement in childhood and were unhappy at school. Nor has one had to be particularly skilful on television (neither Mrs Thatcher nor Mr Foot was an effective performer at the time of election), or very popular with the public (neither Mrs Thatcher nor Mr Foot was the most popular candidate for the party leadership, according to the opinion polls). When they vote for a leader MPs are influenced by various motives—friendship, party unity, prospects of election victory, policy, and calculations about career advancement. On the whole, parliamentary leadership has attracted and rewarded a personality and style in which such qualities as perceived reliability, self-restraint, safety, trustworthiness—in short, 'character'—have figured prominently.

Political selectivity

Political recruitment in Britain is selective, as it is in other countries. There are frequent complaints about Parliament's social unrepresentativeness, particularly its male and upper- and middle-class character (see Table 5.5). Yet surveys and election results do not suggest that voters feel that Parliament should be a social microcosm of the nation. Moreover, the views of many MPs may be quite unrepresentative of those with similar occupational backgrounds to their own; after all, MPs have actually left those occupations to enter politics.

If such features as political interest, political

Table 5.5. Background of MPs, councillors, and Cabinet Ministers compared with electorate, 1987 (%)

	Councillors	MPs	Cabinet Ministers	Electorate
Males	83	94	95	48
Middle-aged (35–54)	41	67	75	33
University education	50	65	85	4
Manual workers	33	5	0	50
Professional, administrative, and managerial	41	84	100	16

Source: Butler and Kavanagh (1988), ch. 9.

awareness, and ambition for a political career are not distributed randomly through the population, then the pool of would-be politicians is bound to be unrepresentative. The more significant selectivity is at the earliest stage when people present themselves for a political career. Thereafter, it is the preferences of the local activists who choose candidates and of voters who elect MPs which shape the composition of the political élite. The process of the election itself—by which a majority select a minority, or even one person, to represent them—may alter the status of relationships between the voter and his representative.

Members of other élites are, if anything, even more exclusive in social and educational background. A study of political and economic leaders in eighteen various élite sectors found that half had been educated at public schools and Oxbridge. Similar selectivity is seen among senior army officers, judges, bishops in the Church of England, and directors of the Bank of England and the 'Big Four' clearing banks.

Patronage

Apart from politicians and civil servants (see Ch. 13), there are other posts and positions of political influence, many of which are part-time and unpaid. The number of such posts has increased in the post-war period with the growth of the Welfare State government intervention in the economy and the decline of local government. Appointments to such bodies enlarge the patronage available to ministers. They include full- and part-time members of various boards, committees of inquiry, working parties, Royal Commissions, 11,000 Justices of the Peace, as well as many other agencies which Conservative ministers have set up since 1979, in the wake of the loss of responsibilities by local government.

Often members of such bodies have achieved some prominence in other walks of life—people like Lord Rees-Mogg (a former *Times* editor), or Lord Armstrong of Ilminster (a former Cabinet Secretary), for example, are members of many such committees. The committees have access to Whitehall and their reports may influence the mass media, the climate of opinion, and the policy process. As W. L. Guttsman observed, 'Officially or unofficially they tender advice, suggest lines of policy and criticise others' (1963: 338). Some individuals are appointed as 'independent' persons, others 'stand for' or represent an interest. These are 'The Great and the Good' and are still largely drawn from the upper middle class (Hennessy 1986); compared, however, with those of twenty years ago there are fewer senior lawyers and academics and more businessmen and accountants, a reflection of the long period of Conservative rule. There is also a tendency for retired ministers and senior civil servants to take up appointments in other élite groups. For civil servants, the move is often to academe (particularly Oxbridge), the City, or industry. Ex-Labour ministers usually found refuge on government committees, Royal Commissions, and the boards of nationalized industries, but such opportunities have narrowed. Conservatives move to industry and finance. Interestingly, the traffic is largely one way, from the Government to other sectors. The British tradition of élite cooption—or election being largely a matter of form—is continued.

The growing role of such *quangos* in the past decade has provoked complaints about bodies which are not elected and only weakly accountable to Parliament and the extent of political patronage. In 1994 they were responsible for nearly a third of all central government spending. They have flourished as local government has lost many functions and former state-owned enterprises and institutes have been privatized. One study reveals that in housing, education, development and training, and health there are nearly 5,000 non-elected bodies taking decisions and allocating public funds. Such bodies make important decisions and allocate funds to schools, hospitals, and police as well as universities. They constitute, according to John Stewart (1994), a new 'magistracy'. Conservative (and Labour) ministers have tried to appoint more politically sympathetic members to the bodies. Ministers may use such appointments as rewards for political supporters, notably from companies which have contributed to Tory party funds, and to ensure that government policy is carried out. Such bodies allow ministers to bypass other, less sympathetic, organizations; displace responsibility elsewhere for unpopular policies or decisions on politically sensitive matters; and to claim that they have consulted with or acted on the advice of an outside group of distinguished people.

There is something of a grey area surrounding such appointments. Local magistrates are appointed by the Lord Chancellor, on the advice of local advisory committees whose membership is a closely kept secret. Royal Commissions are formally appointed by the Crown, on the advice of the Prime Minister, and members of tribunals, operating under the Council of Tribunals, are appointed by the department to which the tribunal is attached. It is known that Whitehall departments search for names of potential appointees and a list is kept by the Public Appointments Unit (PAU) based in the Cabinet Office. The Cabinet Office claimed in 1993 that ministers were responsible for 42,600 appointments to public bodies. The less important appointments are made by the PAU but the more important are made by ministers or senior civil servants. The Prime Minister also has considerable powers of patronage. In addition to appointing all ministers and parliamentary private secretaries, he approves all permanent secretaries and has a say in the appointment of Regius professors at Oxford and Cambridge universities, judges, and bishops of the Church of England. Many retiring MPs (particularly Conservative MPs) are able to anticipate a political honour from the Prime Minister as a reward for political services, at least a knighthood and perhaps a peerage.

Summary

- For most British citizens (approximately 80 per cent) political participation is confined to voting and there are relatively few opportunities for this. The remaining 20 per cent become more involved in other kinds of political activities such as local politics and protest marches. It is from this group that political leaders are recruited, but more specifically from about the 4 per cent who actively involve themselves in party politics.
- Political leaders who emerge from this process are disproportionately middle-class, middle-aged, well-educated males, and, with rare exceptions, members of a political party.
- Candidate selection is the prerogative of constituency parties but the central offices maintain lists of approved candidates and frequently try to exert influence in these matters. The Labour party has succeeded in increasing the number of women selected and elected by insisting on women-only shortlists in some safe seats.
- There is a growing convergence in the social and educational backgrounds of MPs in the two main parties. A meritocratic strain of professional politician is emerging.
- There is a clearly defined ladder of career advancement for MPs—from PPS, to junior minister, to cabinet rank. Preferment to ministerial office is dependent upon several variables, the most important being the electoral fortunes of the party, standing with the party leader, popularity or following within the party, and parliamentary experience.
- Both main parties have similar procedures for electing the party leader. Once elected, the leader of the Conservative party seems to have less security of tenure.
- Except in rare circumstances, leaders tend to be reconcilers who have acquired skills of party management during a longish apprenticeship. In addition, leaders who become Prime Minister have usually had experience of one or more of the great offices of state.
- Margaret Thatcher, John Major, Neil Kinnock, and Tony Blair may represent a trend towards a younger, less experienced kind of politician emerging as leader.
- That politicians are unrepresentative of the population at large is an inevitable outcome of the uneven distribution of political interest, political awareness, and political ambition.
- The increasing number of semi-autonomous executive agencies, quangos, etc. has increased the opportunities for political participation and influence. But those selected to serve on such bodies tend to be unrepresentative and often include former politicians and bureaucrats, and party acolytes.
- Politics remains predominantly a middle-class, middle-aged male activity. Institutional arrangements and selection processes help to perpetuate this.

CHRONOLOGY

1955 Churchill retires from the premiership; Eden emerges as the new Conservative party leader and premier

1957 Eden retires in the wake of the Suez debacle and Macmillan emerges as the new leader of the Conservative party and premier, despite expectations that Butler would become leader

1960 Wilson unsuccessfully challenges Gaitskell for leadership of the Labour party

1963 In the wake of the Profumo scandal Macmillan retires from the premiership on health grounds and Home emerges as the new leader and premier
Wilson is elected leader of the Labour party after the sudden death of Gaitskell

1965 The Conservative party adopts a new electoral system for choosing a leader. Heath is elected

1975 After losing three out of four elections Heath agrees to introduce new rules for challenging the Conservative party leader. Mrs Thatcher successfully challenges Heath

1976 Wilson dramatically retires from premiership and James Callaghan is elected as Labour party leader, becoming Prime Minister

1980 The Labour party adopts new rules for mandatory reselection of MPs. Callaghan retires from party leadership and Foot is elected to replace him

1981 The Labour party adopts a new system under which the election of the party leader is no longer confined to the PLP but the electoral college is to include constituency parties and unions

1983 Foot stands down as Labour Party leader; Kinnock is elected by the new system

1988 Benn unsuccessfully challenges Kinnock for party leadership

1989 Anthony Meyer's 'stalking horse' challenge to Mrs Thatcher's leadership of the Conservative party. The Labour party changes rules on mandatory reselection: only required if favoured by a majority of constituency party members

1990 Heseltine challenges Mrs Thatcher and succeeds in unseating her, but John Major is elected leader and becomes Prime Minister. Rules are amended to make challenges dependent upon support of 10 per cent of Conservative MPs

1992 Kinnock resigns the Labour Party leadership in the wake of the 1992 election defeat. John Smith elected leader

1994 Smith dies suddenly. Tony Blair elected leader of the Labour party

1995 Major challenges his critics to 'put up or shut up'. He wins the leadership contest on the first round. Rules regarding challenges are again to be amended

ESSAY/DISCUSSION TOPICS

1. 'Women only' shortlists for safe seats will solve the problem of the under-representation of women in Parliament. Discuss.
2. What evidence exists to substantiate the view that the Conservative party places a premium on loyalty to the party leader?
3. Party leadership in British politics is synonymous with party management. Discuss
4. To what extent are there signs of convergence between the two main parties in respect of the social profile of their MPs, and in the election of and role of their leaders?
5. The representative institution is unrepresentative. Explain this paradox in relation to the House of Commons.

RESEARCH EXERCISES

1. A foreign visitor asks you what an 18-year-old would have to do if he/she has an ambition of one day becoming Prime Minister. Explain the qualities they would need, the steps they would need to take, and the obstacles they might encounter.
2. What is a quango? Choose two examples of quangos in your locality. Explain their functions, who appoints their members, and discover as much as you can about the backgrounds of the people appointed.

FURTHER READING

On mass participation see G. Parry, G. Moyser, and M. Day, *Political Participation and Democracy in Britain* (Cambridge: Cambridge University Press, 1992). On MPs see P. Norris and J. Lovenduski, *Political Recruitment* (Cambridge: Cambridge University Press, 1994). On political leaders see M. Punnett, *Selecting the Party Leaders* (Hemel Hempstead: Harvester Wheatsheaf, 1992).

The following may also be consulted:

Foley, Michael, 'Presidential Politics in Britain', *Talking Politics*, 6/3 (1994).

Jones, Bernard, 'The Labour Leadership Contest 1993', *Talking Politics*, 5/2 (1993).

—— 'The Labour Leadership Contest 1994', *Talking Politics*, 7/2 (1994/5).

Kelly, Richard,'After Margaret: The Conservative Party since 1990', *Talking Politics*, 5/3 (1993).

Norton, Philip, 'A New Breed of MP?', *Politics Review*, Feb. 1994.

6 ELECTIONS AND VOTING

Reader's Guide

Elections play an important part in legitimizing the exercise of government power. It is important, therefore, that the electoral system is accepted as fair by voters and political actors. It is often claimed that Britain's first-past-the-post system has the advantages of producing strong, one-party, accountable government. But it also results in a disproportional relationship between votes and seats gained, and this has been a source of pressure for electoral reform.

The electoral system is a framework of rules defining such issues as voter and candidate eligibility, constituency boundaries and size, and limits on campaign expenditure. When elections take place they yield a wealth of statistical material on voting behaviour. This is supplemented, at and between election times, by opinion polls. This data is of enduring interest to politicians, political scientists, and media pundits, who seek to determine why electors will vote, or have voted, in a particular way. Voting behaviour is influenced by a combination of long-term and short-term factors. Since the 1970s the extent of, and explanations for, partisan- and class-dealignment have been keenly debated topics in this field.

This chapter examines the electoral system and the likely consequences of proposals for reform. It analyses the factors that influence voting behaviour and, finally, considers the role of elections in British democracy.

GENERAL elections serve several political purposes in democracies. In Britain they reinforce *the legitimacy* of a regime in so far as people are disposed to accept a government which has emerged according to established procedures. They also help to choose a government, though the nature of the party and electoral systems determines whether this is done directly by voters or indirectly by elected members of the legislature; where a coalition is formed, then bargaining between the party leaders determines the composition of the Government. Elections also offer cues for the direction of public policy, particularly where parties and candidates stand on programmes. Finally, they provide an institutionalized and non-violent method for resolving political disagreements and changing governments. The removal van that draws up outside 10 Downing Street, the day after a Prime Minister has lost a general election, contrasts with the violence or bickering that accompanies a change of government in some other states. *Open, competitive elections are a peaceful means of achieving political and constitutional change.*

This chapter examines the electoral system and the likely consequences of proposals for reform. It then analyses the factors that influence voting behaviour and, finally, considers the role of elections in British democracy.

The electoral system

Britain and France are alone in Western Europe in not using proportional representation (PR) to elect MPs. In the British first-past-the-post system, the candidate with the most votes, whether or not it is a majority, wins the constituency. Yet there have been a few exceptions in Britain to election by the single-member constituency, one-person-one-vote principles. For example, a form of PR was used for elections for university seats and the Stormont Parliament in Ulster until 1929, and there were some double-member seats until 1950.

An electoral system includes procedures for

BOX 6.1. THE GENERAL ELECTION OF 28 FEBRUARY 1974

In February 1974 Edward Heath, exasperated by his battle with the National Union of Mineworkers and the general industrial unrest provoked by his Industrial Relations Act (1971) and U-turn on incomes policy (1972), called the 'Who Governs Britain?' election.

The electorate seemed as unsure of the answer as the government.

Labour	301	37% of the votes cast
Conservatives	297	38.2% of the votes cast
Liberals	14	19.3% of the votes cast
SNP	7	
Plaid Cymru	2	
Ulster Unionists	11	
SDLP	1	

Four days elapsed during which Mr Heath attempted, unsuccessfully, to cobble together an arrangement with Mr Thorpe and the Liberals. On Monday, 4 March, Harold Wilson entered Downing Street to form the first minority government since 1929.

translating individual votes into seats in the legislature. Systems may be considered on the criteria of *representativeness* (i.e. producing a proportional relationship of seats to votes) and *majoritarianism* (i.e. producing a government majority). At general elections since 1945 the working of the British electoral system, coupled with the dominant two-party system, has always, except for February 1974, produced a majority of seats for one party and one-party government. It has been less successful in producing a House of Commons in which seats are distributed between parties in proportion to their electoral support. The disparity between votes and seats for the individual parties has sometimes been marked. In 1945, for example, Labour won 48 per cent of the votes and 62 per cent of the seats; in 1959 the Conservatives got 49 per cent of the votes and 58 per cent of the seats. More striking was the 1983 general election in which the Conservatives gained 63 per cent of the seats for 42.4 per cent of the votes. There has been a marked bias (in seats) for the winning party and a smaller one for the opposition, although in 1992 it was the second party, Labour, which gained the most from the bias. The two exceptions, when the party with the most votes did not get the most seats, were in 1951 and February 1974 (Labour and Conservative respectively). Similar disproportionality is evident in the handful of other states—India, Canada, New Zealand, and the United States—which use this 'British' first-past-the-post method.

Yet the price of these disproportional outcomes has widely been regarded as acceptable. Proportionalism was often unfavourably associated with a multiplicity of parties, coalitions, and unstable governments. The fact that most other West European states had a form of PR did not recommend itself when one remembered the political instability of inter-war Germany, post-war Italy, and France. Defenders of the British system often argued that if a choice had to be made between the goals of representativeness and majoritarianism, then the latter was more desirable—the

BOX 6.2. ELECTIONS AND THE CONSTITUTION

Britain's electoral system is the product of history and convenience. Most rules governing the franchise (voter eligibility) are contained in statutes and thus form part of the written constitution.

Parliament

The Parliament Act of 1911 reduced the maximum life of Parliament from seven to five years. Within that five years the Prime Minister, by convention of the unwritten constitution, advises the monarch on the exercise of the Royal Prerogative to dissolve Parliament and sets a date for a general election.

During both world wars the maximum life of Parliament was extended by statute—demonstrating that no special procedure is required to change the British constitution.

Voters

The franchise was gradually democratized by a series of Electoral Reform and Representation of the People Acts (see the chronology at the end of this chapter).

Today there is universal adult suffrage: everyone aged 18 and over can vote, except convicted felons, certified lunatics, and peers of the realm. In 1992 there were 43 million people registered to vote.

Candidates

Any British citizen or citizen of the Commonwealth or the Irish Republic

- resident in Britain
- 21 years or over
- eligible and registered to vote
- who has the signatures of 10 electors
- and £500 deposit—forfeited if less than 5% of the vote in the constituency is received

exceptions
- Anglican and Roman Catholic clergy
- judges
- members of the armed forces
- some local government officers

Elections are part of the legitimation process. Citizens recognize the government's right to rule if it was chosen by known and accepted procedures. An Act of Parliament may be disliked and unwelcome but it is recognized as legitimate if it has been passed by a majority of representatives elected in the accepted manner.

The disproportionate nature of the British electoral system does occasionally raise questions of legitimacy, for example, the Poll Tax, introduced by the Conservatives who were elected on only 42.3% of the vote in 1987.

claims of government, as it were, won out over those of representativeness. A system which gave full power to the winning party and condemned the Opposition party to virtual impotence was accepted by the latter because it hoped to gain the full fruits of office at a later election. *The saving graces of the British system were that it produced strong stable government, provided voters with a choice between clear party alternatives, and ensured that a government with a majority could be held accountable at the next election.*

In general elections between 1951 and 1970 the aggregate shares of votes and seats for the two main parties did not diverge too sharply: the Conservative and Labour parties between them gained an average of 92 per cent of votes and 98 per cent of seats. The Liberal party was penalized because its votes were not spatially/geographically concentrated. The system also exaggerated the amount of change: a swing of 1 per cent in votes either way between Conservative and Labour could transfer approximately twenty seats from one party to the other, so adding a difference of forty to any gap in seats between them. In fact, there was little turnover of seats at general elections and the two main parties concentrated on the fifty or so marginal seats which decided the outcome of a general election.

In the 1970s, however, the electoral system ceased to work in the regular ways defended by its supporters, largely because of the rise of the centre party. It was less successful in producing parliamentary majorities. In the two 1974 general elections the outcomes failed to produce a government majority sufficient to last for a full Parliament, and in February 1974 there was a minority government for the first time since 1929. Declining support for the two main parties has meant that one party may gain a majority of seats with less than 40 per cent of the votes (as Labour did in October 1974), and large majorities with just 42 per cent (as the Conservatives did in 1983 and 1987). Finally, the system has operated with more disproportional results. The combined Labour–Conservative share of the votes in the six general elections from 1974 to 1992 slumped to an average of 75 per cent, but their average combined share

BOX 6.3. THE GENERAL ELECTION OF 1983

The British electoral system does not distribute seats proportionally in accordance with votes cast. The 1983 general election provides a particularly glaring example of a disproportional result.

	Votes	% of votes cast	Seats won as % of total
Conservatives	13,012,316	42.4	61.1
Labour	8,456,934	27.6	32.2
Lib/SDP Alliance	7,780,949	25.4	3.5

Labour with 27.6% of the vote won 209 seats

Alliance with 25.4% of the vote won 23 seats

Each Conservative seat 'cost' 33,000 votes

Each Labour seat 'cost' 40,000 votes

Each Alliance seat 'cost' 338,000 votes

of the seats in the House of Commons was hardly affected, at 95 per cent. In 1983 and 1987 the Liberal SDP Alliance gained 23 and 22 seats respectively instead of the 160 plus which its votes would have gained in a more proportional electoral system. Although changes in electoral behaviour and the fragmentation of the party system have reduced Labour–Conservative support among the electorate, the electoral system has protected the two parties' dominance in parliamentary seats. In 1992 there was some recovery of the two-party share, to 78 per cent, the highest since 1979.

Electoral reform

In recent years there has developed an intellectual critique of the working of the British party and electoral system. Liberal party critics of the two-party system have long complained, understandably, about how the electoral system discriminated against them. In the 1970s they were joined by objectors to the adversarial way in which the two disciplined parties operated, and the tendency for the alternation of Labour and Conservative parties in government to produce sharp and unsettling reversals in policies across many areas. Under a proportional system a party with less than 50 per cent of the vote is unlikely to have a majority of seats and will be pressed to compromise its policies and perhaps enter a coalition with another party if it is to form a durable government. Most critics of the adversarial system favour a proportional electoral system because they believe that it will lead to broader based governments, that is, coalitions and the adoption of policies more representative of public opinion. Interestingly, the concept of political stability in this context refers more to continuity of policy than to durability of government. In fact, in the 1960s and 1970s discontinuities of policy occurred as often within the lifetime of a government as following a change of government.

George Brown, Labour Foreign Secretary (1966–8), reflecting on Labour's surprise defeat in the 1970 general election, clearly saw the way in which social change was beginning to work to Labour's disadvantage:

'In another generation I doubt if "traditional" Labour areas will exist in the electoral sense. They are declining now, and being replaced by new housing estates . . . and at every election Labour will have to fight for the loyalties of the people who live there. They will never become permanent Labour voters — it's our job to prevent them from becoming Tory voters. We've got to learn to understand these people. They are working-class sons and daughters who have become office workers, draughtsmen and the like, and they feel that they have emancipated themselves from working-class backgrounds, that they have come up in the world. It's easy to be irritated by this sort of snobbism, but it's better to try to understand it. In terms of physical amenity they *have* come up in the world, mainly as a result of Labour policies and political pressure since the end of the First World War . . .
In a real sense, the Labour Party is suffering now from its own achievement. The outstanding achievement is to have brought about a degree of prosperity reflected in the standard of living of the bulk of the people in Britain far beyond what we thought possible in 1945, and infinitely far beyond anything we ever dreamed of in the 1930s.

George Brown, *In My Way* (Harmondsworth: Penguin, 1972), 261–2

Discussion of electoral reform is not new. Proportional representation nearly gained a parliamentary majority in 1917, after the Speaker's Conference (1916–17) recommended the adoption of proportional representation in urban seats, and did so in the Commons in 1931. Reformers wanted to prevent the possibility of a government, backed by only a minority of voters, pushing through radical measures; they also wanted to give fair representation to smaller parties. As Vernon Bogdanor (1981) notes, proportional representation was advocated for practical purposes—first, a desire to lessen the powers of party; secondly, the urge to protect minorities.

In recent years, pressures have grown for the adoption of a more proportional electoral system. Europe is a force for change. With its present system, Britain is the odd man out in the elections to the European Parliament; all other states use a form of PR. Since 1983, as the Conservatives gained decisive election victories on a minority of voters, so political calculation has led some Labour MPs to join Liberal Democrats to call for PR. Historically, the Conservatives have gained more than Labour from the first-past-the-post system. Parties have a keen interest in electoral systems because the translation of votes into seats affects their political fortunes. To date neither of the Labour or Conservative parties has considered it in their interest to change the system, for under a proportional system it is unlikely that either would achieve a majority of seats. Labour reformers have been pessimistic, until the mid-1990s, about the party's chances of being in government again unless it was a part of a coalition.

Many of the anticipated merits and deficiencies of the adoption of PR in Britain must remain matters of speculation until it has been accomplished (of which there is little prospect in the foreseeable future) and the consequences assessed. In part, the effects would depend on the type of proportional representation system employed. With the *single transferable vote system*, a voter's first preference in multi-member constituencies is transferred if it cannot be used to help elect the first-choice candidate, either because the candidate has sufficient votes for victory or has no chance of election. In the multi-member constituencies (in which the parties nominate an ordered list of the candidates) the personal link between the members and constituents would probably decline since the member would be more dependent on the party machine. The *alternative vote system* retains single-member constituencies but asks voters to indicate their order of preferences for candidates. If no candidate has a majority of first preferences then the second preferences of low-ranked candidates are distributed until one candidate receives over 50 per cent of the votes. The system is not proportional and would have made very little difference to the outcome of the 1992 election. Another system is the *additional member* system, as used in Germany. Some MPs are elected locally, as under the present system, and another half are allocated at regional level so that a party's total number of seats are proportional to its share of the vote. Again, this system gives influence to the party machine, which draws up the list of 'top up' MPs.

If no one party gained 50 per cent of the votes (and seats) then it is likely that either coalition or minority governments would ensue. This need not result in unstable governments; since 1945 the average length of tenure of British governments is about the median for West European governments, most of which have been coalition or minority governments. Coalitions would probably weaken the influence of any one party's manifesto in the inter-party bargaining over the contents of a government's programme and limit a government party's direct responsibility to the electorate for its record.

Party regionalization

The electoral system in Britain has also been a force over time for a sharper geographical division of political representation. North Britain (Scotland and the North of England) has remained solidly Labour at most general elections between 1951 and 1992 (even at times of national swings to the Conservatives) while the South and Midlands have moved sharply to the Conservatives over that time (see Fig. 6.1). Urban and city centre seats have become more Labour, while suburban and rural seats have become more Conservative. Since the 1979 election Labour seats have outnumbered Conservative by 2 to 1 in Scotland, Wales, and the North of England, but Conservative seats have outnumbered Labour by 4 to 1 in the Midlands and South of England. By 1992 only a handful of Labour MPs represented rural seats and few Conservatives represented seats in heavy industrial areas, inner cities, or the urban North. In 1992 there was a modest reversal of this regional trend, as the South was affected by the economic recession that had earlier damaged the North, and turned away from the Conservatives.

John Curtice and Michael Steed (1980) have suggested three reasons for this *growing regionalization of party strength and declining electoral competitiveness*. One has been the impact of third-party (Liberal and/or Nationalist) intervention, which has tended to draw support from the weaker of the two main parties. A second is the growing importance of the local political 'environment' and/or reference groups which dispose voters, regardless of their class, to support the dominant local party. Finally, there have been variations in the rate of change in the mix of social classes between the North and the South. The middle-class migration from the inner-city centres and the relatively slower growth of the middle class in the North and the conurbations have reinforced the geographical differences in the distribution of social classes. One consequence of these trends is that in the 1980s the two main parties were less able to aggregate interests across regions or produce policies that would be equally appealing to different regions (Conservatives to the north, Labour to the south). The likely effects of the adoption of a more proportional electoral system are that the parties' seats would be more evenly distributed across the country than at present and the parties might be encouraged to pay more attention to aggregating interests across regions.

To ensure that there is an approximate equality in the individual weight of votes across the country, impartial Boundary Commissions (a separate one for each of the four nations of the United Kingdom) periodically review and make recommendations (between every ten to fifteen years) about the size of constituencies, with a view to rendering their electorates more equal in size. The Commissions work out a quota for each nation by dividing the total electorate by the total number of seats. In making recommendations they are permitted to make allowances for sparsely populated constituencies and for the sense of a community in an existing seat. They are also required to respect the boundaries of counties and London boroughs. The Commissions have a reputation for political impartiality in submitting recommendations, but political pressures from parties for the retention of politically 'safe' constituencies also play a part in affecting recommendations. Scotland and Wales were guaranteed a minimum number of seats by the 1944 Redistribution of Seats Act, but they have long been

Fig. 6.1. Regional support for the parties after the 1992 general election (% of votes cast)

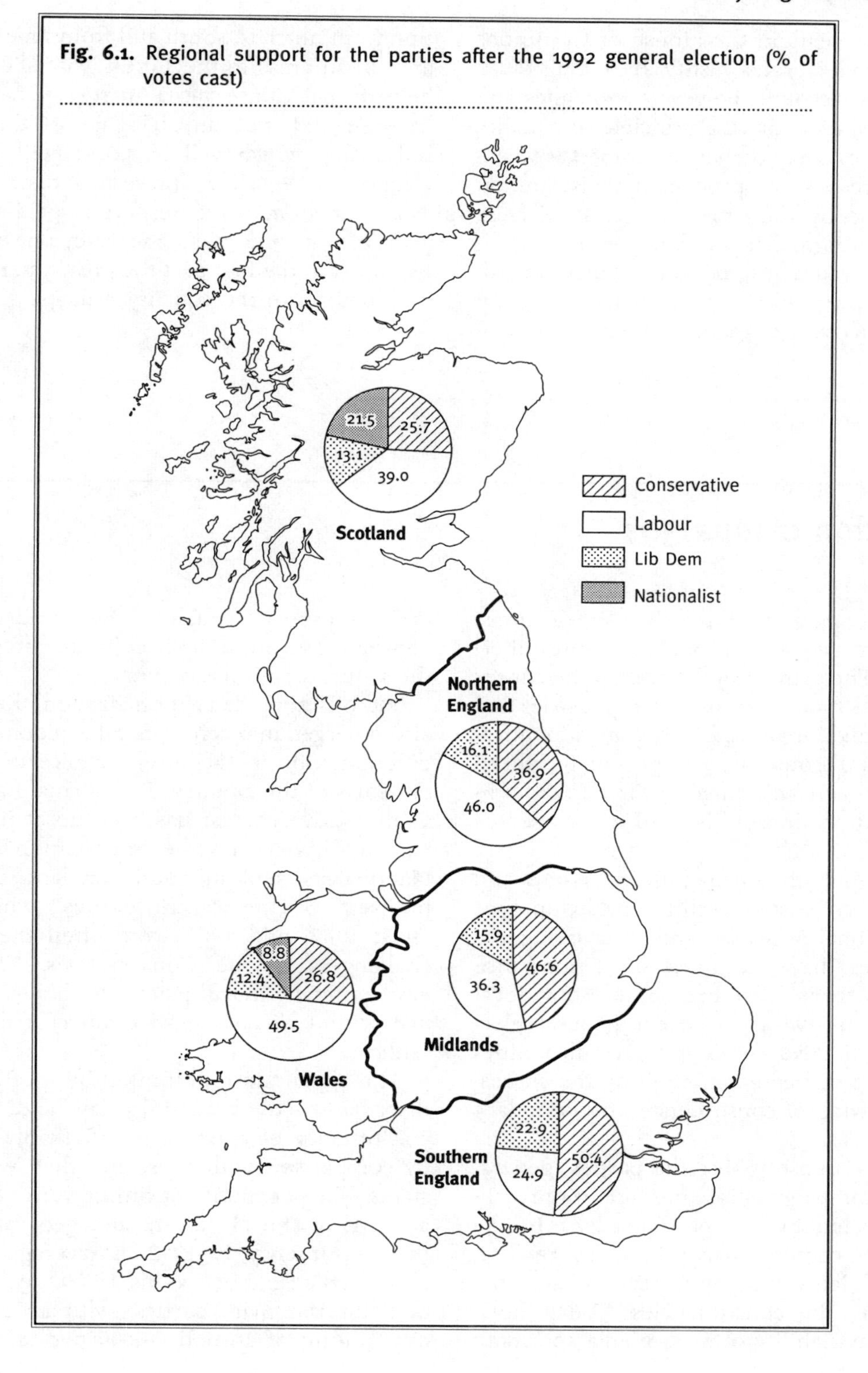

given more seats in the House of Commons than their electorates justify. By taking these factors into account, however, anomalies and inequalities arise and the principle of equality in the size of constituency electorates is weakened. Moreover, the process of redistribution is slow and by the time changes have been made new anomalies will have occurred.

The last redrawing of constituency boundaries was prior to the 1983 general election and took account of the movement of population from north to south and from inner cities and urban areas to the shires. That allocation helped the Conservative party at Labour's expense and will do so again in the next redrawing, which will be completed for an election in 1996. It is interesting that, in this aspect at least, some respect is paid to the principle of 'one man, one vote, one value'. By contrast, the first-past-the-post system pays no attention to the equality principle.

Election campaigns

There is a paradox about British elections: they are more national but also more differentiated. The growth in the twentieth century of the mass media, nation-wide industries and services, and organized, disciplined political parties has produced (except for Northern Ireland) a *'nationalization' of the election campaign*. 'Nationalization' is used in two senses here.

One is that the main political issues and events which shape electoral behaviour are more national in scope, and local influences and issues have declined in importance (although there have been interesting local variations in swing in recent general elections). Distinctive community ties in politics may also have been weakened by the successive redrawing of constituencies by Boundary Commissions.

A second sense is that the parties' general election campaign strategies are more centrally directed by the party leaders. In the nineteenth century, parties had to rely on local candidates and supporters to carry the message in the constituencies. Today, however, the availability of mass media and modern transport enables the leaders to communicate these policies more directly to the voters across the country.

Yet, offsetting this nationalization, there has also emerged in recent general elections more differentiation in the voters' choice in different parts of the country. The normal Labour–Conservative contest has been upset in Scotland and Wales by the intervention of the Nationalists, and in South England by the progress of the centre parties, which in 1983, 1987, and 1992 were often the main challengers to the Conservatives. *There is now less of a national pattern to election results and general elections provide less of a national verdict.*

The expenditure permitted by local candidates is strictly limited by law and British constituency elections are remarkably cheap by comparison with those in other Western states. The permitted maximum varies to take account of the electorate and geography of the constituency. In 1992 the average allowance worked out at around £5,000 per candidate for the main parties, with an average expenditure of around £4,000 per candidate.

There is, however, no limit on expenditure by the party headquarters. In 1992 the three main parties spent a total of nearly £23 million (£11 million Conservative, £10 million Labour, and £2 million Liberal) during the four weeks of the campaign, compared to around £15 million in 1987. In contrast to the position in the United States, the British parties are provided with free broadcasting time on radio and television, and precluded from purchasing advertising time on these media. Another contrast to the United States is that advertising on regional broadcasting media and such other initiatives as the use of telephones for canvassing and conducting private local opinion polls are not legally possible unless the spending limits on local parties are lifted drastically.

There is a ritual element to local campaigns as candidates canvass, deliver leaflets, and address public meetings. But there is some evidence that active local parties can boost voting support for the party (Seyd and Whiteley 1992). In marginal seats a popular candidate and good organization may be worth a few hundred votes, enough to affect the result. Voting by post allows scope for an efficient organization to gain an advantage, for the older the register the more voters are likely to have moved and to be eligible for a postal vote. Half the postal votes go to people permanently registered as infirm or likely to be away on business. In 1992 there were 720,000 valid postal ballots, amounting to 2 per cent of the total vote. Conservatives are usually better than Labour at organizing postal votes and their middle-class supporters are more likely to qualify and apply for them. The party probably owed five seats to these votes in 1992.

There is also evidence that an established MP, because of the media attention he may attract and the constituency services he may provide, is able to build up a personal vote, that is, support himself as an individual rather than as a representative of his party. One may test the personality 'effect' by comparing the 1992 swing in those seats in which the incumbent did not stand with the average swing in the region. Curtice and Steed calculated that the effect of an established MP was to attract between 750 and 1,000 votes which would have otherwise gone to his or her opponent.

Electoral decline of the two-party system

Post-war support for the two main parties divides into two periods, with a sharp break occurring in 1974. The two parties' 'normal' votes for the two periods (1945–70 and 1974–92) are shown in Table 6.1. Each party's 'normal' vote can be compiled from figures on party identification and recent local and national election results (Rose 1992, Crewe 1993: 95). It reflects the party's underlying support, although it can vary due to short-term factors. In the first half of the post-war period, the two main parties gained an average of over 90 per cent of the vote at general elections, and were evenly matched in shares of the vote (45–6 per cent) and in the periods spent in government. But in subsequent general elections the combined average has fallen to 75 per cent, with Labour the main casualty. The Conservative vote has averaged 39.8 per cent, 6.4 per cent higher than Labour's share.

Table 6.1. Party voting, 1945–1970 and 1974–1992 (%)

	Conservative	Labour	Liberal	Other
Mean, 1945–70	45.2	46.1	7.1	1.6
Mean, Feb. 1974–92	40.7	34.4	19.5	5.5
Range, Feb. 1974–92	39.8 ± 4.1	33.4 ± 5.8	19.6 ± 5.8	5.5 ± 1.1
General election result				
1992	41.9	34.4	17.8	5.0

Source: Rose (1992), 453.

In general elections Britain has moved from a *competitive two-party system* to a *dominant one-party system*. But for all the Conservative dominance, one should remember that in the 1983, 1987, and 1992 elections the Conservatives secured office on a lower percentage of the vote than in any previous victory, bar 1922.

There is some evidence that the Labour party has been in the process of long-term decline since at least 1970. The 1992 election was the sixth in succession in which it gained less than 40 per cent of the votes and at each election since 1964 the proportion of Labour identifiers among those entering the electorate for the first time has fallen. The party's traditional electoral advantages over the Conservatives in social class and party identification have weakened, and a number of the policies adopted in the 1980s failed to strike a chord among many of its supporters. The interesting question about a 1996–7 general election is whether a reformed Labour party and a discredited Conservative party can between them reverse recent general election trends.

Minor parties

The main 'third' party in the post-war period has been the Liberal party, now Liberal Democrats. In general elections since 1970 it has gained an average of about 20 per cent of the vote, rising to a quarter in 1983 when it was part of the Alliance with the new Social Democratic Party. It is, however, no nearer to a breakthrough than it was twenty years ago; its 18 per cent share of the vote in 1992 was the same as in October 1974. The party's main electoral support has been in the south-west, but since the 1970s it also made gains in the south-east and in urban areas. The centre-party supporters have constituted a social microcosm of the electorate, being drawn fairly evenly from different classes, and placing the Liberals squarely in the middle of a spectrum of left–right political ideology, equidistant from Labour and Conservative.

The decline of Labour–Conservative dominance has been marked throughout Britain. In England this has been affected by the rise of the centre party, in Scotland and Wales by the rise of nationalism. In Scotland the high point for the Scottish Nationalist Party (SNP) was 1974 (Oct.) when it gained a third of the vote. But as interest in devolution waned, so support for the SNP has fallen from 1974 levels. In Wales, the nationalist (Plaid Cymru) party has had less success and its share of the vote has not exceeded 10 per cent. There has been little popular support for Welsh independence or devolution and the party's support is cultural, mainly confined to rural and Welsh-speaking areas.

In Northern Ireland, the main change has been not so much in voting behaviour as in the line-up of the parties. Until 1972 the dominant Protestant party, the Ulster Unionists, supported the Conservatives at Westminster. In 1972, however, the Conservative government forfeited that support when it suspended the Stormont Parliament, and the Unionist party has since split into different Protestant and Unionist parties. They still won thirteen of the seventeen seats in 1992, with nationalist-backed candidates winning the other four seats. In 1992 the Conservative party, alone of the main British parties, contested some seats in the province.

Electoral behaviour

In an attempt to impose some sort of order on the many factors which constitute forces for stability or change in voting behaviour, we may distinguish between long-term and short-term factors. *Long-term factors* include the social structure and partisanship (or party identification). Because these change only gradually they are forces for continuity in the short term, although, as they change over time, they may provide the basis for an eventual shift in voting behaviour.

But of course electoral behaviour is not stable and in recent elections voters have become more volatile. Some of the change in election results stems from the obvious feature that the composition of the electorate is different from one election to the next. The changes due to deaths and comings of age of voters, and to emigration and immigration, mean that about a tenth of the electorate is new every five years. Nearly a half of those on the electoral register in 1996 or 1997 will not have cast their first ballot before 1974, the last time Labour won a general election.

The changes in party leadership and policy, as well as the events of an election campaign, and the build-up to it—essentially *short-term factors*—may also produce electoral change.

Party identification

Many factors dispose a person to vote as he did at the previous election. *Party loyalty or identification*—the tendency of a voter—to think of himself as a Labour, a Liberal, or a Conservative voter is one important force for stability. As shown in Table 6.2, some four-

Table 6.2. Trends in party identification (%)

	Conservative	Labour	Liberal/SDP
1964	39	42	12
1966	36	45	10
1970	40	43	8
Feb. 1974	35	40	13
Oct. 1974	34	40	14
1979	38	36	12
1983	36	31	17
1987	37	30	16
1992	42	31	12

Source: British Election Study cross-section survey.

fifths of the British electorate identify with a party, though the proportion of electors identifying with the Conservative and Labour parties fell from 81 to 67 per cent from 1964 to 1987, with some recovery in 1992. Over the same period the proportions of 'strong identifiers' with the two parties fell even more. The decline was fairly even across all social groups but was more marked among the young, and each new cohort of voters was less partisan than its predecessors.

Compared with other voters, the strong party identifiers are more likely to agree with their party's policies, cast a vote, participate in politics, and believe in the usefulness of the electoral process. They are 'reliable' supporters and likely to remain loyal, however badly the party is performing. Surveys have indicated that party allegiance tends to harden over time; the reason why many young or first-time voters are so changeable is that they have not 'learnt' party outlooks and are therefore less 'immune' to the influence of contemporary issues and events. Older voters, with a more settled allegiance, are more adept at screening new political issues and personalities, so that existing partisanship is maintained. Understandably, one of the main goals of a party's election campaign is to mobilize traditional supporters.

Class and voting

Social class has been important in studies of voting behaviour largely because of the weakness of other social factors. However, if we again divide post-war general elections into two halves, the decline in class voting is quite marked. In the 1945–70 period the Conservatives regularly gained four-fifths of the middle-class vote and Labour three-fifths of the working-class vote; in the period since, these shares have fallen to less than three-fifths and one-half respectively. In the first period about two-thirds voted with their class-party, compared with less than half in the second period. In 1992, the AB, or middle class, swung from the Conservatives by 5.5 per cent but the C2, or skilled workers, 2.2 per cent to them (see Table 6.3).

It is worth charting the changing electoral importance of social class in the twentieth century. As Butler and Stokes (1969, 1974) argue, it took time for working-class voters to be weaned away from their established loyalties to other parties and a further period for the young to be socialized in Labour-voting homes. They show that the generation that came of age in the 1940s moved decisively to Labour. This generation had the highest level of class alignment and contained the

Table 6.3. Occupational class and party support in 1992 (%)

	AB	C1	C2	DE
Conservative	56	52	38	30
Labour	20	25	41	50
Lib Dem	22	19	17	15
Swing to Labour	4%	1.5%	3.5%	1%

Note: AB = professional and managerial workers, C1 = other white collar workers, DE = unskilled and casual workers. Columns do not total 100 because voting for other parties is not shown.

Source: Denver in *Talking Politics* 5/1 (1992).

highest proportion which regarded politics as the expression of class conflict. The balance of class political loyalties was still moving in Labour's favour up to 1970. As the Conservative-inclined old died and left the electoral register and the Labour-inclined young entered it, so the net effects of deaths, comings of age, and higher birth rates of the working class worked to Labour's advantage. *Labour was the 'natural' majority party if it could mobilize its potential maximum vote.* The Conservatives, to overturn the forces of electoral demography, had to rely on superior performance in office, more attractive leaders and policies, and Labour's shortcomings.

Recent social changes, however, particularly the decline in numbers of manual workers, council house tenants, and trade unionists, and the more middle-class character of society, have eroded Labour's advantages and transformed it into a 'natural' minority. One calculation of the net political effects of social change on the parties' hypothetical 'natural' vote is that between 1964 and 1987 they cost Labour some 4 per cent of the vote and increased the Conservative share by 3 per cent (Heath, Jowell, and Curtice 1991).

There has been a vigorous debate about the exact relationship between social class and voting behaviour, and about how this might explain the sharp decline in Labour's share of the vote in general elections for years. *According to Ivor Crewe (1986), the working class has declined in size and Labour has gained a diminishing share of it.* As the middle class has grown in size so also the Conservatives have gained a smaller share of that vote. *Party dealignment*, or the falling support for the two main parties, is not in doubt. Crewe and others claim that there is also a *class dealignment*, that is, there is less relationship between a person's social class and his or her vote. In recent elections less than half the voters have been voting with their 'natural' class party.

On the other hand, Heath, Jowell, and Curtice (1987) claim that there has been no class dealignment. They agree that class sizes have changed, but claim that *relative* class voting has not; too much attention has been paid to *absolute* levels of party support in the working and middle classes. What matters is whether a party's support has changed *relative* to its support in another; this determines the extent of *relative* class voting. Since 1964, Labour's support has fallen in both middle and working classes and there has therefore been no relative change. They argue that Labour's electoral difficulties have come less from the weakening of class loyalties than from the reduced size of the working class. They suggest that social change (i.e. the smaller size of the working class) explains about half of the decline in Labour's vote since 1964. Politics—choices of leaders, policies, and strategies—explain the other half. Crewe and others, according to this thesis, have confused the decline of Labour with the decline of class voting.

As the two social classes have become more fragmented they have become less politically homogeneous. What Crewe calls the *new working class* (resident in the South, not a union member, employed in the private sector, and as owner-occupier) has been more likely to vote Conservative than Labour. The *traditional working class* (in

Table 6.4. Party choice of the 'old' and 'new' working class in 1992 (%)

	Old working class			New working class		
	North/ Scotland	Council tenant	Union member	South	Owner-occupier	Non-union
Conservative	23	20	23	38	38	34
Labour	52	57	45	36	39	43
Lib Dem	13	12	16	22	18	16
Labour lead	+29	+37	+22	−2	+1	+9

Source: Denver in *Talking Politics*, 5/1 (1992).

opposing categories to those mentioned above) is more likely to vote Labour than Conservative (see Table 6.4). Among white-collar workers there is a division between those in the public and private sector. Less than half of white-collar workers in the public sector voted Conservative in 1987, compared with two-thirds of those in the private sector. Conservative support has been particularly strong among self-employed in both the working and middle classes.

A neo-Marxist perspective emphasizes the division between the 'consumption' and the 'production' classes which cuts across traditional working and middle classes. Thus there can be differences of interest between those employed in the public versus the private sector or those dependent on public services, such as transport and council houses, against those with private cars and owning their own homes. This school emphasizes that internal divisions within the working class are a reason for not voting Labour.

Part of the weakening relationship between vote and class has been a consequence of the growth of third-party support in the 1970s and 1980s. The rise of the Alliance in the 1980s in particular helped to undermine the simple model of a two-class, two-party system. A large part of the disagreement arises from rival definitions of class voting and whether it is measured absolutely or relatively. Yet some features remain clear. *The Conservative party has been a cross-class party,* drawing some 40 per cent of its support from the working class and 60 per cent from the middle class. For all that *Labour* is an unsuccessful working-class party it *still draws two-thirds of its support from that class*. Labour's electoral decline in the past twenty years has been faster than changes in social structure or values. It is worth noting that the Socialist share of the vote has hardly changed over the past two decades in Austria, Scandinavia, France, and Italy, and it has increased in Australia and New Zealand. The good news for Labour from a comparative survey is that such features as wider home ownership, affluence, and the embourgeoisement of the working class are not necessarily electorally adverse. Democratic socialist parties can still thrive in prosperous societies. A good deal of the electorate's disenchantment with the Labour party in the 1970s and 1980s must be explained in political rather than sociological terms.

Issues

Issues have to pass a number of tests if they are to shift voters from an established loyalty. These include:

- public opinion is unevenly divided on the issue
- a voter feels strongly about the issue
- a voter identifies the parties with different positions on the issue and believes that one of them can do something to improve matters

These criteria greatly limit the scope for issues to shape voting. Many issues which deeply concern political activists or commentators may not interest voters at all. Unless voters think an issue is salient, they are not likely to hold considered and stable views on it. Often the typical voter's thinking is of the nature that unemployment, inflation, and strikes are 'a bad thing' and that 'something' should be done about them. A voter may agree with a party's policy on an issue but not believe that a party might be able to make much of a difference.

For a number of years Labour managed—because of class or party loyalty—to attract support among voters who disagreed with many of its policies. The disagreement was particularly sharp in 1983, when only a third of Labour supporters agreed with the party's promises to extend public ownership, increase spending on the social services, and protect existing trade union rights and immunities, and in 1987 when many disagreed with the party's defence policy. Labour's difficulty in the 1970s and early 1980s was that the psychological (party identification), social (class), and ideological (policy) sources of support had all been weakened.

The view that such factors as leadership and party image have become increasingly important in determining election results is illustrated by Denis Healey's reflections on Labour's 1983 election defeat:

'Labour started the election with enormous handicaps. Michael Foot was a kindly and cultured man, as well as a brilliant orator, but he simply did not look like a potential Prime Minister; he failed to command public respect as Leader of the Opposition . . .

' . . . The election was not lost in the three weeks of the campaign but in the three years which preceded it. In that period Labour managed to lose about twenty percentage points in the opinion polls. In that period the Party itself acquired a highly unfavourable public image, based on disunity, extremism, crankiness and general unfitness to government.'

Denis Healey, *The Time of My Life* (Harmondsworth: Penguin, 1990), 499 and 502

In the 1987 and 1992 elections, however, Labour was preferred over the Conservatives on three key issues—jobs, health, and education—among those who regarded the issues as important, while the Conservatives led on the question of which party could deliver prosperity. More significant than particular issues as an influence on the vote is the *party's image*—perceptions of the strength of its leadership, competence, and trustworthiness—and Labour trailed here. Memories of Labour's record in office, particularly the winter of industrial disruption in 1979, impressions of the record cultivated by opponents, and internal party divisions in the early 1980s fostered this sense of distrust, and was still present during the 1992 general election.

Short-term factors

The events during and preceding an election campaign can also change votes. The miners' strike in January 1974, the industrial disputes and disruption in the winter of 1979, the recapture of the Falklands in 1982, the honeymoon period of John Major's first year as Prime Minister (1991) were all given extended coverage by the mass media and influenced voters' perceptions of the parties.

Another short-term factor causing voters to shift allegiance is the state of the economy, or rather the combined effects of the voters' perceptions of how the economy has performed, changes in voters' economic circumstances, and their personal expectations for the future. Survey evidence shows that voters who feel that their economic circumstances have improved are likely to 'reward' the party in government by switching to it, and to defect if they feel that their conditions have worsened. Governments of both parties, aware of the voters' mood, have often tried to manufacture short-term economic 'booms' during the run-up to a general election. Most recent general elections have been held in years which saw living standards rise faster than in non-election years. The perception that the Conservative party was the best party to guard the increased prosperity was significant in influencing voting behaviour in the 1983 and 1987 elections. Voters who believed that they had prospered economically over the lifetime of the Parliament were particularly likely to vote Conservative. In 1992, a crucial number of voters calculated that Labour would do a worse job than the Conservatives in improving their living standards. Policy voting favoured Labour but pocketbook voting favoured the Conservatives (see Table 6.5).

Table 6.5. When voters made up their minds (%)

	1983	1987	1992
Long time ago	78	81	73
2–3 weeks before election	14	12	13
Last few days of campaign	8	7	14

Source: Denver in *Talking Politics*, 5/1 (1992).

In the 1950s and 1960s it was conventional wisdom to claim that election campaigns did not decide elections. Voting loyalties were formed over the long run, were fairly stable, and net changes in voting support during campaigns were virtually self-cancelling. The old party-identification model of voting behaviour assumed that an individual's party identification hardened over time and was passed on to the children: partisanship was more or less self-perpetuating and made for the continuity of the party system. Since the 1970s, however, there has been great electoral volatility between and during election campaigns. The stabilizing factors, particularly party loyalty, have weakened, and give more scope for the influence of short-term factors, including the events associated with the build-up to the election, such as the Falklands war in 1982, tax cuts and prosperity in 1987, and the replacement of Thatcher by Major before 1992. In April 1992 the Conservative party finished 7.6 per cent of the vote ahead of Labour. Three years later it trailed Labour in the opinion polls by over thirty points.

Opinion polls

Opinion polls are almost inseparable from the conduct of general elections in Western states (Kavanagh 1995, Broughton 1995). The mass media, which publish polls, are concerned to make a story about which party is in the lead and is likely to win an election. The political parties also sponsor private polls to help them plan their campaigns. The major polling organizations such as Gallup, NOP, or MORI have been established for some years now and they supply most of their political polling to the mass media. By interviewing a sample of 1,000 or so voters, drawn randomly from the electorate, the polls claim to be correct within a margin of plus or minus 3 per cent in 95 per cent of cases. The polls have a good record in predicting election winners, though they have been spectacularly wrong on three occasions, 1970, February 1974, and 1992. In 1992 the final forecasts of the five major pollsters averaged a 1.3 per cent Labour lead compared with a 7.6 per cent lead in votes for the Conservatives on polling day. The average error in the final forecasts of the five major polls was less than 2 per cent in 1979, 1983, and 1987.

It is frequently alleged that the polls, by making forecasts, influence the election result, for example, by creating a 'bandwagon' of support for the leading party. Since 1964 the number of opinion polls has increased and their findings have become more publicized. More voters are now aware of what the polls are saying and are potentially more open to influence. But there is no evidence that the polls do create a 'bandwagon'. In general elections since 1964, except for 1987, the party ahead in the final survey did worse than predicted on polling day. There has been clear evidence in by-elections, however, that voters use the reported findings of constituency polls to 'punish' one or other of the two main parties, and for the Liberal candidates often to benefit from this tactical voting. And there is no doubt that the findings of polls influence the morale of the politicians and how a campaign is reported by the media (Kavanagh 1995). In 1970 commentators and bookmakers wrote off the possibility of a Conservative election victory because Labour was so far ahead in the polls. In 1992 few anticipated a Conservative majority because all the polls pointed either to a Labour victory or to a hung Parliament with Labour as the largest party. Inevitably, commentators wrote of a brilliant Labour campaign and a poor Conservative one, until the result was known (Kavanagh 1995).

Manifesto and mandate

British political parties are programmatic. They fight general elections on manifestos and promise, if elected, to carry them out. Leaders regard the promises of a manifesto as a 'contract' between the party and its voters, and on the whole parties in government do carry out a great many of the specific legislative promises. Manifestos are also designed to persuade the voters. In addition to making proposals for specific policy areas, the tone of a manifesto is important in colouring a party's image as, for example, 'national', 'sectional', or 'radical'. They are final statements of party policy and within the party may be regarded almost as a draft legislative programme for the new Parliament. When governments introduce policies which are sharply at variance with the promises, or which were not foreshadowed in the manifesto, the Opposition is apt to cry, 'You have not got a mandate!' or 'You have broken your promises!'.

Newly elected ministers invariably choose to regard an election victory as conferring a mandate on their policies. The idea of the electoral mandate has changed over the years and even today it has no agreed meaning. Essentially, it refers to the authority of a government to carry out its election promises—which may or may not be enshrined in a manifesto. In some general elections, a particular issue may have predominated and the main parties' positions may have been so clearly differentiated on it—and recognized as such by many voters—that the outcome of the election is widely regarded as having clear implications for the subsequent direction of policy. Occasionally it has been argued that a government should not introduce a major policy, or one concerning constitutional matters, without putting the issue clearly before the voters. Conservatives argued along these lines over Liberal proposals to reform the House of Lords in 1910–11, and some Labour leaders objected to Mr Heath taking Britain into the EC without first putting it to the test of a general election.

Another and a related meaning of the mandate is that the election enables the voters to indicate their preferences between broad party programmes rather than for a particular policy. Since 1945 the manifestos of the main parties have become longer and more detailed. As central government has extended the range of its activities and responsibilities, and more pressure groups have organized to press a policy on government, so the number of issues on which parties are expected to take a line has grown. Governments now claim to have a mandate for a battery of items in the manifesto, however picayune each may be. As pledges are fulfilled, so the Government spokesman claims that he has kept faith with the electorate, regardless of the popularity of the measure. But voters for a party simply cannot be expected to be aware of all the proposals, let alone agree with them.

Surveys have dented the idea that voting decisions are largely determined by specific policy considerations. They have shown the substantial disparities which may exist, particularly in the case of the Labour party in the 1970s and 1980s, between a voter's preferences on policies and the positions of the favoured party. A voter's decision is based on many considerations, including traditional allegiance and perceptions of the parties' records, competence, and leadership, as well as policy preferences. It is unrealistic to expect most voters to be aware of or, if aware, even agree with all of them. In 1987, for example, many Conservative voters disapproved of the party's proposals on water privatization, health reforms, and the poll tax. Parties present a broad package of policies and voters have no way of voting for or against particular items in a programme at general elections.

What then remains of the idea of the mandate? One can certainly make out a case that it is used by party leaders when they find it expedient to do so. Yet parties in office do carry out the great majority of their election promises when these can be cast in legislative form. A clear policy commitment strengthens the hand of a minister with his civil servants if he wants to initiate a new policy. A government is also likely to be at its strongest *vis-à-vis* public opinion, pressure groups, and the Opposition in the first year or so of office when the memory of its electoral success and 'mandate' is still fresh. Governments

BOX 6.4. PRESSURE FOR CHANGE: ELECTORAL REFORM

1. During the 1970s the system seemed to be failing to deliver its supposed advantages of strong, one-party, accountable government.
2. Unfairness seemed more marked when an increasing share of the poll but not the seats was won by third parties.
3. The Liberal party has understandably been a keen advocate of reform.
4. Europhiles see the electoral system as yet another incompatability between Britain and her European partners.
5. Labour, fearing permanent opposition after four election defeats in a row, began to flirt with the idea of electoral reform, and in 1992 set up the Plant Committee to investigate the matter.

Changes Advocated

1. List System. The whole country is treated as one constituency and the electorate votes for lists of party candidates. Parties are allocated seats in proportion to the votes they gain.

Advantages
- direct proportional relationship between votes and seats
- easier for women and ethnic minorities to get elected

Disadvantages
- forfeits link between MPs and local constituency
- can result in proliferation of small parties and short-lived coalitions, less accountable government

2. Single Transferable vote (favoured by the Lib/Dems). Multi-member constituencies. Electors mark candidates in order of preference. The following formula is used for the quota each candidate requires to be elected:

$$\left(\frac{\text{number of votes cast}}{\text{number of seats +1}}\right) +1$$

Advantages
- retains MP—constituency link
- voters can choose between different candidates of the same party

Disadvantages
- complicated

3. Alternative Member System (favoured by Labour's Plant Committee). This would reduce the number of locally elected MPs by about 50% (and double the size of constituencies). Each voter has two votes—one for a constituency MP elected according to the traditional simple majority system, one for party lists which elect MPs proportionately.

Advantages
- some proportionality but also a constituency link

Disadvantages
- complicated
- larger constituencies
- two classes of MPs, one accountable to party rather than constituents

Chances of Change	**Slim**, for the following reasons:
	1. Any single-party government in a position to introduce change has no incentive to do so.
	2. Labour's interest in change has waned as its election hopes have waxed under Tony Blair.
	3. Low voter interest in the issue.
	4. 1992 saw some revival of the two-party share of the poll.

which break promises, particularly if these are popular, are likely to get into trouble not only with the Opposition but with their own supporters, both inside and outside Parliament—as John Major's government discovered when it sharply increased taxes after its 1992 election triumph. The mandate, in its modern form, means that most voters broadly approve (or prefer in comparison to anything else on offer) a party's general sense of policy direction rather than its specific policies, and agree with its right to carry out policies which it laid before the country. In fact, voters are more interested in manifesto aspirations (such as full employment, lower inflation, better public services and safer neighbourhoods, etc.) than specific policy promises.

Conclusion

Modern political representation assigns an important role to political parties and elections which provide the opportunities for popular choice. In another, less institutional, sense representation also has to do with a high degree of trust and identification between voters and politicians. The government's authority or right to claim obedience from citizens rests in large part on its representativeness (that is, it has emerged through a competitive election).

Many features of representation and the electoral system have become matters of political dispute in recent years. Electoral reform (i.e. the adoption of PR) has become urgent as governments have been elected by a smaller share of the electorate and as the centre-party voting support has increased impressively but with meagre representation in Parliament. In the first seven post-war general elections, between 1956 and 1970, the party in Government was elected by an average of 47.5 per cent of the votes. But in the next six general elections, between 1974 and 1992, this average fell to 41.1 per cent. With more than two major political parties the present electoral system works less predictably and it may produce even more disproportional results than before. In 1983 the Conservatives actually had a smaller share of the vote than in 1979, but they trebled their overall majority in seats to 144. Questions about representation were raised when in 1981 nearly thirty Labour MPs defected to the SDP without any reference to constituents who had returned them as Labour MPs two years earlier. The referendum has been introduced as a supplementary channel of electoral opinion, but with little debate about its constitutional implications. Finally, the 'mandate' concept has been attacked, as more interventionist governments have liberally interpreted their election as a blanket approval for their policies.

The introduction of a more proportional voting system, the shift to multi-partisanship, and the formation of coalitions would affect the conduct of party politics, the role of Parliament, and relations between government and Opposition. What the trends undermine is a dichotomous view of British politics and society—a political dimension of Left and Right which correlates with two political parties and two social classes.

We can certainly point to a dealignment of the Conservative–Labour duopoly over the electorate and more specifically to a decline of the Labour party. Compared with the 1950s and 1960s the Conservative average share of the vote in the six elections (February 1974 to 1992) has dropped by 3 per cent, Labour's by 10 per cent. Since 1979, thanks in part to the electoral system, Britain has had a dominant Conservative party system. With just over 40 per cent of the vote, it has won handsome Parliamentary victories, as Labour and the centre parties have divided the non-Conservative vote.

Summary

- Elections play an important part in the process of legitimization and political communication in Britain.
- Britain's first-past-the-post electoral system has traditionally been defended on the grounds that it produces strong, stable, and accountable governments; voters have a clear choice between party alternatives; and it is accepted and easily understood.
- Since the 1970s this view shows signs of breaking down. Third parties have succeeded in capturing a larger share of the vote. This produced weak, unstable governments in 1974. In the 1980s it produced strong, stable, accountable, one-party government but questions of legitimacy are raised about an electoral system which delivers one party large parliamentary majorities, enabling it to introduce radical and often unpopular policies, based on a minority of the votes cast.
- Pressure for electoral reform is widespread but not strong. Different groups prefer different alternative systems. Each of the alternatives brings advantages but also disadvantages; the main opposition party is at best ambivalent; the electorate is disinterested and, crucially, those in a position to introduce reform, the government of the day, have no incentive to do so.
- The media, the nature of the issues, personalities, and the fact that there are no legal limits on national campaign expenditure have all contributed to a growing nationalization of election campaigns.
- Since the 1970s, partisanship has declined, social classes have changed size and become more differentiated, and class loyalty has weakened. These trends have helped to undermine a dichotomous view of British politics and society—a politics of Left and Right which correlates with two political parties and two social classes.

- Voting behaviour has become more volatile and less predictable. We now have to look at the image, style, and perceived competence of party leaders, and the events which take place at or near election time, in order to explain the shifting patterns of voter preferences.

- Manifestos and the concept of the mandate still play a part in the electoral and legitimation process, but the nature of the system and the complexity of factors that determine voter preferences make it difficult to be precise about the significance of either.

- This changing pattern raises many questions about the next election (1996/1997). Will the Labour party be able to capitalize on any advantages it has in short-term determinants of voting behaviour — personalities, events, the present government's record — to overcome the long-term erosion of its traditional base of support? Will there be a continuation of the trend, discernible in 1992, of a return to the two-party system? Will the case for electoral reform once again be marginalized along with its third-party advocates?

CHRONOLOGY

1832 Reform Act: middle classes can vote, with property qualification

1867 Reform Act: vote extended to urban skilled workers

1872 Secret Ballot Act: eradication of corrupt practices

1884 Reform Act: vote extended to rural and urban labourers

1911 Parliament Act: maximum life of Parliament reduced from 7 to 5 years

1918 Representation of the People Act: votes for men 21 and over and women 30 and over

1928 Representation of the People Act: voting age for women reduced to 21 years

1949 Representation of the People Act: abolition of plural voting (businesses, universities)

1969 Representation of the People Act: votes for 18 and over

1873 Referendum in Northern Ireland on the border question

1975 Referendum on continued membership of the Common Market

1979 Referendums in Scotland and Wales on devolution

1985 Representation of the People Act: those living and/or holidaying abroad can apply for a postal vote

ESSAY/DISCUSSION TOPICS

1. Why has pressure for electoral reform increased during the last twenty years?
2. Have social changes condemned the Labour party to be a permanent opposition party?
3. How democratic is the British electoral system?
4. If the case for electoral reform is self-evident, why has reform not been introduced?
5. In what sense can the 1970s be described as a watershed in British politics?
6. Is social class still a decisive factor in determining the outcome of elections in Britain?

RESEARCH EXERCISES

1. When, why, and with what results have referendums been held in the United Kingdom? Examine the case of those arguing for, and those arguing against, the holding of a referendum on the question of Britain participating in the European single currency.
2. What party do you intend to vote for at the next election? Which of the factors used by political scientists to explain voting behaviour apply in your case? If possible, extend this exercise to others who are willing to participate.

FURTHER READING

For a good introduction see D. Denver, *Elections and Voting Behaviour in Britain* (Oxford: Philip Allan, 1989). For the 1992 General Election see A. Heath, R. Jowell, and J. Curtice (eds.), *Labour's Last Chance* (Aldershot: Dartmouth, 1994), D. Butler and D. Kavanagh, *The British General Election of 1992* (London: Macmillan, 1992), and the special issue of *Parliamentary Affairs* (1992). On new techniques of campaigning, see D. Kavanagh, *Election Campaigning: The New Politics of Marketing* (Oxford: Blackwell, 1995) and M. Scammell, *Designer Politics* (London: Macmillan, 1995).

The following may also be consulted:

Conley, Frank, 'The 1992 General Election: The End of Psephology?', *Talking Politics*, 5/3 (1993).

Connelly, James, 'The Single Transferable Vote', *Talking Politics*, 5/2 (1993).

Crewe, Ivor, 'Voting and the Electorate', in P. Dunleavy *et al.*, *Developments in British Politics 4* (London: Macmillan, 1993).

—— 'Changing Basis of Party Support, 1979–1992', *Politics Review*, Feb. 1993.

Denver, David, 'The 1992 General Election', *Talking Politics*, 5/1 (1992).

Eatwell, Roger, 'Opinions Polls in 1992', *Talking Politics*, 5/2 (1993).

Rose, R., and McAllister, I., *Voters Begin to Choose* (London: Sage, 1986).

7 POLITICAL PARTIES

Reader's guide

The party system in Britain is in constant flux. Survival for any party depends upon its ability to adapt to the changing political climate. Britain's first-past-the-post electoral system favours a two-party system, but, over time, not necessarily the same two parties. In the early years of this century the Conservative and Liberal parties were dominant, but the Labour party was able to capitalize on working-class suffrage, a change in attitudes towards, and expectations of, the state, and a divided Liberal party in order to replace the latter as a dominant party. The current upheavals in the Labour party, both in terms of policy reviews and constitutional and structural change, can be seen as a response to the signals of four consecutive electoral defeats — adapt or decline. The electoral success of the Conservative party throughout the twentieth century is a testament to its adaptability. It has responded flexibly to changes brought by war, economic depression, universal adult suffrage, the rise of the welfare state, and the loss of the empire.

Parties recruit political leaders, they mobilize the mass electorate, and they present voters with a choice of policies. Given the vital role of parties, it is important to examine the nature and effectiveness of the British party system. What kind of party system does Britain have? What has been the impact of origins and history on the present-day organization, structure, and cohesiveness of the parties? What is the secret of the Conservative party's success? What kinds of difficulties are encountered by third parties? Where is the locus of power in British political parties — with the members, the elected representatives, the central organs, or the leadership? How are party unity and discipline maintained? What is meant by party government?

This chapter addresses these questions. It also examines the contemporary debate about whether parties do in fact make any difference to government policy. It concludes by exploring some of the ways in which declining party support and membership is changing the nature of the parties and forcing them to seek alternative ways of performing some of their traditional functions.

THE conduct of democratic politics today is almost inconceivable without organized political parties. By nominating candidates for election, parties play the main role in political recruitment; they also mobilize many thousands of people into political activity, influence the political preferences and outlooks of many voters, and, by aggregating policies into programmes, enable people to make choices at elections about how they will be governed.

British government is frequently described as parliamentary and/or Cabinet government. But we may also speak of it as *party government*. In some ways the most important characteristic of the modern British Parliament and Cabinet is their relationship to the parties. Since 1945, the Cabinet has been formed by members of one party, and for all but a period of thirty-eight months one party has had an overall majority in the House of Commons.

This majoritarianism has combined with other features to provide Britain with *responsible party government* (Birch 1964). The term 'responsible' refers to the ability of the electorate to elect a government and hold it accountable for its record in office at the next election. Such a direct link between the voter and government is less easy to achieve in a political system which has a separation of powers and different parties may control the executive and legislature (as in the United States), or which has coalition governments (as do most West European states). The fact that the British parties stand on programmes or manifestos and that MPs are highly disciplined in the division lobbies also encourages this sense of responsibility.

BOX 7.1. KEY CONCEPTS: POLITICAL PARTIES

Political Parties	Multi-interest groups which organize in order to win government office and enact their policies
Structure	Some parties are cohesive disciplined organizations held together by a strong commitment to an ideology, e.g. the Communist party in the former USSR. Some are loosely cobbled electoral machines, e.g. parties in the USA. British parties are somewhere between these two models: • fairly cohesive and disciplined • with some ideology • recruiting electoral candidates and party leaders from within the party
Functions	• political recruitment • aggregate interests • educate and mobilize the electorate • provide opportunities for participation • make liberal democracy work by taking up policy positions and presenting the electorate with a choice
Types of party system	• **one party dominant**: there are several parties in the system in terms of votes cast and seats in the legislature, but only one party has any real prospect of governing with an overall majority or being the dominant partner in any coalition. Four election victories in succession (1979 to 1992) have raised questions about whether the Conservative party has achieved this kind of dominance in Britain. • **two-party**: there may be several parties but two share dominance, either being capable of gaining an overall majority and of forming the government. This has traditionally been the description applied to the British system. • **multi-party**: more than two parties, none likely to achieve overall majority; therefore coalition governments are the norm.

The party system

Party systems are often defined according to the *number of significant political parties, measured in terms of votes or seats*. Whether Britain is regarded as having a two-party system depends on which criterion is applied. We may talk of a two-party system in Parliament, since the Labour and Conservative parties regularly collect over 90 per cent of the seats. Such features as Cabinet dominance of the Commons, stable government, and the duality of government and opposition have been linked to the two-party system. But among the electorate, support for the two parties is much lower and has, over time, declined. It is also worth noting that the perception of a two-party system being 'normal' in Britain has been based on a short period of recent political history. Before 1914 the Liberal–Conservative hegemony was challenged by the Labour and Irish Nationalist parties. In the 1920s there was effectively a three-party system, as Labour gradually displaced the Liberals as the second party. Since 1970 the Labour–Conservative dominance has been challenged by the growth of political nationalism in Scotland and Wales, of the Liberals in England, and of the Alliance in the 1980s. Thus the term two-party system is more accurate for the 1945–70 period only (Kavanagh 1994*b*).

Similar qualifications apply to the 'norm' of majority one-party government. Between 1910 and 1915 the Liberal government depended on the support of other parties for a working majority in the Commons. There were also two periods of minority Labour government (10 months in 1924 and 1929–31) and some thirteen years of coalition government between 1915 and 1945.

While no party system is immune to change, the *British party system has been more stable than those in most other West European countries*. In every general election since 1918 the two largest parties in votes and seats have been Conservative and Labour. That general election witnessed the replacement of the Liberal party by Labour as the second largest party and the withdrawal of the Irish Nationalists. It was the last realignment of the British party system. Since 1918 the effect of the electoral system (which penalizes third parties, whose vote is dispersed) and the adaptability and opportunism of the two established parties have prevented the breakthrough of a viable third party. Any discussion of the possibility and form of a new realignment should take note of the forces which precipitated change in the past. Historically, a shift in the balance of strength between parties or the emergence of a major new party has been associated with:

- *a split in a dominant party* (in the Liberal party, for example, in 1886 over Home Rule for Ireland which helped the Conservatives, and again in 1916 which helped Labour, and the exit of over a score of Labour MPs to form the Social Democratic Party in 1981)
- *the emergence of a major new issue* (the Irish demand for Home Rule in 1880 and the rise of the Irish Nationalist party, or political nationalism in Scotland since 1974)
- *a change in the composition of the electorate* (the wider enfranchisement of the working class in 1918 which helped the rise of Labour)

It is striking how 'natural' the two-party system has seemed to many British voters. In Chapter 6 we discussed an institutional explanation of the two-party dominance in Parliament—the *first-past-the-post electoral system*. But this influence is subject to qualification: the electoral system penalizes parties whose support is spread across constituencies and may encourage voters not to 'waste' a vote on such parties. Significantly, however, it did not hold back the Irish Nationalist party, which concentrated its votes in Ireland. A second possible explanation is *social class*, leaving aside the third of the working class that regularly votes Conservative. The relative weakness in the twentieth century of other social differences has probably lessened the opportunity for significant third parties, based for example on language, race, or religion, to emerge.

Conservatives

The Conservative party has electorally been the most successful right-wing party anywhere in the twentieth century. It has managed to thrive without a large peasant or rural vote, or exploiting religion or nationalism, sources which have been important for right-wing parties elsewhere. The party's electoral dominance began when the Liberals split over Irish Home Rule in 1886. In the 110 years since then it has been in office, alone or in coalition, for seventy-six years. The party has gained from coalition and from splits in the other parties. But it is also a tribute to its political skill and opportunism that it has managed to win so many general elections this century in a largely working-class electorate (Seldon and Ball 1994).

Conservatism

It is difficult to discern a Conservative ideology or coherent set of principles (Norton and Aughey 1981). Indeed the importance attached to winning or retaining office has required a certain flexibility and opportunism. The need to appeal to a mass electorate has precluded the party from speaking for a narrow section, or a party of reaction. Disraeli's claim that the Tory party was 'a national party or it is nothing' has been echoed by many of his successors, and during the twentieth century it has extended its support from the middle class to a large minority of the working class. The party has accepted many outcomes which it once stoutly opposed, such as the weakening of the House of Lords, self-government for Ireland, votes for women, nationalization, a large Welfare State, high levels of public expenditure and taxation, loss of empire, and so on. For such an electorally successful party it has championed many lost causes.

There have been two important strands in twentieth-century Conservatism. The 'neo-liberal' element defends the values of limited government, low taxes, individualism, self-reliance, and free enterprise. It does allow the state a role, but largely for 'essential' services like maintaining law and order, defence, and stable prices. Margaret Thatcher and the late Sir Keith Joseph have been the most politically influential advocates of these ideas. In John Major's Cabinet it is right-wing ministers like Peter Lilley and Michael Portillo, various free-market think-tanks, and the Centre Forward group who advocate big cuts in income tax and state spending, and the encouragement of market forces.

There is also a 'Tory' strand, which is prepared to welcome the constructive interventions of the state. It advocates a paternalist and pragmatic role for government in regulating the market, correcting instances of market failure, and providing welfare and other public services which the market might not provide for all. After the party's heavy defeat in the 1945 election, many Conservatives accepted a positive role for the state in pro-

BOX 7.2. IDEOLOGY AND THE CONSERVATIVE PARTY

It is difficult to identify a Conservative ideology or a coherent set of principles. **To win elections** seems to have been the guiding principle, together with an acceptance that this requires **opportunism and flexibility.**

Conservatism is traditionally associated with

- **pragmatism rather than doctrine**
- **gradual reform rather than radical change**

Nevertheless, two strands coexist in the Conservative party which, although sharing some core values, can be distinguished by attitudes towards economic and social policy. The policies pursued and advocated by the party will depend upon which strand is dominant at the time.

	ATTITUDE AND ADVOCATES	ECONOMIC POSITIONS	SOCIAL TRAITS	SHARED POSITIONS
NEO-LIBERAL	*More doctrinaire* Lady Thatcher Michael Portillo John Redwood Peter Lilley	Libertarian Limited government Free enterprise Low taxation Low government spending Sound money, strong pound	Authoritarian Individualist Self-reliant Individual responsibility Respect for authority Accepts inequality	**Law and Order**
'ONE NATION'	*Pragmatic* Kenneth Clarke Douglas Hurd Michael Heseltine	Welcome state intervention to regulate market failure	Paternalist — the fortunate have the duty to govern in the best interest of all classes Recognize links between social deprivation, unemployment, and crime Individual and society share responsibility	**Strong defence** **Patriotism**

moting full employment and welfare, not least because they calculated that this was essential to win elections in a largely working-class electorate. By 1951 the party was back in office and prepared to accept a good deal of the previous Labour government's work—in extending public ownership, establishing the National Health Service, and maintaining welfare and full employment. All post-war Conservative leaders until Mrs Thatcher have shared this outlook and regarded it as good electoral politics. Its most prominent advocates today are Kenneth Clarke, Chris Patten, and Douglas Hurd, as well as the Tory Reform Group.

Mrs Thatcher, however, broke with this tradition and espoused a different set of values and policies. She and her supporters claimed that by 1979 collectivism had gone too far and that the policies of post-war governments (Conservative as well as Labour) had pursued social democratic politics and thereby had damaged the qualities of self-help, self-reliance, and thrift.

Thatcherism

The term 'Thatcherism' has often been interchanged with a number of other terms, including the following:

(1) *Monetarism*, or the claim that an increase in the supply of money above the rate of production would produce inflation: a case of too much money chasing too few goods
(2) *The new right*, including the *libertarians* who advocate a free economic market and the *authoritarians* who believe in a strong role for social authorities and the state; Thatcherism was a blend of both
(3) *Style*, associated with her direct 'no-nonsense', even abrasive approach to Cabinet colleagues, opponents, pressure groups, trade unions, the USSR, and EC leaders. She claimed that she was a 'conviction' as opposed to a 'consensus' politician. She was not interested in compromises but in doing what she considered to be 'right'
(4) *Values*. Her values and beliefs included the following:

- *Sound money*. This probably reflected her early background as a shopkeeper's daughter and her hatred of government borrowing, public spending deficits and inflation
- *The free market*. She believed that the free market was inherently more efficient than state planning, and that it was morally superior because it gave people the opportunity to make their own choices. This belief reflected itself in the policies of privatization of state-owned industries, reduction in subsidies to industry, sale of council houses, changes in industrial relations, including the abolition of the 'closed shop', and the refusal to impose controls on incomes, prices, and dividends
- *Reducing the economic role of the state*, particularly as a spender, taxer, and provider. The more the state does, the less scope there is for enterprise
- *Reducing income tax*, shifting from direct to indirect taxes. Lower taxes would improve incentives to work as well as promote freedom
- *Cutting public spending* as a share of GDP. Mrs Thatcher was hostile to the public sector, with the exception of what she regarded as the essential services of defence and police. She regarded much of the public sector as inefficient because it lacked competition and the threat of bankruptcy, and was cushioned by 'tax-payers' money'
- *The rejection of egalitarianism* as a goal of social policy. A notable collection of her speeches, delivered before she became Prime Minister, was titled 'Let our Children Grow Tall'. Too often, she held, the pursuit of policies of redistribution and fairness held back talent and discouraged hard work. Freedom meant the freedom to be different, the freedom to be unequal
- *Reversing Britain's relative economic decline*. She sensed that too many of her predecessors were prepared to preside over Britain's decline, rather than take the tough measures necessary to reverse it.

Mrs Thatcher also believed in what one of her friends, the late Shirley Letwin, called the 'vigorous virtues'. These include a commitment to strong defence, a firm stand on law and order (reflected in her support for capital punishment and a sharp increase in police numbers, pay, and equipment), and an assertive foreign policy. This assertiveness was often seen in negotiations with the European Community, particularly over Britain's budgetary contributions and in resisting proposals that smacked of greater integration or gave more initiative to the European Commission or Parliament. Mrs Thatcher was more willing to look towards the United States, under Reagan, than the European Community. She was a nationalist in her attachment to a sovereign British Parliament and not *communautaire*.

Labour

The first steps towards the creation of a Labour party were taken in 1900 with the establishment of a Labour Representation Committee; the party's name was adopted in 1906. Before 1914 most Labour MPs were concerned with a narrow range of issues, for example, the bargaining rights of trade unions, an eight-hour working day, and benefits for the unemployed. The origins of the party lay less in a desire to advance a socialist ideology than in a pragmatic calculation about the best way to advance the interests of the organized working class.

In 1918 Labour adopted a new programme, *Labour and the New Social Order*. This called for the public ownership of the major industries, full employment, and a financial and economic policy which would redistribute wealth to the working class, a programme which enabled Labour to distinguish itself from the Liberal party. The Liberal party had split in 1916 and Labour now had the opportunity to become both the second-largest party in Parliament and the official opposition. Although the extension of voting rights to all adult males (in 1918) and then women (in 1918 and 1928) made the electorate predominantly working class, Labour enjoyed only two short spells of minority government (1924 and 1929–31) in the next twenty years. It entered the wartime coalition in 1940 and formed its first majority government in 1945. Since 1951 Labour has been in government for only eleven of the subsequent forty-five years (1951–96).

Labour's rise was associated with several broader political and social changes. One was the growing acceptance of the ideas of social and economic equality and the entitlement of poor people to better conditions. Another was the idea of positive government, the belief that state intervention, by either regulating or replacing the free market and by providing welfare, could be used to promote social and economic goals. Public ownership of major enterprises and economic planning were not new ideas in 1918, but they were promoted more vigorously by Labour. It was the only party which wholeheartedly advocated a greater social and economic role for the state. The party's rise also provided an opportunity for people of humble social origins to enter politics. The House

BOX 7.3. IDEOLOGY AND THE LABOUR PARTY

The Labour party has traditionally been regarded as more doctrinaire than the Conservative party. It was born from cooperation between trade unions, socialist societies, Fabians, and the Cooperative Movement. All four strands have influenced the content of Labour party ideology, which has been:

- collectivist and cooperative
- egalitarian
- championing the working class, minorities, and the underprivileged
- for state provision of health, housing, education, and welfare
- for redistribution of wealth through taxation
- for state ownership; nationalization of key industries has formed a major part of party policy

There is something of a cyclical pattern of dominance in the Labour party: Labour usually gets into office when the right/centre-right of the party is dominant. Loss of office has usually seen the left in ascendance until a prolonged period in opposition allows the right to recapture control over policy-making within the party.

Left Wing

- Public ownership — litmus test of socialism
- Equality of outcome
- Against nuclear weapons
- Cool to NATO and USA
- Doctrinaire in attitude
- Advocates: Tony Benn Ken Livingstone
- Traditionalists — Old Labour

Right Wing

- Ambivalent or negative about public ownership
- Equality of opportunity
- Accept nuclear weapons
- Pro NATO and USA
- Pragmatic in attitude
- Advocates: Tony Blair Gordon Brown
- Modernists — New Labour

of Commons became less socially homogeneous after 1918, as Labour MPs from working-class backgrounds and with little formal education were elected (see Ch. 5).

Labour and socialism

The Labour party has been ambiguous about the meaning of socialism. As a coalition of interests and values there have been different interpretations of the concept. It may be useful to distinguish four strands since 1918.

1. The sectional interest of the *trade unions*. The unions have been prepared to accept economic planning and controls over the economy, but exclude wage bargaining. They have paid lip-service to redistribution but have been lukewarm in practice about achieving greater equality of incomes; powerful unions have clung jealously to craft distinctions and wage differentials. The unions have worked for the election of a Labour government because they thought they could influence its general economic policy and improve their bargaining position *vis-à-vis* employers.

2. The *Fabian* belief in public ownership and the efficacy of gradual change. The founders of the Fabian Society in the late nineteenth century were intensely practical people; in so far as the Webbs, George Bernard Shaw, and others were socialists it was largely because they thought that state management and control of economic activities would be more efficient than unregulated capitalism. They saw the state as neutral, not exploitative, and rejected Marxist notions of the inevitability of class conflict and revolution.

3. The *socialist* strand, represented originally by the Independent Labour party, and now by the left wing, has looked to the gradual transformation of capitalism into a classless egalitarian society.

4. *Marxism*, with its class analysis, emphasis on economic structure as the determinant of political behaviour, and hostility to capitalism (Miliband 1961, Coates 1975). But its rejection of parliamentary methods for achieving the dictatorship of a class has had little following in the Labour party. More recently, some members of the left wing, particularly those associated with the Trotskyist Militant Tendency, have supported 'direct action' and rejected exclusive reliance on parliamentary methods to achieve reforms.

By the end of the twentieth century each of the strands has declined. Labour has been affected by the worldwide decline of communism and diminished confidence in the efficacy of state ownership and central economic planning. Labour is now a social democratic party, like its West European counterparts. As such it supports constitutional reform, environmental protection, minority rights, the Social Chapter and public–private sector partnerships.

Other parties

For the past seventy years the *Liberals*, now Liberal Democrats, have been the major third party in the United Kingdom and qualified the accuracy of the label 'the British two-party system' (Bogdanor 1983*b*). Until 1918 it was one of the two parties of government and still in office in 1916. In that year it split between the supporters of Asquith, the Prime Minister, and Lloyd George, his successor. Thereafter, the party steadily lost support among a larger and working-class electorate. Since the 1960s, however, the party had a number of by-election successes and has seen its general election vote rise to a fifth of the total, although the number of MPs has never exceeded seventeen (in 1983).

The creation in 1981 of the *Social Democratic Party* posed the most formidable challenge yet to the post-war two-party system. A number of right-wing Labour MPs became

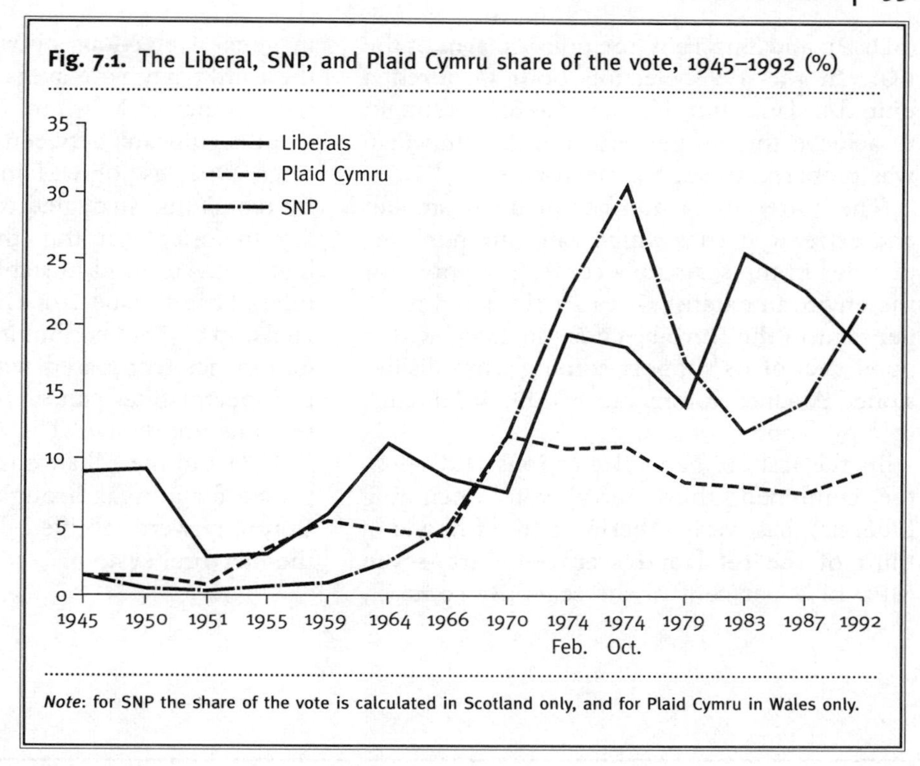

Fig. 7.1. The Liberal, SNP, and Plaid Cymru share of the vote, 1945–1992 (%)

Note: for SNP the share of the vote is calculated in Scotland only, and for Plaid Cymru in Wales only.

increasingly disillusioned with the left-wing drift of the party's policies and the constitutional changes (particularly those which gave the trade unions greater power in the election of the party leader), and formed the new party. The SDP broadly agreed with the Liberals on many policies, including British membership of the EC, proportional representation, industrial co-ownership, incomes policy, and constitutional reform. In the 1983 general election the Alliance of the two parties gained over 25 per cent of the vote—coming close to overtaking Labour—but fell back slightly in the 1987 election. But on both occasions the operation of the electoral system ensured that it gained only a handful of seats. After the 1987 election leading figures in both parties pressed for outright merger. This was accomplished in 1988 and the new successor party eventually called itself the *Liberal Democrats*. It has, however, never recaptured the level of support gained by the Alliance. In the mid-1990s the prospects of a powerful centre party and a realignment of the party system remain as distant as ever.

The *Scottish National Party* emerged as a major electoral force in Scotland in the 1970s but its subsequent fortunes declined. In 1992 it gained a fifth of the vote and has three of Scotland's seventy-two MPs. In Wales, *Plaid Cymru*, or nationalist party, has made less headway. In 1992 it gained 10 per cent of the vote and returned three of the thirty-six Welsh MPs. The nationalist parties favour greater autonomy for their nations, leading to independence in the case of Scotland.

In Northern Ireland, the once-dominant—and largely Protestant—Unionist party has splintered into rival successor parties since 1974. The *Official Unionists* and the *Democratic Unionists* parties returned fourteen of Northern Ireland's seventeen MPs in 1987. On the other side are the—mainly Catholic—Social and Democratic Labour Party

(SDLP), and Sinn Fein, the political arm of the IRA. In the 1992 election both favoured a united Ireland, but whereas the SDLP sought to achieve this by peaceful means, Sinn Fein was prepared to support terrorism.

There are also a number of other smaller and extreme parties which gain tiny numbers of votes in the seats they contest. Support for the environmentalist *Green Party* soared to 15 per cent in the Euro-elections in 1989, with a good deal of its support coming from disillusioned Alliance voters, but in 1992 it fell back to 1 per cent.

In the last six general elections (1974–92), the combined 'third party' vote (including Liberals) has varied between a fifth and a third of the total and averaged thirty-seven MPs, or 6 per cent of the seats. By contrast, in general elections between 1950 and 1970 the third-party vote averaged 10 per cent and the number of MPs ten. The number proved to be significant between 1974 and 1979 and since 1992, as John Major's government, hit by rebellions, struggled to find a parliamentary majority. But the 'third force' covers a highly diverse and unstable set of parties and offers little prospect of changing the political landscape. The continuity of the Con–Lab dominance, compared with most other West European states since 1945 or even 1918, remains impressive. The Scottish Nationalists in 1974 and the Alliance in 1983 seemed to be poised for a breakthrough, but on each occasion they were rebuffed, largely as a result of the electoral system.

Constituency parties

The tasks of local parties are to fight elections, nominate candidates, raise funds, and propagate party policy. Many constituency activists may be more interested in local politics and elections than in what happens at the next general election. The most important role of the constituency party is the selection of parliamentary candidates, particularly in those seats which the party has a good chance of winning. In all the main parties candidates are selected according to national guidelines and subject to approval by the party headquarters. This is rarely refused, although the Labour leadership has assumed the power to impose candidates in by-elections.

All the main parties have had problems attracting and retaining members. Compared with the 1950s, the Labour and Conservative parties now have considerably fewer members. A similar decline is occurring in many parties in other Western states also—leading some observers to wonder if the age of mass political parties is over—but contrasts with the increase in membership of many voluntary groups. Members of parties are affiliated at the constituency level or, in the case of Labour, they may be affiliated through a trade union. Labour's individual membership in mid-1995 was about 330,000, giving an average constituency figure of about 500, with an average of ninety 'activist' members in each constituency (Seyd and Whiteley 1992). The Conservative party estimates its national membership to be about 500,000, an average of about 800 per constituency. Some strong Liberal parties have substantial memberships and compare favourably with any party in the country, but most are centred on a small group or an enterprising individual and may come to life only with the stimulus of a general election.

Parties need active members to raise funds, fight elections by canvassing and addressing and delivering leaflets, and represent the political party in the local community. It is also from such members that local party officers and candidates for local elections are drawn. The decline over the years in the number of political activists amounts to what Seyd and Whiteley call a 'de-energising' of grass-roots politics. Contrary to stereotype, they report that the typical Labour members are middle-class professionals who own their own homes, and their views are only slightly to the left of ordinary Labour voters. Tony Blair's campaign in 1995 to reform Clause Four of the Labour party's constitution was backed overwhelmingly by party members (see below, Box 7.7). Of 441 constituency ballots, only three opposed change: individual members voted 85 per cent in favour of change, and 15 per cent against. The typical Conservative member is middle-aged (average age 62), inactive, and, again contrary to stereotype,

Table 7.1. Social characteristics of Labour and Conservative party members and voters (%)

	Labour members 1989	Labour voters 1992	Conservative members 1992	Conservative voters 1992
Gender				
Male	61	48	51	45
Female	39	52	49	55
Age				
25 and under	5	11	1	9
26–35	17	21	4	20
36–45	26	18	11	18
46–55	17	15	17	18
56–65	16	16	24	14
66 and over	19	18	43	21
Social class				
Salariat	49	16	55	38
Routine non-manual	16	21	17	24
Petty bourgeois	4	3	14	12
Foreman and technicians	5	6	6	5
Working class	26	54	8	22
House tenure				
Own property	75	58	91	86
Rented from council	18	34	3	7
Other rented	6	8	5	7
Household income (£ p.a.)				
Under 10,000	38	49	26	25
10,000–15,000	18	18	19	14
15,000–20,000	15	13	15	16
20,000 and over	30	19	42	44

Source: Adapted from Seyd, P. and Whiteley, P., 'Labour and Conservative Party Members Compared', *Politics Review*, IV(3), Philip Allan (1995).

not Thatcherite in his views—for example, 8 in 10 believe the government should spend more to get rid of poverty. However, the Tory activists are more Thatcherite—and unrepresentative of the membership at large (Whiteley *et al.* 1994) (see Table 7.1).

Organization and authority

Power within the British political parties has effectively been centralized for many years. Model rules, drawn up by the party headquarters, govern the operation of the constituency parties which can be disbanded on the decision of the centre. In this respect the 'top-down' formal structure of the parties resembles that of the British political system. By tradition, the main task of the parties outside Parliament is to support, or be a handmaiden of, the party in Parliament. But this view is coming under challenge. In 1976 the Liberals decided to elect the party leader by the mass membership, as did the SDP in 1982. The more significant and controversial changes were in the Labour party, which, after all, is a potential party of government. Demands by activists for more 'accountability' and 'participation' resulted in the party, between 1979 and 1981, accepting the mandatory reselection of MPs and the election of the party leader by an electoral college in which the votes of MPs were outnumbered by the trade unions and constituency parties. Within the Conservative party the 'Charter' movement campaigns for more party democracy, including direct election of the party chairman by members, fuller reporting of party accounts, and a greater policy-making role for the annual Conference.

Labour

There are three major institutions in the Labour party.

First, there is the *Party Conference*, which meets annually. According to the party's constitution, the work of the party is under the direction and control of the Party Conference. Over a thousand elected delegates vote on issues during the week-long Conference.

Secondly, there is the *National Executive Committee (NEC)*. Twenty-seven of its twenty-nine members are elected by all or part of the Conference: the executive represents the different sections of the party and according to the constitution 'is, subject to the control and directions of the Party Conference, the Administrative Authority of the Party'. The members include twelve elected by the trade unions, a Treasurer, and five women's representatives elected by the entire Conference, seven elected by the constituency parties. In addition the Young Socialists and the socialist and Co-operative societies have one member each, and seats are reserved for the leader and deputy leader of the party.

The third major institution in the party is the *Parliamentary Labour Party (PLP)*, consisting of Labour MPs.

We may discern *different types of representation in the party* (Kavanagh 1982). There is *functional* representation in the trade union section of the NEC, *social* representation of women on the NEC, and *elective* representation, operating variously through mandates and the direct election of Conference delegates and the party leader and deputy leader. Conference delegates may claim to represent the memberships of their organizations, the NEC to represent the Conference delegates and, via them, the members of the party, and Labour MPs to represent the voters.

Labour and trade unions

Labour differs from other British parties in that a major producer interest—thirty-eight trade unions—affiliated forty-one members to it. As an organization the party is weak in terms of finance and members, while the trade unions have long been strong in both. The unions also dominate the policy-making bodies of the party, the NEC, and the Conference. In 1995 the unions controlled 70 per cent of Conference votes, elected twelve trade union members to the twenty-nine–member NEC, and determined the election in all of eighteen of the twenty-nine places. Union influence was reinforced by new arrangements for electing the leader (1981) which gave the unions 40 per cent of the electoral college vote, later reduced to one-third. The party's dependence on the unions is a product of history, and 'peculiar' when compared with party–union relations in other countries (Crouch 1982). Trade union power in Conference has rested on the block vote, whereby the majority in a delegation decide how the entire union vote is cast. In the 1970s the five largest unions or, more precisely, majorities in them could together cast a majority at the Conference. The result was that if party leaders and big unions agreed, then they could 'manage' Conference. Not surprisingly, the key to leading the party involved keeping the unions on one's side.

Yet for all this 'power' of the unions, it is misleading to see them as manipulating the party. The unions are not of one mind on many issues and have traditionally left 'political' or non-industrial issues to the parliamentary leaders. The unions' effective power was usually concentrated on the few issues which they regard as priorities. On such occasions the unions become involved in drafting Conference resolutions and NEC policy statements and even defeating the platform in Conference votes. Hence the importance of events between 1967 and 1969, and again in 1978–9, when the Labour governments transgressed the traditional separation of spheres by 'invading' trade union areas of wage bargaining and industrial relations (Minkin 1991).

Just as Labour responded to electoral reverses in the 1980s by changing policies and trying to alter its image, so it has also reformed its institutions. In spite of significant union opposition the Labour leader John Smith persuaded the 1993 Conference to adopt 'one member one vote' (OMOV) for elections at all levels of the party from Conference to selection of candidates in local parties. This had the effect of ending the union block vote. It may be that reliance on the unions for 'indirect' members and for funds has weakened the incentives for Labour to become a mass party. In 1993 the party reduced the size of the union vote at Conference to 70 per cent, with the intention of providing more incentive for constituency parties to recruit members, and in 1995 decided to reduce it further to 50 per cent. Some of the unions, including the largest, the TGWU, which opposed Blair's revision of Clause Four in 1995, did not ballot members—a poor advertisement for union democracy. Unions are being kept at arms' length in the modern Labour party.

BOX 7.4. LABOUR AND THE TRADE UNIONS: A MARRIAGE OF INCONVENIENCE?

There are historical, constitutional, and financial links between the Labour party and the trade unions.

Origins

- The Labour party was formed as the political wing of the labour movement, of which the trade unions were the industrial wing.

Constitutional

- 38 trade unions are affiliated to the Labour party, representing a membership of 4.6 million.
- Trade union votes account for 70% of Conference votes.
- 12 NEC seats are allocated to the trade unions, and in all, 18 out of the 29 NEC seats are controlled by union votes.
- Levy-paying members of affiliated trade unions account for a third of the votes in the electoral college used for choosing the Party leader.
- Levy-paying members of affiliated unions account for a third of the votes when the 'one member one vote' system is used in constituency party matters.

Financial

- Trade unions account for approximately 75% of Labour party funds.
- Approximately 50% of Labour MPs are sponsored by trade unions.

Pressure for Change

1. Trade unions themselves are coming under increasing financial restraints, partly as a result of falling membership and partly arising from the need to direct some of their lobbying efforts to the EU.
2. The '*new*' Labour party is anxious to substantiate its claim to be a national, non-sectional party, and the traditionally close relationship between the Labour party and the trade unions appears to be an impediment to this. Trade unions pay the piper; therefore it is assumed that they call the tune. Folk memory lingers on of 1969 (In Place of Strife); 1974 (the Social Contract) and 1978–79 (Winter of Discontent)
3. Support for the Labour party among trade unionists is falling: in 1974 55% of trade unionists voted Labour, in 1992 only 46% did so.

Changes to date

1. The trade union block vote has been reduced to 70% and Tony Blair now seems intent upon reducing it further to 50%.
2. Tony Blair is publicly resisting trade union pressure to accept a commitment to a minimum wage.

Conservatives

The Conservative structure also has *three main institutions*. Yet relationships are much simpler than in the Labour party, largely because the ultimate authority of the party leader is clear. The *parliamentary party* consists of Conservative peers and MPs, and the latter elect the party leader. In opposition the leader selects members of the Shadow Cabinet or so-called Consultative Committee. The party leader appoints the chairman and major officers of Central Office which directs the *party organization*, including the Research Department. Finally, there is the *National Union*, a federation of the constituency parties, which organizes the party's annual Conference.

Conservative Conferences differ from Labour's in three main respects. First, the Conservative members attend as representatives of constituency associations and organizations and not as mandated delegates. Secondly, the Conference resolutions are only 'advisory', expressions of opinion which the party leaders may or may not act on. Thirdly, it is the Conference of the National Union, not of the party. This separation was reflected in the tradition of the party leaders arriving and addressing the delegates at the end of the Conference. Edward Heath broke with the tradition in 1965 by attending the entire Conference, and his example has been followed by leaders ever since. In terms of formal structure the Conservative party in Parliament is quite independent of the extra-parliamentary organization and in practice the National Union has supported the party leaders on major issues. Richard Kelly's (1989) study of Conservative Conferences argues that the support shown by representatives at Conferences has less to do with deference

BOX 7.5. THE CONSERVATIVE PARTY AND BUSINESS: A CLANDESTINE AFFAIR?

The Conservative party succeeded in convincing a once predominantly working-class electorate that it is a national rather than a sectional party. Nevertheless, there are close links between the Conservative party and the captains of industry and financiers of the City. These links are, however, less formal and therefore, less obvious than those between the Labour party and the trade unions.

In 1994 the question of political funding was made the subject of a report by the Parliamentary Select Committee on Home Affairs. The Labour members of the committee submitted a minority report criticizing Conservative party funding. Controversy centred around the scale and secrecy and the source of individual donations, and on business contributions.

Conservative Party Finances 1988–1992 (£ million)

	Income	Constituencies	Business	Unspecified
1988	15	1.2	4.5	9
1989	8.6	1.2	3.5	3.2
1990	9.1	1.2	2.9	4.2
1991	13	1.2	2.8	7.7
1992	22	1.3	2.93	17

Source: adapted from the *Guardian*, 13 March, 1993.

than with the leadership correctly anticipating and responding to the concerns of members. When dissent breaks out it is usually because the leadership has been out of touch. Conferences have proved troublesome over immigration (1979), Rhodesia (1980), law and order (1981), and Europe (1992). In each case the grass-roots dissenters gained in strength because their concerns were reflected by Tory MPs.

Power in the parties

Observers have long disagreed about the distribution of power in the parties, particularly in the Labour party. Labour has caricatured the Conservatives as élitist, with the party leader and his colleagues being relatively immune to grass-roots pressure and the extra-parliamentary organizations being supportive and deferential. Conservatives, by contrast, have praised the support and loyalty shown by the members and the opportunities this provides for the exercise of firm leadership. *Labour's doctrine of intra-party democracy*, according to which policies are made by the Party Conference and the parliamentary leadership is accountable to the membership, has long been defended by party supporters and attacked by Conservatives. Robert McKenzie wrote a classic book, *British Political Parties* (1963) in which he claimed (like many Conservatives) that Labour's intra-party democracy and the sovereignty of the Party Conference are incompatible with the British constitution. He posed the question: to whom are Labour MPs responsible—the Party Conference or Parliamentary leadership—when the two bodies disagree? He also pointed to the impracticability of extra-parliamentary bodies instructing the party in government; the claim for the autonomy of the party leader and his front-bench colleagues rests on the fact that they are in government or are prospective office-holders and have a wider responsibility than to the Party Conference.

But, according to McKenzie's reading of the party histories, the actual distribution of power in the parties, when in government, is similar. He showed that Labour leaders managed, by various stratagems, compromises, accidents, and through the willingness of the large trade unions to back the parliamentary leadership, to avoid the strict regime of Conference control. The impact of Conference resolutions which go against the wishes of the parliamentary bodies was softened by other provisions in the party constitution which allowed the PLP to decide 'the timing and application' of the policy in Parliament, and that only Conference resolutions carried by a two-thirds majority automatically become party policy. Labour party leaders since 1935 have been reasonably secure in office—indeed, their average tenure has been longer than that of Conservative leaders and none has been pushed out. The Conservative party outside Parliament has been more supportive of the leader, particularly when the party is in office. Loyalty to the leadership has been a more reliable rallying cry than loyalty to something as debatable as Conservative principles. Yet Conservative MPs have been willing to withdraw support for the leadership, as in the cases of Austen and Neville Chamberlain, Sir Alec Douglas-Home, Mr Heath, and in 1990, Mrs Thatcher (see Ch. 5).

Disappointment with the record of the

Labour governments (1974–9) and their failure to implement a number of left-wing Conference resolutions encouraged the left to work for a change in the formal structure of the party. What was the point of winning the battle over policy, critics argued, if Labour ministers and their civil servants watered down the policies? Between 1979 and 1981 the reformers wrested from the MPs their exclusive right to elect the party leader and deputy leader, made the reselection of MPs mandatory for all local parties, and won a series of Conference votes for left-wing policies. MPs were made more accountable to local activists, the PLP to extra-parliamentary pressures, and the leader's election was decided by a largely extra-parliamentary electorate. Many of the changes were made in the name of making the party more democratic. *These developments appeared to refute McKenzie's thesis about the autonomy, in practice, of the PLP. There was no doubt that the extra-parliamentary party was in the driving seat at this time* (Minkin 1980, Kogan and Kogan 1982, Kavanagh 1985).

The victory was short-lived. Successive election defeats encouraged assumptions that the party (a) had to be more centrist in policy, that is repudiate many of the Conference-backed left-wing policies, and (b) required a strong leader to compete with Mrs Thatcher, one who was clearly not beholden to the trade unions or Conference. Mr Kinnock, in spite of the constraints of the new rules, managed to get his way on most issues and undo the effects of the constitutional changes. Labour leaders appear to be as strong as ever in relation to the extra-party structure. For example:

- After the 1992 election defeat Labour decided to give policy-making to a 190–member Policy Forum which worked in seven policy commissions. The commissions report to Conference and give a larger role to parliamentary figures than the old system did.
- For election campaigning purposes Mr Kinnock appointed a number of groups, including a hand-picked Campaign Management Team, which to some extent bypassed the NEC. The Shadow Communications Agency, consisting of volunteers from the communications industry, operated with a good deal of independence during the 1987 and 1992 general elections.
- The post-1987 policy review was also conducted independently of the NEC, although the policy groups contained NEC members. The groups' reports effectively presented Conference with a *fait accompli*.
- In 1990 the force of mandatory reselection of MPs was diluted when Conference decided that reselection should only take place if approved by a majority of balloted party members. All constituencies now use secret ballots to choose Conference delegates and leadership and deputy leadership candidates in the electoral college.
- At a Special Conference in 1995 Tony Blair appealed directly and successfully to members, over the heads of activists and union leaders, to change Clause Four.

The surge of reform in the late 1980s and 1990s has been carried out in the name of making the party more democratic, just like the previous surge in the early 1980s. But if the reforms in the early 1980s were promoted and designed to strengthen the left and local activists and to limit the autonomy of the leadership, the more recent reforms have been promoted by the leadership with the aim of increasing the independence of the party leaders and weakening activists—assumed to be left-wing and unrepresentative of Labour voters (Smith 1994, Hughes and Wintour 1990, Seyd 1987).

The power relationships in each party have varied over time with the issues, personalities, and events, and according to whether or not the

BOX 7.6. POWER IN THE PARTIES

	CONSERVATIVE	LABOUR
Structure	• Hierarchical; top-down pyramid. • Authority is concentrated in the leadership and PCP	• Egalitarian; not pyramidal but linear. • The leadership and PLP are only one wing of the party. PLP CLP TUs COOPs Fabians Conference Elected NEC
Leader	• Free hand in selection of Cabinet or Shadow Cabinet. • Dominant influence on policy. • Dominant influence on campaigns. • Appoints party chairman.	• Traditionally has greater freedom if in government; more restricted by constituent parties when in opposition. • Tony Blair, however, has become exceptionally dominant whilst in opposition. • Shadow Cabinet elected by PLP. • Policy determined by leadership, plus conference and NEC. • NEC elected by conference.
Conference	• Rally: assembly of parliamentary and extra-parliamentary party. • Traditionally key-note unity, although conferences have become more truculent; less easily managed. • Generally supportive of leadership.	• Assembly of party. • Elects NEC to manage party between conferences. • Proposals receiving 2/3 majority become part of the party programme, but the leader plus the NEC decide what will be included in the manifesto.
Constituency parties	• Supportive role: – raising funds – recruiting members – canvassing for support – organizing electoral support • Main power: candidate selection subject to Central Office approval.	• Supportive: – raising funds – recruiting members – canvassing for support – organizing electoral support • Main power: candidate selection subject to NEC approval.

party is in office. It is clear that the centre of power in the Conservative party has always lain with the parliamentary party, and that this is now so for Labour. The main challenge to the authority of the Conservative leader has come from parliamentary colleagues; to the Labour leader from the NEC, leaders of the major trade unions, and parliamentary colleagues who can count on the support of these groups. In recent years, the balance within the Conservative party in Parliament has shifted between the leader and the MPs, as the former is now elected and may be dismissed by the latter in the annual election. In the Labour party the constitutional changes shifted the balance from the parliamentary leadership and the PLP to the extra-parliamentary organs. Yet, in spite of the organizational changes, Mr Kinnock, Mr Smith, and Mr Blair have got their way on the important issues and, although in opposition, have wielded as much power as previous Labour premiers. The reason has largely to do with the party's sense of frustration at being out of office for so long. The spectacle of Conference defying the leader was not calculated to enhance Labour's electoral appeal. The desire to support the leader, provide an image of a united party, and adopt policies to please the electorate rather than Conference have mattered more in strengthening the leader's position.

Party structures and how they operate

Some conclusions about the party structures and how they operate are warranted. *First, the parties outside Parliament should be seen as pressure groups on the parliamentary leadership.* This is more apparent with Labour because of the position granted to the trade unions in policy-making bodies like the annual Party Conference and the NEC.

Secondly, *the different histories and origins of the parties have been important.* For most of the twentieth century, the Conservative leader has been Prime Minister and his colleagues Cabinet ministers, while Labour leaders have spent more time in opposition. Since 1922 all ten Tory leaders have been Prime Minister at some time: this is so for only four of Labour's eleven leaders, as of 1995. Possession of office tends to produce a pattern of authoritativeness: deference in the party is more marked than when it is in opposition. Labour, unlike its rival, was created by an extra-parliamentary body and during its first twenty years the trade unions were a more substantial and influential body than the small number of Labour MPs. The party and movement acquired a marked extra-parliamentary character early on, one that it has never entirely lost. The historical facts are also reflected in the different tone and atmosphere of the two parties' annual conferences. Delegates to Labour's Conferences have believed that they *should* have a major say in making party policy, and it is expected that party leaders should explain and defend themselves. There have been no such expectations in the Conservative organization.

Yet it is also clear that long periods in opposition or government can produce reactions. The loss of four consecutive general elections and the excesses of party—really activist—democracy in Labour triggered a reaction which ended by strengthening the party leadership. The long period of Conservative rule and sharpening divisions in the party over Europe and the post-Thatcherite direction of the party also produced a reaction, one which left the leader weaker.

It is clear that Labour's *structure, with its separate centres of decision-making, has facilitated political divisions and factionalism in the party.* The parliamentary wing is only one element in the Labour movement; MPs defeated on an issue in the PLP can carry their case to the NEC and Conference and try to overturn the majority view of MPs. Tony Benn, in the late 1970s and early 1980s, was also skilful in using his powerful position in the policy-making forums of the NEC to push policies that most of the PLP did not favour. The political and structural divisions in the Labour party mean that the leaders face a complicated task in managing the diverse strands and keeping the party together, particularly when majorities in different institutions are in disagreement.

Party finance

Both Labour and Liberal Democrats favour state funding of political parties. The Conservative party, largely because it calculates that it benefits more than the other parties from the *status quo*, is opposed.

British parties, locally and nationally, struggle to run effective organizations. Both Labour and Conservative parties depend heavily on contributions from interest groups. But whereas trade union donations to Labour are openly declared, some business support for the Conservatives is kept secret.

The sources of Conservative finance are (a) quotas from constituency associations, (b) contributions from business and, (c) other funds, routed through such bodies as Aims of Industry or British United Industrialists, or secret 'one-off' gifts. Yet the loss of members and the effects of the recession on business has hit those sources. By the end of 1994 the party was £18 million in debt.

Labour depends on the affiliated trade unions, who contribute 55 per cent of the party's annual running costs and, in 1992, over 80 per cent of its £11.4 million election fund. In addition the unions sponsor more than a half of the party's MPs, so providing most of the candidates' election expenses and contributing to the local party.

Convergence and divergence

Study of the parties' election manifestos shows that *the parties have differentiated themselves on many policy matters*. For example, there have been clear and persistent differences between Labour and Conservatives on such issues as taking firms and industries into public ownership (more or less), council house building (more or less), and the priority accorded to increases in public expenditure versus reductions in income tax. Labour party rhetoric refers to 'equality', 'redistribution', and 'social justice'; Conservative party rhetoric refers to 'personal responsibility', 'freedom', 'incentives', and 'enterprise'. The forces making for differentiation arise from party tradition and ideology, calculations about electoral support (particularly among existing and potential supporters), reactions to what the other party in government is doing, and the need to justify a party's established policy. In many other areas, such as foreign policy, defence (apart from Labour's support for unilateralism in the 1980s), Northern Ireland, and constitutional matters, the parties have not sought out distinctive and well-publicized policy lines. *But much of the pressure for divergence or convergence depends on which policy group is dominant in the party.* A Thatcher-led Conservative party presented a sharper contrast to Labour than did the party of the 1950s and 1960s, and when the left is stronger in the Labour party, as in the early 1980s, it promises to break with the policies of consensus and continuity.

Voting returns, the parties' self-images, and the media presentation of policies as the struggle of the government versus the opposition may suggest that the two parties are monoliths. This has rarely been the case and several issues in recent years have proved especially troublesome for the two parties. Entry to and attitudes to greater integration in the EC, the holding of referendums, trade union reforms, incomes policies, defence, and the poll tax have all triggered major divisions within one or both of the major parties.

The continuities in many policies between the two parties when in office, even when they have promised differently, and the U-turns which governments before 1979 made in mid-term, encouraged the suspicion that a good part of the difference between the parties is between the 'ins' and 'outs', that is, that a party's policies depend largely on whether or not it is in government. In the 1960s and 1970s the party in opposition usually criticized the government of the day, particularly for controls on incomes, social and economic policies, and public expenditure priorities. Yet when that party was in office it usually pursued policies which were not dissimilar to those it criticized. Proximity to, or occupation of, office appears to be a powerful educator. Of course, a government may claim that a problem is never exactly the same as it appeared when it was a party in opposition. In office the pressures and influences are different, and the 'governmentalists' in the party have greater sway. (We exclude the minor parties from this argument because they have been able to advocate policies with little concern that they will be held responsible for implementing them). It is also the case that a series of election defeats can force a party to move to new common ground. After the defeat in the 1983 and subsequent elections Labour accepted a growing number of Conservative policies which it had earlier opposed.

The differences within the parties are sometimes as important as those between them. In many West European countries a multi-party system encourages some parties to emphasize their differences from rivals, and be more 'pure' in the ideology they advocate and the interests they defend. In France and Italy, for example, the political left was for long divided between Socialist and Communist parties. But in Britain the two largest parties have usually tried to be 'catch-all' parties, appealing to a wide range of interests in search of votes and playing down ideological or sectional appeals (Labour's left-wing platform in 1983 and the party's worst election result for over fifty years only confimed the lesson). The existence of two large parties has also meant that the parties themselves have been embryonic coalitions, embracing their own left and right wings.

Left and right factions

There have been clear and sometimes bitter differences between Conservative MPs, as in the 1950s and 1960s over the pace of decolonization and, from 1975 to 1983, between supporters of the 'monetarist' approach associated with Mrs Thatcher and advocates of a more interventionist set of industrial policies associated with Mr Heath, and more recently over Europe. The 92 Group, on the right, favours a 'tough' line on law and order and immigration, advocates big cuts in public spending and income tax, and resists any further dilution of British sovereignty in Europe. On the left the Tory Reform Group favours a more moderate stand on all of the above issues.

Disagreement on policy among Labour MPs for long expressed itself in organized factions. The left has divided into the 'soft' *Tribune* and 'hard' *Campaign* groups. The political right, for long a majority in the PLP, has been slower to organize formally. It has differed from the left over attitudes on such issues as:

- the extent of public ownership
- American foreign policy
- nuclear weapons
- membership of the EEC

In the 1980s the right in the Labour party lost ground, partly because of its support for causes unpopular in the party—the EEC and incomes policy—and partly because the formation of the SDP involved the loss of many of its leading figures and placed some of those who remained under suspicion of being potential defectors. Factionalism implies a degree of coherence or constraint in a group's position on issues: we expect certain policy positions to go together. For example, many Labour left-wingers have usually been in favour of more public ownership, unilateral nuclear disarmament, and British withdrawal from the EC. There has also been a larger centrist grouping of MPs which has often proved crucial in resolving differences between the two wings of the parliamentary party and in electing Labour leaders.

The strengths of the left and right have also varied across the different party institutions. Since 1945 the right has usually been in a majority in the Cabinet and Shadow Cabinet, in the PLP, and, until the late 1960s, in the trade unions and the NEC. The left has usually been stronger among the constituency activists and, until 1982, in the NEC. Between 1979 and 1983 the PLP shifted to the left, and this was reflected in the election of Mr Foot as leader and changes in party policy on defence and the EC between 1979 and 1983. The party decided to give up nuclear weapons and withdraw from the EC (see p. 74). Under Mr Kinnock, although he was elected as a man of the left, the policies shifted to the centre. By 1987 the party was in favour of membership of the EC and by 1992 abandoned support for unilateralism, the closed shop, and public ownership. Like most western social democratic

BOX 7.7. THE LABOUR PARTY AND THE REFORM OF CLAUSE FOUR

Old Clause Four

To secure for the workers by hand or by brain the full fruits of their industry and the most equitable distribution thereof that may be possible upon the basis of the common ownership of the means of production, distribution and exchange, and the best obtainable system of popular administration and control of each industry and service.

New Clause Four

The Labour party is a democratic socialist party. It believes that by the strength of our common endeavour, we achieve more than we achieve alone, so as to create for each of us the means to realise our true potential and for all of us a community in which power, wealth and opportunity are in the hands of the many not the few, where the rights we enjoy reflect the duties we owe, and where we live together, freely, in a spirit of solidarity, tolerance and respect.

To these ends we work for:

- A dynamic economy, serving the public interest, in which the enterprise of the market and the rigour of competition are joined with the forces of partnership and co-operation to produce the wealth the nation needs and the opportunity for all to work and prosper, with a thriving private sector and high quality public services, where those undertakings essential to the common good are either owned by the public or accountable to them;
- A just society, which judges its strength by the condition of the weak as much as the strong;
- An open democracy, in which government is held to account by the people; decisions are taken as far as practicable by the communities they affect;
- A healthy environment, which we protect, enhance and hold in trust for future generations.

Labour will work in pursuit of these aims with trade unions . . . voluntary organisations, consumer groups and other representative bodies . . . On the basis of these principles, Labour seeks the trust of the people to govern.

parties, the 'new' Labour party advocated constitutional reform, partnership between the private and public sectors, and integration with Europe. This policy thrust continued under John Smith and climaxed in Tony Blair's successful campaign in 1995 to redraft Clause Four and its commitment to common ownership (nationalization) of the means of production, distribution, and exchange (see Box 7.7).

Labour's left and right have also differed on electoral strategy. This disagreement is an old one and has always plagued socialist parties to some degree. *Revisionism* was associated in the 1890s and 1900s with the German socialist Edward Bernstein, who saw that, contrary to Marxist predictions, the workers were not becoming materially poorer or revolutionary in outlook, and capitalism was not breaking down. In the Labour party, the revisionists claimed that the socialist goal of a more equal society could be accomplished without the public ownership of major industries and services. The left, on the other hand, has regarded public ownership as the litmus test of socialism. After Labour had lost three successive general elections in the 1950s, 'revisionists' argued that the party was doomed to continued electoral decline unless it changed its policies and its image (Abrams and Rose 1960). Drawing on survey research, they claimed that nationalization was unpopular, the manual working class was diminishing, workers and their families aspired to own more material goods, and a more classless or less proletarian appeal was needed. Virtually the same analysis was made after Labour again lost three successive elections in 1987, and with even greater force when it lost again in 1992. Few were prepared to argue that four successive Conservative election victories and a large vote for the centre party proved that voters were turning away from Labour because it was not left-wing enough. The combination of electoral sociology and practical political considerations suggested a shift to the political right.

In the 1980s, the party factions were described in new terms. In the Conservative party a 'wet' was somebody who still clung to the 'One Nation' style of conservatism, believed that the costs in increased unemployment of Mrs Thatcher's policies were too high, and favoured a more positive role for government and public expenditure in assisting economic recovery. A 'dry' Conservative favoured a fundamental shift in resources to the private from the public sector, 'privatization' of many public services and state enterprises, income tax cuts, and a reduction in state spending for welfare. During the 1990s the division is more between *consolidators* versus supporters of further Thatcherite policies, particularly more privatization and cuts in taxes and public spending. The more bitter division has been over Europe, between those who resist further measures of integration and would even entertain Britain's withdrawal, and those who see Britain's future in Europe, including membership of a common currency. The 'right' is further divided between the free market or individualist wing, which favours economic competition and consumer choice, and a more authoritarian wing, which favours hierarchy, traditional values, the authority of figures such as parents, teachers, and the independence of the UK *vis-à-vis* Europe. Its values are expressed in such outlets as the *Telegraph*, *Spectator*, and *Salisbury Review*.

In the Labour party most members of the Tribune Group are now regarded as part of the 'soft' left. The 'hard' left, which traditionally favoured unilateralism, much more public ownership and planning of the economy, withdrawal from the EEC, and the removal of British troops from Northern Ireland, has been marginalized. The difference now is between 'new' Labour and 'old' Labour, with the latter clinging to the traditional positions on Clause Four, class politics, the trade union connection, and high spending and taxing.

Party unity and discipline

British parties have long been noted for their remarkable unity when it comes to voting. *Party loyalty* has been crucial to the executive's dominance of the House of Commons, to its secure tenure of office until the next election, and to the idea of responsible, programmatic, party government. The commitment of most MPs to a party programme and the likelihood that one party will have a majority in Parliament enables the British electorate to choose between two potential parties of government at general elections.

Party discipline is one reason which induces MPs to follow a common line when it comes to voting in the House of Commons. A party line is agreed after soundings and discussions between the Chief Whip and MPs, and the final decision is expected to be binding on all MPs. A member who defies a three-line whip (the strongest form of instruction to follow the line) runs the risk of having the party whip withdrawn; this is tantamount to expulsion from the party. The Standing Orders of the Labour party state that '. . . membership (of the PLP) involves the acceptance of the decision of the Party Meeting', although abstention is allowed if an MP claims that the policy offends his or her conscience.

The Conservative leader always appoints the Chief Whip, the Labour leader only when he is Prime Minister; in Opposition Labour's Chief Whip is elected by the PLP. The tasks of the Chief Whip and his assistants are to try and ensure that enough members are present in the House of Commons or nearby to vote the appropriate way in divisions, and to act as intermediaries between the leaders and backbenchers. Whips try to gauge the mood of members, assess how they will express their unhappiness with party policies, and cajole, bully, or conciliate the potential dissident. Expulsion is very rarely used if the number of likely dissidents is large, because it may then be counter-productive. In 1969 the Labour Chief Whip effectively aborted Mr Wilson's plan to reform the trade unions by reporting to the Cabinet that the bill could not get a parliamentary majority. In 1971, when nearly a third of Labour MPs defied the whip and voted for membership of the EC, there was little the whips could do. In 1986 the Conservative government did not proceed with the sale of Land Rover to an American company when the whips reported that many of the party's MPs would vote against it. In 1995 the withdrawal of the Conservative whip from nine MPs who had persistently defied the whips on European issues was hardly a success. Not only were most of them supported by their local associations, but the government's majority in the Commons was threatened. Invoking disciplinary sanctions reflects the failure of the normal process of decision-making in the parliamentary party.

We therefore have to look for more *positive reasons* for the cohesiveness of parties, for the loyalty of MPs in the division lobbies. In contrast to the United States, people vote at general elections for politicians because they are representatives of a party, not independents or celebrities in their own right or as local spokesmen. An MP's party label is more widely and better known to his constituents than his personality and political ideas. And most MPs know this: if they want to be re-elected by the voters then a party label is virtually a prerequisite. Moreover, they will have built up loyalty to the party and to the parliamentary colleagues and local supporters with whom they have been associated for many years. MPs also know that a divided and faction-ridden party is not likely to be well regarded by the electorate.

The increase in the number of government appointments in the hands of the Prime Minister has also helped the executive; around a hundred MPs receive some kind of appointment, from Cabinet ministers down to parliamentary private secretary. As all are bound by collective responsibility, a Prime Minister is virtually assured of the support of a third of his party in any vote. Office, or the hope of it, helps loyalty. But again, this is not a complete explanation, for many MPs have no hope or expectation of office. The threat of dissolution is hardly a sanction to wield against dissident MPs. The likely outcome is that a divided government would be defeated in the election, the Prime Minister lose office, and many MPs in marginal constituencies forfeit their seats.

Because a party is not monolithic, followers will have different views about its principles. Although Conservative and Labour MPs may have differing values and assumptions, the differences are not of rival ideologies. The past convergence in several policies—*over time*—between governments of either party may point to the strength of outside circumstances and/or the weakness of party principles. Some so-called party principles are rationalizations of what a party has done. The Labour deputy leader Herbert Morrison once claimed that 'Socialism is what the Labour Government does'.

Party government?

A party in government, which has an assured majority in the House of Commons, some control of the parliamentary timetable, and is not hamstrung by constitutional checks and balances, is well placed to get its legislation through. Compared with other Western democracies the British system of government provides formidable resources for a party in office. Critics and defenders of this system agree that, given the lack of formal constraints on the legislative majority, the government is usually able to carry through its major legislation. Yet if the ability to pass measures through Parliament reflects the government's strength, then the failure of the government to realize its broader policy goals *vis-à-vis* society and the economy may correlatively be taken as a sign of its weakness. Advocates of responsible party government in Britain, and representative government in general, usually assert that parties should have an impact—an intended one. After all, parties are elected to office by voters, and for parties to play a minimal role in government weakens the influence of elections and the electorate. Yet parties in office often disappoint their supporters.

There is no completely satisfactory way to test for the effects of a party in government (Rose 1984*a*). One may compare the promises and statements of intent of governments with actual outcomes at the end of their terms. But one then has to speculate about what might have happened if a different party had been in control. Perhaps, given the circumstances, another party might have acted similarly. Governments may have other effects—on public opinion or patterns of public spending, for example—than those which were planned, or their policies may be modified in response to changing events or shifts in public opinion.

Two rival views about the effects of parties have been propounded by academics and practising politicians. *One is that parties are too strong* because they can wield the power of the elective dictatorship. The first claims that the 'adversary' form of the two-party system, combined with the all-or-nothing nature of one-party government, produces abrupt reversals of policy when one set of partisans replaces another in government. *This view claims that parties do influence government, though, regrettably, for short-term and ideological motives* (Finer 1975*b*). Critics of the adversarial system recommend the adop-

BOX 7.8. THEORY AND PRACTICE: RESPONSIBLE PARTY GOVERNMENT

Theory

- The party that wins a majority of Commons seats becomes the government.
- Its majority enables it to control the Commons' timetable and to enact the policy programme on which it was elected.
- At election time the governing party is responsible for its record.
- The governing party may not have a good working majority, e.g. 1951, 1964, 1974, and 1977–9; even John Major's majority of 14 in 1994–5 was precarious on some issues, such as Europe and VAT on fuel.

Practice

- A party may promise much before being elected only to discover that the reality of holding office is different. Opposition parties, except for a few courtesy meetings in the run-up to an election, are cut off from Whitehall advice, facts, and figures.
- Civil servants may have a vested interest in the continuity of departmental policy and have many strategies for blocking radical initiatives.
- Much policy is 'inherited' commitment; for example, in 1990 75 per cent of public spending went on policy commitments made before 1939.
- Party promises may disappear in the face of international events, e.g. the quadrupling of Arab oil prices in 1974.
- There may be incompatibility between party policy on some issues, for example, the rejection of the Social Chapter and commitment to EU membership.
- National governments have no control over globalized economy, technology, and information and communications networks.

For party government to be a reality, special circumstances may need to exist:

- political will
- a large majority
- a weak opposition
- a long period in office
- a 'fair international wind'

tion of proportional representation as a means of limiting the impact of party government. In the 1970s the main argument for electoral reform shifted from achieving electoral justice to promoting efficient government by slowing the abrupt reversal of policies with changes of government. This last argument has lost its force since 1979 because of the continuity of the Conservative party in office, and the first regained its relevance because of the size of the Alliance vote in 1983 and 1987.

A different view is that parties are weak. Parties usually are poorly prepared in opposition, whereas civil servants are well entrenched and have many strategies for frustrating ministers who pursue policies which the civil servants regard as impracticable. Moreover, the same recurring problems and similar constraints, or the sense of 'ongoing reality', force parties, regardless of ideological leanings, into following a broadly similar set of policies (Rose 1984*a*). A good part of 'reality' comes in the form of inheritance of public policy and government spending from previous governments. In 1990 almost three-quarters of public spending went on policy commitments made by governments before 1939 (Rose and Davies 1995). Another limit on national government is the impact of events beyond its borders, for example the quadrupling of Arab oil prices in 1974, or of policy decisions taken elsewhere, for example from the European Union. If one tries to relate party incumbency to economic conditions—rates of inflation, unemployment, growth, and prosperity—it is striking how modest have been the differences associated with party control of government.

The preceding observations do not, however, completely refute the adversarial analysis. If, as the adversarial critics point out, changes in party control have led to discontinuities in policy, it is also the case that the trends towards convergence have occurred *despite* the parties in office attempting to pursue different policies. Between 1945 and 1975 there was a change in government policy on prices and incomes every thirteen months on average, and between 1958 and 1974 a change in corporation tax every two years on average. But in those years changes were as likely to occur within the lifetime of a government as with a change of party control. In 1970, for example, the new Conservative government reversed its predecessor's policies on the reorganization of secondary education, aid to industry, prices and incomes controls, and pensions. In 1974 Labour introduced new policies for each of these, as well as for the EEC, housing, finance, and trade union law. Yet by the end of their terms of office each of the governments had moderated its original policies. By 1973 the Conservatives had moved to a statutory prices and incomes policy and massive state intervention in industry, the reorganization of secondary schooling along comprehensive lines proceeded, and the trade union legislation under the Industrial Relations Act was effectively non-operational. By 1979 Britain, under Labour, was still in the EC, the government had an incomes policy of sorts, had curbed the growth of public expenditure, and private education remained. What is important to note, therefore, is that governments have often abruptly or gradually changed their policies, often around the mid-term of Parliament. The continuities occurred in spite of the parties' attempts to try different policies. This may have less to do with ideology than with political 'learning' or having to cope with events. *The adversarial critique may explain the early stages of a government's record, but Rose's 'moving consensus' (that is, the initiatives of one party in government are gradually accepted by the other party) is more relevant to an understanding of the long-term trends.*

The Thatcher governments were an exception to the consensus thesis. They have been committed to shift the balance from the state and collectivist values to market forces and individualism. Although it took a decade to reduce state spending as a share of GNP from the 1979 level, it did shift the amount of resources directed to particular programmes, increasing that on defence and law and order and cutting that on industry and, drastically, housing. Many nationalized assets were sold off to private investors, services and goods provided by local and central government were either contracted out or 'privatized', and, by 1990, over a million council houses had been sold to tenants. It shifted the balance from direct to indirect taxation. Its industrial relations legislation, aided by changing patterns of work and high unemployment, severely weakened the trade unions. It would be difficult to argue, in the light of this experience, that a determined government still 'makes no difference' (Kavanagh 1990).

It may be that if a party is determined, has a large majority, is faced with a discredited or feeble opposition, and is in office for a long period (say ten or twenty years), then it may have a much greater impact on society and the economy. In Sweden and the United States in the 1930s and 1940s, and in West Germany during the 1950s and 1960s, the dominant parties (Socialist, Democrats, and Christian Democrats respectively) had such an impact. The 1992 Major government provides an interesting contrary case. Its tiny majority has made it vulnerable to defections, divisions on the key issue of Europe have further weakened it, the post-Kinnock Labour party is widely seen and approved as 'new Labour', and there is no widespread perception of a 'Major agenda' (Kavanagh 1994).

A changing system?

For much of the 1974–87 period the two-party system was under stress, and for a time the rise of the Alliance threatened to displace Labour as the second party. Since 1987 and

BOX 7.9. BRITISH POLITICS TODAY: THE POLITICAL PARTIES

Current trends and pressures for change include:

1. The future of the link between the Labour party and the trade unions. Whilst '*new*' Labour might wish to shrug off its sectional image and perceptions that it is 'in hock' to the trade unions, this poses a major question for party finances.
2. Financial worries are not confined to Labour. Even the Conservative party, the wealthiest British party, has struggled with persistent deficits since the 1980s, having an accumulated deficit of approximately £18 million by 1994.
3. A question on the agenda is whether Britain, like most other Western democracies, should provide state funding for political parties. There is a small amount (approx. £2 million) of funding 'Short Money' available to the opposition parties for their parliamentary work, but at present all funding of election campaigns has to be raised by the parties themselves.
4. The question of funding has been exacerbated by dwindling party membership. Over a forty-year period from 1953 to 1993 Conservative party membership fell from 2.75 million to .50 million. Labour membership over the same period fell from 815,758 individual members and 5.5 million trade union affiliated members to 279,000 individual members plus 4.5 million trade union affiliated members.

Falling membership also brings associated problems of:

- narrowing the field of political recruitment
- reducing the effectiveness of parties at constituency level

the collapse of the Alliance this prospect has receded.

The introduction of electoral reform and coalition government would fundamentally alter the relationships between and within the main political parties (see Ch. 6). Some of the objections to coalitions—instability, weakness, discontinuity of policy—may have been valid for Germany under Weimar and in France during the Fourth Republic. They are, however, less applicable to the present regimes. Disraeli claimed that 'England does not love coalitions', a dislike that often shades into suspicion of such other allegedly 'un-British' phenomena as proportional representation and multi-partyism. But the 'norm' of one-party majority government has prevailed for only about a third of the period since 1900, and the introduction of proportional representation has been seriously considered in the past.

If one examines the historical record of coalitions or informal party cooperation in Britain since 1914, then the one-party model has been relaxed in one of the following circumstances:

- *wartime* (e.g. 1916 and 1940) and its aftermath (e.g. 1918–22)
- *economic crises* and lack of a one-party majority (for example, the 1931 coalition which followed the collapse of the minority Labour government)
- *a minority government* which refuses to dissolve Parliament; this has led to cooperation with another party (e.g. Labour and Liberals 1930–1) or a pact (e.g. Labour and Liberals 1977–8)

Some critics of the results of one-party governments and the adversary two-party format advocate proportional representation (PR), not necessarily to manufacture a coalition (though this is likely), but to ensure that government also has the backing of a majority of the electorate. Another view is that coalitions, regardless of whether or not they are an improvement on one-party government, will be a logical outcome of the continued fragmentation of the party system and the probable failure of one party to gain a majority of seats. The introduction of PR—perhaps as a response to a series of deadlocked general election results—may then consolidate the fragmentation. Finally, the long period of Conservative rule, on the basis of some 42 per cent of the vote, has converted some in the Labour party to the merits of electoral reform. Historically, the Conservatives have gained disproportionately from the British electoral system. PR and a possible Lib–Lab alliance might improve the chances of having a non-Conservative government.

The emergence of formal coalitions would challenge a number of British political and constitutional assumptions. It would weaken stronger interpretations of the electoral 'mandate', since the parties forming a government would have to bargain about policies and Cabinet seats. Presumably it would also limit the political freedom of the Prime Minister to control ministerial appointments and dissolve Parliament, as he or she would have to consult another party leader over the allocation of posts. Coalitions have in the past tended to break up, or suffer strains, because the parties' leaders tended to be more coalescent than the backbenchers or party activists. Finally, the role of the monarchy, in responding to a Prime Minister's request for a dissolution, or in asking a party leader to form an administration, would probably become more political, more significant, and, constitutionally, more controversial.

Discussions about the desired effects of government and party raise a question of values—how much power should a party have over society? One view, found on the left wing of the Labour party, is that a government must take greater powers if it wants to achieve radical change in the society and economy. Checks, for example, from interest groups, the EU, or the House of Lords, the reformers might point out, are restraints on the elected government and the electors who voted for it. The free-market proponents of the Conservative party are suspicious of and view government activity as more likely to reduce individual choice and freedom than to increase it.

Conclusion

The concern in the United States about the decline of political parties has echoes in Britain, to where the era of the mass political party may be drawing to a close. Declining levels of strong party identification and falling levels of party membership and activity indicate the marginal role of parties in the lives of most people. A MORI poll conducted in January 1993 found that very few people were willing to carry out demanding political activites for the party they supported. For example, only 2 per cent expressed willingness to canvass by telephone for the party or stop strangers in the street and discuss their party's merits, and a mere 4 per cent would speak out for it at another party's meeting. The widespread agreement among voters on many values means that the Labour and Conservative parties do not represent sharply contrasting values or issue preferences among their supporters. A MORI survey in 1988 found that a third of Labour voters held what the researchers called Thatcherite values and a quarter of Conservatives supported socialist values. In some respects the parties have become less ideological, a response to the growth of a more middle-class electorate and the tendency for more voters to have mixed social class characteristics. Interest groups are reluctant to associate closely or openly with political parties, and the Labour party's connections with the trade unions are being loosened. No party has a direct connection with a national paper, and parties cannot purchase broadcasting time (Kavanagh 1994*b*, Mulgan 1994).

Parties are adapting to their broader changes by contracting out some of these functions. Before and during election campaigns they recruit technical experts to assist with publicity, media presentation, opinion polling. For most voters television is the main source of information about politics and this is increasingly taken into account by party strategists for their annual conference. Parties increasingly have turned to independent or sympathetic think-tanks for ideas about policy, Labour to the Institute of Public Policy Research (IPPR), the Conservatives to the Social Market Foundation (SMF), Adam Smith Institute (ASI), and Centre for Policy Studies (CPS). Parties have also drawn on cross-party sources for policy advice. The IPPR Commission on Social Justice and the Welfare State included experts from outside the Labour party.

What is clear is that any diagnosis of the defects of British government cannot ignore the conditions of British political parties: British government, for good or ill, is still party government, and the health of one is inseparable from the health of the other.

Summary

- Parties perform essential functions of recruiting political leaders, mobilizing voters, and presenting voters with policy choices.

- Responsible party government depends on programmatic disciplined parties, which allow the electorate to reflect upon the governing party's record and choose between competing programmes for the future.

- In terms of voter preferences Britain has a multi-party system, but in terms of parliamentary seats and chances of forming a government, Britain has a two-party system.

- The Conservative party's electoral success is a testament to its adaptability, the political skill of its past and present leaders, and a divided opposition. It stands for certain core values, but has two discernible strands that are distinguished by their views on the appropriate role of government in the economy and the appropriate relationship between the state and the individual.

- Thatcherism is a complex of values, style, and policies associated with Mrs Thatcher's period of leadership. It includes monetarist and free market economic policies — sound money, reductions in taxation and government spending; social authoritarianism — self-help, thrift, individual responsibility, respect for authority; and foreign policy of Atlanticism.

- The Labour party began life as a clearly sectional party, originating from an extra-parliamentary group organizing to press for a better deal for the working class. It has been socialist in rhetoric but social democratic in reality, demanding a bigger share of the cake for the less well-off but recognizing that it is a capitalist cake. It has always accepted that the state has a significant role to play in delivering economic and social well-being to citizens.

- The Labour party began life as a political wing of the labour movement, the trade unions being the industrial wing. The close association between the trade unions and the Labour party has been a source of criticism and serves to undermine the Labour party's claim to be a national non-sectional party. This relationship is being challenged and changed by the current Labour party leadership.

- With the notable exception of the nationalist votes concentrated in the Celtic fringe, third parties are disadvantaged by the electoral system and the fact that the major social division in Britain is into two broad classes. The brief heyday of the SDP/Liberal Alliance in the early 1980s is a testament to these systemic disadvantages.

- Three distinct levels are discernible in the structure of the two main parties — constituency level, central organizations, and parliamentary party. Despite constitutional differences, power tends to be centralized in both parties, the parliamentary wing being supported in terms of funds and canvassing by the constituency parties, whose main power lies in candidate selection. Attempts to reinvigorate internal democracy or activist democracy in the Labour party in the 1980s has been reversed in the wake of four election defeats.

- The exact balance of power within the parties shifts over time in relation to issues, personalities, and electoral fortunes. Some conclusions can be drawn from an examination of the internal structure of the parties: (1) the extra-parliamentary wing acts as a pressure group on the leadership; (2) history and origin influence structure, power relationships, and expectations; (3) electoral success strengthens the centre; and (4) in Labour the plurality of power centres has encouraged division and factionalism.
- Since 1983 two main parties have appealed to floating voters in the centre of the political spectrum, and this inevitably leads to a degree of convergence between them. There is some divergence in respect of the role of government in the economy and this becomes more or less marked in accordance with which wing is in control in each party.
- Party unity and discipline is partly explained by the whipping system and expectations of constituency parties, but ambition linked with patronage and shared values also play a significant role.
- Britain has party government: the government is formed from a majority in the House of Commons, which controls the timetable and, with few exceptions, can rely on its majority to enact its programme. Some question whether parties really make a difference, arguing that beneath the adversarial rhetoric, despite changing the party in power, there is continuity rather than change of policy.
- The weakening of partisanship among the electorate has had an impact throughout the party system, reducing activists and potential recruits, etc. This has forced parties to revamp their images and to contract out some traditional functions. If events, for example the introduction of PR or the continued decline in support for the two main parties, lead to coalitions, then this is likely to bring about other constitutional changes.

CHRONOLOGY

1900 Labour Representation Committee formed

1906 Labour party formed. Liberal general election landslide; the first Labour MPs are elected; electoral pact with Liberals

1916 Lloyd George becomes PM; splits Liberal party

1918 Lloyd George Coalition wins 'Coupon Election'. Labour party constitution: commitment to socialism, Clause Four

1922 Conservatives form the 1922 Committee, leave the Lloyd George Coalition, and win election

1924 MacDonald forms the first Labour government; in the minority, it survives for ten months until the Liberals withdraw support

1929 General election: Labour becomes the largest party for the first time. MacDonald forms government

1931 Cabinet splits over public expenditure cuts. MacDonald resigns but then forms a coalition, denounced by the Labour party. Henderson replaces MacDonald as leader of the Labour party

1932 Lansbury becomes Labour leader

1935 Baldwin replaces MacDonald as PM. Attlee becomes leader of the Labour party. Conservatives win the general election

1940 Churchill forms National Government, includes Labour Cabinet ministers

1945 Labour landslide (144 majority), leading to its first majority government. Churchill appoints Lord Woolton chairman of Conservation party with brief to modernize the party organization

1950 Labour has a narrow election victory

1951 Attlee calls an election and loses to Conservatives, who accept much of Labour's welfare and nationalization legislation; post-war consensus begins

1955 Churchill resigns and is replaced by Eden. The Conservatives win the election. Attlee is succeeded by Gaitskell as Labour leader

1957 Eden resigns, and Macmillan 'emerges' as leader of the Conservatives and PM

1959 Conservatives have a third successive election victory; Labour becomes more 'revisionist'

1963 Macmillan resigns, Home 'emerges' as the new Conservative leader and PM. Harold Wilson is elected Labour leader after Gaitskell's death

1964 Labour wins a narrow election victory

1965 Home resigns, Heath is elected Conservative party leader

1966 Labour wins a 96 majority at general election

1970 Heath wins a Conservative election victory on Selsdon policies which reject post-war consensus

1972 Health U-turns back to incomes policy and increased economic intervention

1974 In February a Labour minority government after the 'Who Governs' election. Revival of Nationalists and Liberals. In October Labour secures a small majority

1975 Conservatives amend the leadership election procedures and Heath loses to Thatcher

1976 Wilson resigns and Callaghan is elected Labour leader and becomes PM

1977 By-election losses; Labour becomes a minority government depending on Lib–Lab pact

1979 Government defeated in Parliament; Conservatives win general election

1980 Callaghan resigns Labour party leadership, Foot is elected to replace him

1981 Labour party Special Conference adopts electoral college for choosing Labour leader. SDP formed. Labour NEC investigates Militant Tendency

1983 Conservatives win the election. Lib/SDP alliance polls nearly as many votes as Labour. Labour NEC expels five members of Militant. Kinnock elected Labour party leader

1986 Labour party begins disciplinary action against Militant in Liverpool

1987	Conservative election victory. Lib/SDP alliance is less successful. Kinnock launches policy review
1988	SDP and Liberals merge
1990	Conservatives depose Thatcher, Major becomes leader and PM
1992	Major wins election. Kinnock resigns, Smith is elected Labour party leader
1994	Smith dies, Blair is elected Labour party leader. Tory Euro rebels vote against party
1995	Major resigns Conservative leadership but regains it in contest. Blair succeeds in bid to redraft Clause Four, a clear signal that the right has recaptured the Labour party

ESSAY/DISCUSSION TOPICS

1. Why is the Labour party more prone to splits and factions than the Conservative party?
2. Why has the Conservative party been so successful electorally?
3. Can the Labour party claim to be more internally democratic than the Conservative party?
4. Does Britain have party government?
5. Did Mrs Thatcher bring an end to consensus in British politics?
6. Britain has a two-party system. Discuss.

RESEARCH EXERCISES

1. Write brief notes on each of the following:

Bruges Group	Tory Reform Group
Militant Tendency	No Turning Back
Manifesto Group	Centre for Policy Studies
Campaign Group	Institute for Public Policy Research

2. How is party policy formulated in the Conservative party and in the Labour party?

FURTHER READING

A good introduction is R. Kelly and R. Garner, *British Political Parties Today* (Manchester: Manchester University Press, 1993). On party ideas and practice see D. Kavanagh, *Thatcher and British Politics: The End of Consensus?* (Oxford: Oxford University Press, 1990) and A. Gamble, *The Free Economy and the Strong State*, 2nd edn. (London: Macmillan, 1994). On the impact of parties see R. Rose, *Do Parties Make a Difference?*, 2nd edn. (London: Macmillan, 1984), Gamble, *The Free Economy*, and D. Marsh and R. Rhodes, *Implementing Thatcher's Policies: Audit of an Era* (Milton Keynes: Open University, 1992).

The following may also be consulted:

Blackmore, Hilary, Pyper, Robert, and Robins, Lynton (eds.), *Britain's Changing Party System* (London: Pinter, 1994).

Ingle, Stephen, 'Political Parties in the 1990s', *Talking Politics* 6/1 (1993).

Kelly, Richard, 'Power in the Tory Party', *Politics Review*, Apr. 1995.

Lennieux, Simon, 'The Future Funding of Political Parties', *Talking Politics*, 7/2 (1995).

8 INTEREST GROUPS

Reader's guide

In the 1970s commentators were questioning whether Britain was becoming ungovernable. Sectional interest groups seemed able to check the power of governments elected to rule in the national interest. In February 1974, Edward Heath called a general election on the question 'Who Governs Britain?', the implied choice being 'the unions' or 'the elected government'. The electorate seemed as unsure as the government about the answer. The result was a hung Parliament, with no party having a clear mandate to rule.

Burgeoning interest-group activity was partly encouraged by, and partly a response to, interventionist government that had become the order of the day in the post-war welfare state. Interest groups provided governments with expertise, extended the opportunities for political participation, and provided channels for a continuous dialogue between government and governed. But some argued that they posed a threat to parliamentary democracy.

Are there different types of interest groups? What techniques do they employ to press their cause? Why do governments tolerate, and in some cases actively encourage, interest-group activity? What enables some groups to achieve 'insider' status? This chapter addresses these questions. It also explores the concept of the corporatist state, examining in some detail the relationship between the government and the main producer-group organizations. Mrs Thatcher rejected not only the policies of the post-war consensus, but also its consensual style of policy-making predicated upon consultation with affected interests. This chapter discusses the deterioration in relations between government and interest groups during her period in office. It concludes by reflecting upon some of the drawbacks associated with interest-group activity.

IF people in Britain were free to influence the government only at general elections, then Rousseau's eighteenth-century gibe about the British being slaves between elections might still have some validity. In the twentieth century, however, two major channels for representing popular opinion have developed. The first is *the party electoral one*, in which a person votes for a party as a member of a territorial constituency, and the second is *the interest group-functional one*, in which a person joins a group consisting of people who have shared attitudes or occupations. Most individuals both possess a vote and belong to a group. But whereas a vote can only be used periodically, perhaps every four years or so in the case of general elections, pressure groups allow citizens a continuous opportunity to affect policy in particular areas.

Interest groups and democracy

Interest groups, which are associated with sectionalism and the possible exercise of sanctions, appear to be inimical to ideas of reasoned discussion and the general welfare. The first major study of pressure groups in Britain was S. E. Finer's *Anonymous Empire*, published in 1958. It demonstrated the existence and operation of such groups within the political parties and Parliament and how they influence policy. Finer concluded that the groups were (as far as the general public was concerned) faceless, voiceless, and unidentifiable—in brief, anonymous. Yet in practice, liberal democracy should also allow for checks on government and limits to majority rule. The opportunity for pressure groups to operate in the political arena depends, therefore, on the existence of such freedoms as those of association, assembly, and speech, and on the acknowledged legitimacy of viewpoints different from those of the government. Recognition and toleration of the diversity of interests in society are necessary conditions for pressure-group activity.

BOX 8.1. KEY CONCEPTS: INTEREST GROUPS

pressure group	A generic term for any kind of group seeking to influence the government to adopt or change particular policies.
types of group	• **promotional**: advocating a cause, e.g. Child Poverty Action • **interest**: representing those engaged in a particular sector, e.g. trade unions, Automobile Association • **peak**: representing a set of interests, e.g. TUC • **episodic**: these spring up to pursue a particular project, e.g. a proposed site for nuclear waste
'insider group'	A group with direct access to the corridors of power, e.g. National Farmers' Union and the Department of Agriculture.
polyarchy	A term coined by Robert Dahl (an American political scientist) to denote a benign political system in which competing groups exert checks and balances and thereby ensure stability and political freedom.
clientelism	Close identity of interest between a government department and client group.
corporatism	Pressure groups incorporated inside the governmental process.
tripartism	A form of corporatism associated with Britain in the 1970s when much economic policy was the outcome of negotiations between the government, the Trades Union Congress (TUC) and the Confederation of British Industry (CBI).

More than half the adult population are subscribing members of at least one organization (such as a trade union) and many belong to a number of groups. (The Royal Society for the Protection of Birds has more members than all of the British political parties put together!) They also belong to other communities of interests in their roles as homeowners, parents, pensioners, car owners, and so on, even though these interests might not be formally organized. In contrast to the opportunities provided by groups for representing specific interests, voting at a general election may not be very effective for expressing and weighing individual views on the issues. Elections provide a rough and ready verdict on the policy packages of the parties, while groups supplement and qualify the representational role of the election by providing for the expression of views on specific issues.

Groups and parties

In so far as groups are voluntary associations, contain like-minded members, and attempt to influence policy, they resemble political parties. But parties differ from groups in three important respects:

1. Parties run candidates at general elections and try to capture political office directly. Although some groups support candidates for a political party at elections, only the trade unions sponsor candidates on a considerable scale.
2. Parties, as would-be governors, develop comprehensive programmes of policies to appeal to a majority of the electorate: in doing so they have to aggregate and strike a balance between the demands of various interests. Groups, by contrast, seek to articulate a sectional interest—even though they may find it good tactics to present it as identical to the public interest.
3. A party in government has to accept responsibility for coordinating and implementing a wide range of policies, whereas a group seeks to influence the policy-makers in the area that concerns it.

Critics may reasonably propose qualifications to the above distinctions. Such parties as the Greens or the Scottish Nationalists have more in common with pressure groups—mainly interested in a single issue—although they operate as a political party. And some trade unions play a dominant role in the Labour party. But the broad distinction is still useful: *the political party seeks office, general influence, and, ultimately, responsibility; the interest group primarily seeks access to decision-makers and to exercise sectional influence.*

The reasons for the *growth of group activity* in this century are not hard to find. As the government's role as an employer, economic decision-maker, taxer, and distributor of benefits has expanded, so it has come to impinge more on the population; and as the responsibilities of government have extended to many areas of life so it has become dependent on many groups for cooperation and compliance. Without continuous consultation and cooperation between groups and government departments, the formulation and implementation of policy would grind to a halt or at least become more difficult. Governments prefer to negotiate with authoritative and representative spokesmen for interests. Ministers may have policies for housing or education but they do not build houses, set mortgage rates, recruit teachers, or build schools. The policies of government consist of initiatives, resources, and decisions which they hope will lead groups and other decision-makers to behave in desirable ways. Cooperation is more usual than conflict.

Types of groups

We may distinguish three types of groups which try to influence government without themselves holding office. First, there are the *promotional or attitude groups*: these advocate a cause and their potential membership is, in theory, coextensive with the entire population. Among such groups are the Abortion Law Reform Association and the Howard League for Penal Reform. We might also include the 'think-tanks' that have emerged in recent years, promoting ideas and policies, for example, the free market Adam Smith Institute and Institute of Economic Affairs or the constitutional reform group *Charter 88*. Second, there are the *interest groups*, usually based on an occupation or economic interest, like professional organizations, trade unions, and business groups.

The distinction between the two groups is not always clear-cut. Some promotional groups may contain supporters who have an economic interest in the cause, and interest groups try as a rule to identify their campaign with a wider cause. Because promotional groups have few tangible rewards to offer their supporters and few sanctions to wield, they appear to be weaker than interest groups.

Finally, there are the *peak groups* which speak on behalf of a set of interests—the Trades Union Congress and the Confederation of British Industry are prominent examples. Some have suggested a fourth type, *latent groups*, in which a common interest exists but has not been an organized group—the homeless, for example, existed long before Shelter was established in 1966. It is worth remembering that very few groups are primarily oriented to the national political process and their interest in the political arena may be episodic.

Context

Group activity in Britain, as elsewhere, is shaped by *the political context*. Some years ago, an American political scientist noted the tendency for the formal structure and approach of pressure groups to resemble that of the national political institutions (Eckstein 1960). In the United States, for example, the dispersal of power between the courts, Congress, the Presidency, and the state governments was mirrored by a dispersal of group activities. In France the concentration of power in the Presidential office and the senior Civil Service has made these the primary targets of groups. In Britain formal decision-making is concentrated in the Whitehall departments; groups therefore regard access to these as crucial. Other political contacts and arenas (such as Parliament and the media), may be used in support or, if there is little access to the higher echelons, as substitutes. But the ability to press a case directly upon a minister—or, more usually, upon his senior civil servants—is a good measure of a group's political standing and its influence with the government.

Consultation with affected and recognized interests is a cultural norm in British politics. An established group expects to be consulted, almost as of right and certainly as a courtesy, about the details of forthcoming government legislation and any administrative change that is likely to affect it. A minister courts trouble and lays himself open to charges of deceit if he neglects this 'rule' of policy-making and political etiquette. Consultation is valuable to both the department and the interest. The former gains information, the opportunity to reassure potential critics, and the chance to amend plans and win understanding and even cooperation from the group. Civil servants may be professional administrators but often they lack specialist or practical knowledge in areas for which they have responsibility. They turn for this, naturally enough, to the representatives of the interests. In return, the group gains official recognition, an opportunity to influence policy at an early stage, and an insight into the thinking of the department. Consultation with some groups has become almost quasi-official and some Acts of Parliament lay down a statutory requirement for consultation, as in meetings between officials of the agriculture department and representatives of the National Farmers' Union (NFU) to negotiate the annual price review for farmers' products. But traditional usage and form are as important as legal requirements in leading to consultation. Some groups have acquired such a status and influence in Whitehall that they have assumed a role similar to that which Bagehot attributed to the monarchy: 'the right to be consulted, the right to encourage, and the right to warn'.

Contacts between group spokesmen and departmental officials may be formal, as on official departmental committees and working parties, or informal, for example, meeting for lunch, or chatting on the phone. In many cases the official and interest spokesman may have developed working relationships over a number of years. Both sides gain from having good personal relations, maintaining contact, and acquiring an accurate understanding of each other's thinking and sticking-points. Interests know that a civil servant cannot commit a minister and that the latter, moreover, cannot commit the Cabinet, which will be the final arbiter if an issue is controversial or politically important. No minister cares to appear before Cabinet colleagues with the reputation, whether deserved or not, of being a captive of a lobby. It is the case, however, that once a decision in principle has been reached by the minister it tends to develop a momentum and, if supported by the Prime Minister, may be difficult to overturn.

A British political party enters government with certain policies worked out in some detail, perhaps even to the stage of having draft legislation. Policies are drawn up with an eye to achieving various goals: maximizing electoral popularity, balancing internal party demands, having an affordable programme, and anticipating the likely reactions of affected groups. Apart from the role of trade unions in the Labour party it is striking how little a party consults with interest groups in drawing up a manifesto. Parties are rarely free in practice to adopt and abandon policies as the mood takes them but are constrained to some degree by their inheritance of interests, ideals, and traditions. Conservative manifestos, for example, regularly favour a lesser role for the state, lower direct taxation, a tight rein on public expenditure, and encouragement for free enterprise, while Labour's favour more expenditure on social services, redistribution of income and wealth, and, until 1992, more state intervention and control over industry.

Where a government claims to have a mandate for a particular measure, then the group may only have the opportunity to discuss details of the proposed legislation. The industrial relations legislation of the 1970 Heath government and the 1979 Thatcher government was formulated and passed with little discussion with the unions. Once an issue becomes politicized in the sense that the major parties assume rival positions, then a group's influence is likely to depend on its proximity to the party in power.

Few groups are able to exploit their members' votes to reward or threaten a minister. People often vote for reasons other than their group interest. They may simultaneously belong to many groups and have rival roles and interests. Affiliation to different groups may not only weaken an individual's identity with any one, but may also provide contrasting cues for voting. For all the trade union leaders' almost monolithic public support for Labour at general elections, less than half of union members have voted for the party in recent general elections. Groups, as well as governments, have to take account of rivals' interests. The lobby for railways is countered by a lobby for road transport, and both may be opposed by environmental groups; the support of the British Medical Association (BMA) for pay beds has been opposed by other health service unions which wish to see them eliminated. The role of the minister, caught in the crossfire of rival lobbies, may resemble that of a broker trying to accommodate different views and arrive at a compromise acceptable both to the various interests and to the party in Parliament.

A final factor is the degree of support for the group's aims provided by the political culture and public opinion. The growing climate of tolerance in the 1960s and 1970s helped organizations campaigning for a relaxation of sanctions against divorce, abortion, and homosexuality. In comparison with the early decades of the century, such groups as the temperance movement (which opposed the sale of alcohol) and the Lord's Day Observance Society (which wanted to keep Sundays free from sport and entertainment events) have declined in influence.

The rise of green issues in the 1980s illustrates the importance of the climate of opinion. Environmental groups such as Greenpeace and Friends of the Earth have been among the fastest growing groups in the past two decades in Britain, as in much of Western Europe. In 1995 Greenpeace claimed a British membership of 400,000 and an income of £9 million, both more than the Labour party could command. Yet little more than a decade ago these groups were widely regarded as eccentric. Their case has been helped by environmental disasters like the Chernobyl explosion, Torrey Canyon disaster, and publicized reports about acid raid, pollution, and the greenhouse effect (McCormick 1991). Leaders in all political parties, business, and trade unions are now concerned to present themselves as 'green', and develop policies to curb the harmful environmental effects of industry and transportation.

Techniques

Groups may exercise influence and exert pressure throughout the *cycle of policy-making*: promoting awareness of a problem, defining alternative policies, trying to influence the content of parliamentary debate, media coverage, and legislation, and then reacting to the implementation of the resulting law, regulation, or circular. The most effective and persistent technique is to lobby the civil servants behind the scenes. Established departmental advisory committees and agencies provide opportunities for such contacts. The importance of a group's ability to present relevant, accurate, and timely information for policy-makers cannot be overestimated. In *negotiation*, a policy emerges out of bargaining

BOX 8.2. A SIMPLIFIED MODEL OF INTEREST GROUPS AND THE POLICY-MAKING CYCLE

Stage	*Group activity*
1. Awareness of need for action: placing the issue on the agenda	• creating a favourable climate of opinion • arousing media and the public • lobbying MPs, parties, think-tanks
2. Formulation: detailed drafting of proposed legislation, regulations, circulars	• attempt to gain access • information, expertise, and cooperation in exchange for consultation and negotiation
3. Legislation: debates and committee stages in both Houses	• lobby MPs, peers, and committee members for amendments • private members' bills
4. Implementation: Whitehall; local authorities; executive agencies	• provide feedback for renewed policy process

EUROPE has added a new dimension to the policy cycle. Many pressure groups now direct some of their lobbying efforts and resources towards Brussels for those policy sectors falling within EU competence.

between a department and the interest. In *consultation*, the views of the group about a proposed line of policy are 'merely' solicited. Even after a policy has been decided against a group's advice the group may still be influential in its implementation.

Access

The chances of a group gaining access to the relevant department or decision-makers depend on various factors. A group's authority is enhanced where it is representative and has a high-density membership, that is, a high proportion of actual to potential members, as in the case of the British Medical Association. A group's leverage, in the shape of the department's reliance on it for information and administrative cooperation, and functional indispensability influence a government's willingness to grant access. A reputation for discretion, responsibility, and confidentiality is also important, not least for distinguishing between 'insider' and 'outsider' groups. Insider groups are likely to have regular access to the department and exchange information—something which is unlikely to take place without mutual trust. A group's adherence to norms of 'responsible behaviour'—are its

BOX 8.3. INSIDER STATUS

In 1989 Wyn Grant categorized groups as 'insider' and 'outsider' (*Pressure Groups, Politics and Democracy in Britain* (London: Philip Allan, 1989)).

What is an insider group?	A group having direct and regular access to the government department responsible for its particular policy sector.
Why do groups seek to become insiders?	In Britain's centralized system of government groups enjoying direct access to Whitehall, where detailed policy is drawn up, have the best chance of influencing policy outcomes.
How do groups become insiders?	• they have information and expertise the government needs • they speak with authority for their sector and have a high density of membership • they have leverage/sanctions • their aims are compatible with the policy agenda of the government of the day
Do all groups seek to become insiders?	No, it brings constraints on a group's freedom of speech and activity and may involve an expectation of consultation in exchange for the group delivering members' cooperation in the policy that results. In short, constraints are the informal 'rules of the game'.

leaders reliable to deal with, will they keep confidences, will they avoid controversy?—increases the likelihood of its access. Outsiders have few rights of access because they have not established trust and/or have little relevant information to provide. The Howard League for Penal Reform has access to the Home Office; RAP (Radical Alternatives to Prisons), which favours the abolition of prison, does not. Arms manufacturers have long had access to Ministry of Defence officials, but the Campaign for Nuclear Disarmament (CND) did not. It is easy to see how, over time, civil servants in a department and group leaders may come to share a similar outlook about 'sensible' policies.

Courting outside opinion

Attempts to court outside opinion occur on two levels. The first is *informed opinion*, represented in the quality newspapers or opinion-forming journals and magazines like *The Times*, *Guardian*, *Daily Telegraph*, *Independent*, and *The Economist*. There are also specialist publications such as the *Times Educational Supplement* for teachers, or *Lancet* and the *British Medical Journal* for doctors. The opinion pages of such usually loyalist papers as the *Daily Mirror* or *Daily Mail* will also carry weight with Labour and Conservative ministers respectively. A group (and indeed the department) seeks a favourable disposition among informed observers, and decision-makers are likely to be aware of, and sensitive to, the judgement of such people.

The second level of influence is *public opinion*. In the nineteenth century public opinion expressed itself through mass petitions, public meetings, and demonstrations to sway the House of Commons. Attempts to persuade public opinion today are done largely by public relations techniques and reliance on the mass media. Public campaigning may create a favourable climate of opinion in which the group can press its demands—media coverage of patients' lengthy delays in receiving hospital treatment, or of the poor conditions of roads, or financially pressed schools laying off teachers, for example, may help the Health, Transport, or Education departments respectively in their Whitehall battles for more resources. A group's leadership may also campaign as a means of demonstrating to members that it is active. Finally, a group may draw on favourable opinion-polling evidence to promote its cause. At all times, a group is careful to promote not merely its sectional interest but to identify the interest with the public welfare. The recruitment of more teachers, improvements in their pay and career prospects, and greater provision of in-service training and facilities are presented as means of providing 'better education', or a British Medical Association campaign is not just for a better deal for doctors but for the health service. Public opinion rarely decides issues, but the climate of opinion affects the standing of a group and is a factor that elected politicians will want to take on board.

Whether or not a group 'goes public' is likely to depend on two conditions. The first is whether the balance of public opinion is favourable. Advocates of the restoration of capital punishment for murder often favour a referendum on the issue because opinion polls show support. Lobbies for 'minority' causes, like penal reform or one-parent families, tend to rely on established and less public channels. The Howard League for Penal Reform has avoided public campaigns, not only because public opinion has generally been opposed to its more tolerant line but also because it calculates that its good relations with the Home Office might be damaged as a result. Civil servants prefer 'behind the scenes' negotiations.

The second condition is whether the group still has access to Whitehall. As a rule, groups holding rallies, marches, and demonstrations usually either lack access or have found it unproductive. Rallies by CND and marches of the unemployed or the National Front are usually a reflection of a failure to gain a hearing in the appropriate departments. In 1995 animal rights groups took direct action to prevent the live export of cattle from Britain to the Continent. More episodic groups sprout up to protest at proposed motorway and bypass schemes or the dumping of nuclear waste. These groups present more direct challenges than other groups with established Whitehall connections.

Lobbying

Finally, a group may lobby friendly MPs. Apart from trade union sponsorship of some Labour MPs, a number of MPs are paid retainers by interest groups, to represent their case in the House. This does not mean that MPs can be 'bought'. They are obliged, if they speak in a debate in the Commons, to declare any financial interest they may have in the matter before the House, and since 1975 there has been a register in which MPs are required to declare their interests and outside earnings. There are also strict rules about bribery: attempts to bribe are a breach of privilege and punishable by the House of Commons. In 1994 and 1995 concern over MPs receiving payments for asking questions in the Commons and working for lobbying groups led to the appointment of the Nolan Committee to report on the standards of conduct in public life. One recommendation, accepted by the government, prevented MPs from doing paid work for lobby companies.

MPs cannot legislate on, or propose, measures involving public expenditure. But the opportunity for private members' legislation allows an MP to sponsor a bill; the abolition of capital punishment (1967) and reform of the abortion law (1967) were both achieved this way. Such reforms are only carried out if the government of the day is willing to make legislative time available, and it will only do this if the minister tacitly approves of the measure. More often the MP's usefulness is that he can present the group's views to the House, inform other MPs about its aims, and provide political advice. Groups may provide authoritative and up-to-date information and assistance to the MP which he would not otherwise get. Parliamentary criticism of the minister for not pursuing a particular policy (which he may privately support) may also help him in his Cabinet battles.

Strength of feeling on an issue in the House of Commons can influence a minister. In 1984 the Education Minister, Sir Keith Joseph, had to abandon plans to increase steeply the parental contribution to support children in higher education, in the face of widespread complaints from Conservative backbenchers. In 1986 similar pressure influenced the Cabinet to decide against the sale of Land Rover to an American company and vote down a measure to allow shopping on Sundays. In 1989 the Minister for Trade and Industry, Lord Young, showed initial sympathy for a report from the Monopolies and Mergers Commission to reduce the number of public houses owned by the major breweries. The brewers fought back, employing a firm of consultants and using press and poster campaigns ('Be vocal, it's your local') and, crucially, by lobbying Conservative MPs—ninety-three of whom signed a supportive House of Commons motion. In the end Lord Young gave way. In all of these cases it was dissent on the part of the government backbenchers that was decisive in advancing the group's case.

BOX 8.4. A SIMPLIFIED MODEL OF PRESSURE-GROUP ACTIVITY

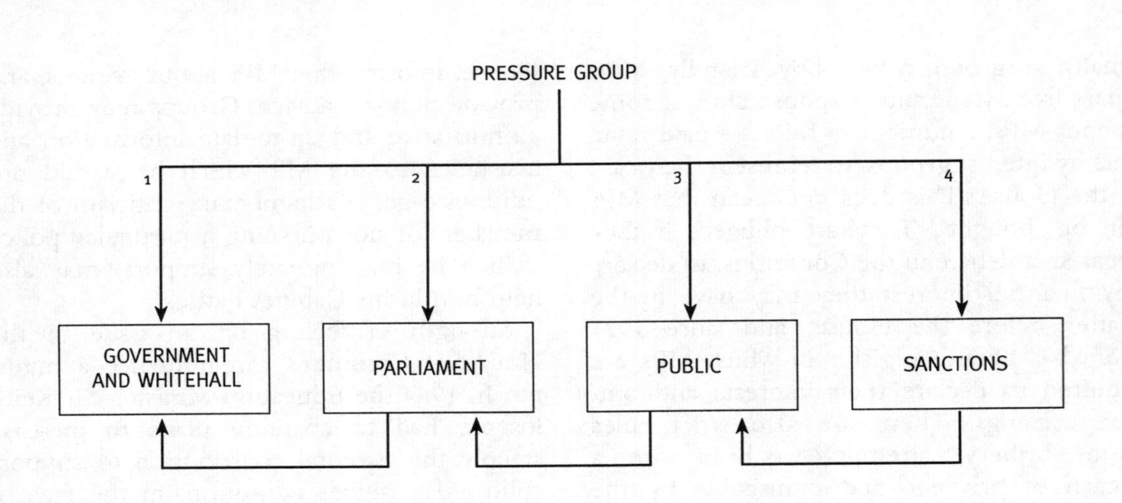

1. Groups attempt to influence government directly and privately; if this fails to bring the desired result . . .
2. Groups lobby MPs, who:
   - question ministers
   - write to civil servants
   - raise issues in party meetings
   - table amendments
   - table 'early day motions'
   - introduce private members' bills

   If this fails . . .
3. Groups may turn to a public campaign:
   - posters
   - mail
   - media advertisements
   - favourable articles in the quality press
   - marches

   Publicity may then persuade MPs;

   if this fails . . .
4. Sanctions may be used:
   - strikes
   - boycotts
   - violence

   aimed at affecting the public, leading to calls for government action

Groups may exert pressure through one or more of these channels simultaneously.

Pressure groups' view of power in the political system

Groups were asked to rank various offices and institutions in terms of their perceived influence over public policy in general. The results were as follows:

	% of groups placing this institution/ office 1st	Overall/ weighted ranking*
Prime Minister	58	2
Cabinet ministers	23	1
Media	13	5
Senior civil servants†	9	3
Junior civil servants‡	3	6
Junior ministers	1	4
House of Lords	1	9
Political parties	1	8
Backbench MPs	–	7

Notes

* Overall rankings were calculated by weighting each group's top ranking by one, its second ranking by two, and so on. The office or institution with the lowest overall score therefore came top of this ranking.

† Senior civil servants = permanent secretary, deputy secretary, and under secretary grades.

‡ Junior civil servants = assistant secretary and principal grades.

Source: Rob Baggot, 'The Measurement of Change in Pressure Groups', *Talking Politics*, 5/1 (1992).

Exercising pressure

Groups may also rely on sanctions or the threat of sanctions. A group can always make this threat as a last resort, but it is a last resort and signifies the failure of alternative approaches. In fact, most groups for most of the time are trying to persuade, inform, and keep in touch with the government, and pressure is used rarely. Nevertheless, it is the threats and pressures that often make the headlines.

Producer groups have the most *leverage*. By withdrawing their labour or even working to rule they may be able to deprive the public of a key service. The effectiveness of sanctions depends on a group's leverage and its willingness to exploit it. Withdrawal of labour in the 1970s by key groups such as coal miners, power workers, or dockers disrupted ordinary life. The action may be aimed at the employer, or at the government where (as in the public sector) it is the employer, but the effect is to coerce the community also. As

Arab oil prices increased in 1973–4, so industry and households in Britain became heavily dependent on coal. This in turn greatly strengthened the bargaining power of the miners and eventually they broke the government's statutory wages policy. Industrial action by other groups, such as civil servants, teachers, and social workers, may be more drawn out and less effective because the public is less immediately dependent on their services. Since 1979, however, strikes and the threats of industrial disruption have been less successful.

A simplified but useful model of pressure-group activity may see the 'typical' group moving through various stages in its attempts to influence government: if that fails it moves to Parliament, then to public campaigning, and, if all else fails, to the threat and use of sanctions.

Trade unions

At the end of 1992 some nine million workers (36 per cent of the work force), were members of trade unions. Since its formation in 1868, the Trades Union Congress (TUC), to which many unions are affiliated, has had no challenge to its position as the peak organization of organized labour, and more than 90 per cent of all trade unionists are in unions affiliated to it. In France and Italy, by contrast, the labour movement has been divided along political and religious lines.

The trade unions differ from other interest groups in that many of them are affiliated to the Labour party and a number also sponsor (or provide financial assistance to) Labour MPs, although the latter are not an important vehicle for union influence. For much of the post-war period the unions were able to negotiate with governments of any political persuasion. Since 1979, however, the union–government contacts have been scaled down, as Conservative governments abandoned incomes policies and pursued economic policies anathema to the unions (Marsh 1992).

In spite of the affiliation of many unions to the Labour party the union movement as a whole has been sensitive to a distinction between its political and industrial roles. Union general secretaries prefer to sit on the TUC's monthly General Council rather than on Labour's NEC. As the voice of organized labour the TUC has to speak on various issues to the government of the day. Although it has obvious sympathies with the Labour party it also has interests which have to be preserved independently of the party and, in contrast to many of its member unions, it is not affiliated to the Labour party.

The dominant position of the unions in the structure of the Labour party has been discussed in Chapter 7. They clearly have a major opportunity to shape the party's policies, particularly when the party is in Opposition. The exercise of that power requires the unions to have a clear idea of what they want and that a clear majority of them are of one mind, two conditions that are rarely achieved. Unions have been most concerned and most united in seeking to preserve a free hand for themselves in the conduct of industrial relations. They have generally been willing to allow the political side of 'the Labour movement' to go its own way. But tensions over industrial relations and incomes policies have strained the party–union connection during periods of Labour government in the 1960s and 1970s. Since 1983 successive Labour leaders have tried to distance the party from the unions (Minkin 1991).

The power of the trade unions has waxed and waned with the political and economic circumstances. After the failure of the 1926 General Strike, the unions abandoned political action and, faced by predominantly Conservative governments in office and high unemployment, found that membership declined and that their bargaining strength was weakened. This has also been the position since 1979. Nevertheless, several factors enhanced the unions' post-war position in bargaining with employers and governments, compared with the earlier period. First, until 1979 governments were fairly sympathetic to the unions' goal of full employment. Secondly, in contrast with most other groups and individuals, trade unions were immune from civil liabilities for losses or damages which arose from actions taken in the cause of a trade dispute. Thirdly, the dominant position of the unions in the Labour party enabled it to influence the party's policies. Finally, British trade unions have a relatively higher proportion of the workforce organized than in many other West European states, although the figure has declined since 1979.

In 1969, the unions and their supporters in the parliamentary Labour party forced the Wilson government to abandon a proposed reform of industrial relations. They also frustrated the operation of Mr Heath's Industrial Relations Act (1971) and eventually secured the repeal of those parts which they found objectionable. In 1974 the economic damage caused by the miners' strike, against the government's statutory incomes policy, convinced the Conservative ministers that they could not carry on without a general election—which they lost. In 1979 the Labour government's claims that it could count on the loyalty and cooperation of the unions were shattered by the official and unofficial strikes that broke out in the so-called 'winter of discontent'. The connected themes of Britain being 'ungovernable' and 'union power' were prominent in British politics in the 1970s.

By 1979, if not earlier, both major parties had come to regard the power of the unions as a problem for the conduct of economic policy and the authority of government. A commitment to manage the economy at, or near, a level of full employment strengthened the hand of the unions in wage bargaining. The potentially inflationary effects of this policy had been recognized in the famous 1944 White Paper on Employment. Governments have tried to cope with the problem variously by:

- *The use of law*, as in the abortive *In Place of Strife* proposals (1969), the failed Industrial Relations Act (1971), and the successful Employment Acts (1982, 1984). These measures have tried to formalize the industrial relations 'jungle' by providing for pre-strike ballots, curbs on picketing and secondary action, sanctions against unofficial strikes, elections of union executives, and limits on the operation of the closed shop
- *Incomes limits*, both statutory and voluntary
- *Conciliation*, or forms of partnership, such as the National Economic Development Council (NEDC) or Labour's Social Contract (1974–9)
- *Unemployment*. The abandonment since the mid-1970s of full employment as a central objective of economic policy and growth of unemployment has weakened the bargaining power of some trade unions

The unions have been in retreat for at least the past decade and a half. Rising unemployment, particularly in manufacturing and changes in work patterns, has reduced the membership over that period by a third. The government and employers have managed to

BOX 8.5. TRADE UNIONS AND POLITICS

In the twentieth century governments have accepted responsibility for a wider range of economic activity. This inevitably politicizes the role of economic interest groups such as trade unions. It is possible to discern a rise and fall in the political power of organized labour. Prior to 1824 trade unions were illegal combinations. In 1900 they were instrumental in the formation of the Labour party. By 1945 they were described as the 'fourth' estate. In the mid-1970s trade unions were incorporated into much of the economic policy process. By 1979 over 50% of the workforce was unionized in approximately forty-eight unions. By 1995 trade unionists accounted for only 37% of the workforce, trade unions had lost many of the legal immunities they previously enjoyed, and union representatives inhabited the outer circles of power. The Labour party has made no promise to restore unions to their pre-1979 status, it refuses to make a specific commitment to a minimum wage, and is in the process of weakening the constitutional position of the trade unions within the Labour party itself.

1824 Repeal of Combination Acts which had outlawed TUs

1875 Conspiracy and Protection of Property Act effectively legalized strikes

1900 Trade unions instrumental in the creation of the Labour party
Taff Vale Judgement: trade unions could be sued by employers for losses resulting from strikes

1906 Trade Disputes Act: reversed the Taff Vale Judgement

1910 Osborne Judgement: prevented the use of TUs' funds for political purposes, undermining the financial basis of the Labour party

1913 Trade Union Act: reversed the Osborne Judgement

1926 General Strike: temporarily weakened the TU movement

1962 Creation of NEDC: forum for TUs and government consultation on economic policy

1964 Frank Cousins, member of the Transport and General Workers Union becomes Minister of Technology

1969 Union pressure forces Harold Wilson and Barbara Castle (Secretary of State for Employment) to back down over 'In Place of Strife' legislation to curb unions

1971 Industrial Relations Act

1974 Heath government brought down by trade unions

1974 Social Contract between TUs and Labour government: social legislation and repeal of Industrial Relations Act in exchange for wage restraint

1978 Winter of Discontent contributes to fall of Callaghan government, 1979

1979 Mrs Thatcher promises 'No more beer and sandwiches at No. 10' (for trade unionists)

1980 Employment Act: curbed picketing and use of the closed shop

1981 TUs acquire 40% vote in electoral college for electing Labour leader

1982 Employment Act: further restrictions on closed shop; TUs liable for members' actions during disputes; prohibits political strikes

1984 Trade Union Act: internal democracy for TUs; secret ballot before strikes and for political levy. Government bans TUs at GCHQ

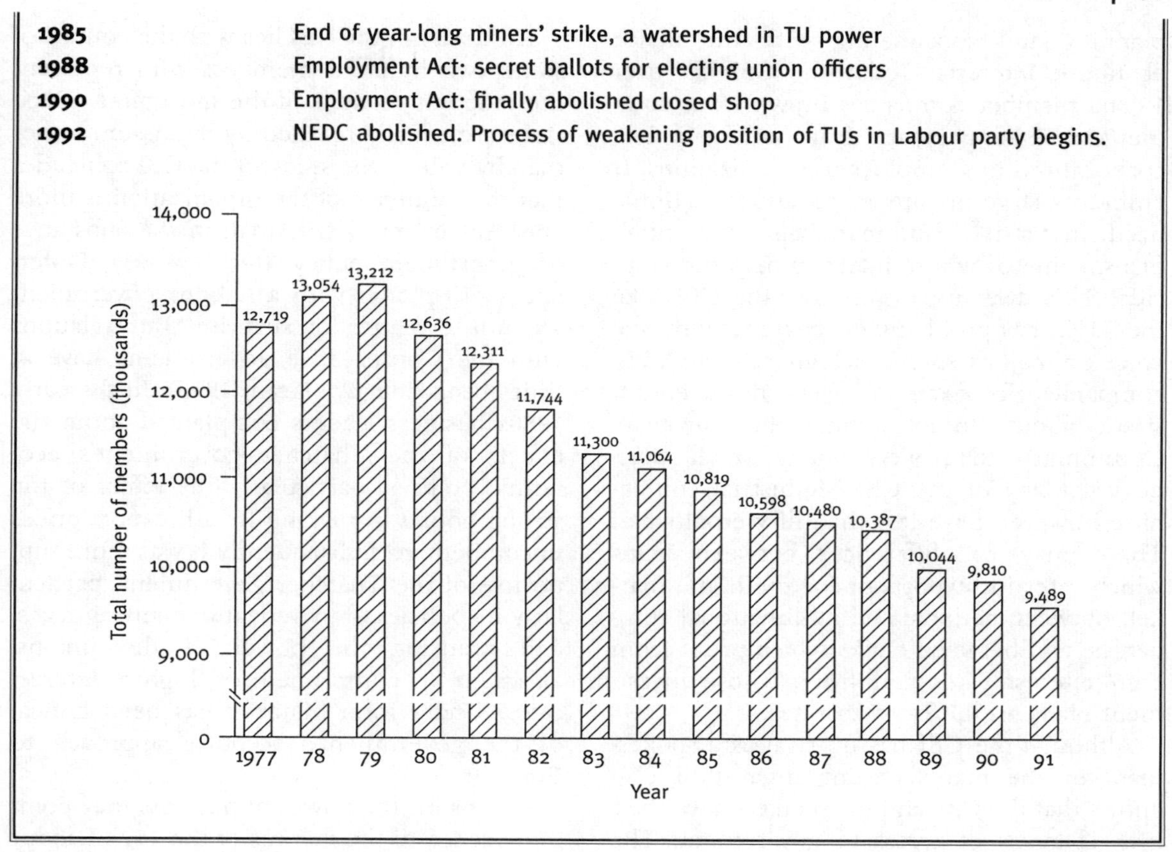

beat most strikes, at times using the government's recent industrial relations laws. The legislation of the Thatcher government has also helped to shift the balance of influence to employers and to individual union members as against union leaders. The government has had some success in extending performance-related pay and local pay bargaining in the public sector. The opportunities for unions to consult with the government have been greatly reduced as the latter has not operated an incomes policy and downgraded and then abolished NEDC.

Business

The major 'peak' organization representing business interests is the Confederation of British Industry (CBI). It was formed in 1965, with government encouragement, out of three existing groupings. At that time the Wilson government was keen on economic planning and consultating with the major economic interests. Today the CBI has over 12,000 member companies drawn from commerce and industry and from among public corporations and employers' organizations. It embraces large private corporations, nationalized industries, multinationals, and small firms, some of whose interests may not coincide. This diversity means that the CBI, like the TUC, has problems in speaking with one voice on major issues which affect capital. Big companies, for example, were more enthusiastic about Britain joining the European Community than were many small businesses. Many of the CBI's industrial companies, however, have less than fifty employees. There may be differences between firms which cater for exports or for the home market, between national and multinational companies, and between banks which profit from high interest rates and firms whose investment plans are hit by such rates.

Although the CBI has no rival as a spokesman for the manufacturing interest, it also knows that the City and the retailers have their own channels of access to government. The Bank of England, for example, acts as a voice for the City to government and vice versa, and there is a Committee of London Clearing Banks, and a City Liaison Committee drawn from the clearing and merchant banks, stockbrokers, and insurance brokers (Moran 1981). Such bodies as the Retail Consortium, the Society of Motor Manufacturers, and the Engineering Employers' Federation exist as separate organizations. Indeed, most large industries have a separate trade association, and big firms like ICI or Shell also enjoy rights of access to government, either directly or through their own associations. In recent years, the Institute of Directors has emerged as a vigorous spokesman for free enterprise, and differs from the CBI in wanting much less state intervention, more privatization, and free-market policies and has been consistently hostile to the idea of a single currency in the EU.

Authority in the CBI lies with the council of some two hundred members who represent the different sections of the movement. Decisions are usually reached by discussion rather than by votes. One study of the CBI concludes that the influence of the organization is more apparent in the details than in the substance of government policy. (Grant 1993). Differences of opinion with a Labour government are not a matter of surprise. But relations with the Conservative governments have at times been difficult (Grant 1993). In the early 1980s business leaders complained about the effects of the Thatcher government's economic policies, particularly the effect of the strong pound in making British export prices uncompetitive. Subsequently it was more supportive of the Thatcher government, particularly its policies of privatization, cutting taxes, and reducing the power of the unions, although it complained of higher interest rates. Under John Major it has been critical of the government's sceptical approach to Europe.

Claims for the power of business may point in evidence to the failures of the 1974 Labour government's plans to take a share in leading companies by 'planning agreements' and for worker participation on the boards of companies. Under CBI pressure, both were gradually watered down; by 1979 only one firm had made a planning agreement and the worker participation scheme had lapsed. Business decisions on investment, production, and employment are crucial to the performance of the economy, and a government has to take this into account. Multinational companies can invest elsewhere if they find the political and economic climate less than favourable. The Thatcher and Major governments have often pointed to the high level of overseas investment in Britain, particularly from the USA and Japan, as a vindication of their industrial relations legislation and opt-out from the EU social chapter.

On the other hand, it might be claimed that

the close relationship between business and government equally reflects the dependence of the former on the latter. A government's specific purchasing and investing decisions and its general economic policy shape the environment within which business operates. This mutual dependence of the state and the large corporations cuts across both the simpler free-enterprise and Marxist models of relations between the two. In the 1970s governments of both parties imposed limits on dividends, prices, and salaries in the private sector as well as the public. The lack of success of much of British industry since 1945 has hardly been a token of a politically dominant big business or of a bond of sympathy between business and government.

Critics of the thesis of business power may also point to the separate interests of the financial sector, the banks, and the insurance and finance houses. The shorthand term for this interest is the City, and it has an authoritative spokesman in the Governor of the Bank of England. British governments have been sensitive to this interest because London has for long been an important international money market. Until recently the pound was a major international currency—and its value is largely determined by the confidence of foreign holders of sterling. For many years the interests of the manufacturing sector have been sacrificed to those of the City whenever a choice had to be made between the two. In the inter-war years, the Treasury orthodoxy was for balanced budgets and, until 1931, for remaining on the Gold Standard, regardless of the damaging effects on industry and unemployment. In the early 1980s Mrs Thatcher's government sacrificed full employment and a competitive exchange rate for the pound to the goal of reducing inflation, and this was consolidated while Britain was a member of the ERM between October 1990 and September 1992. The record of post-war economic policy does not provide convincing proof of the power of private business over government. Crises of sterling and the balance of payments have usually produced economic policies aimed at restoring confidence among financiers despite being damaging to industry (Grant 1993).

The corporatist phase

The concern to produce a faster rate of economic growth and improve government contact with the mass producer interests led the Conservative government to create a National Economic Development Council ('Neddy' or NEDC) in 1962. In common with other West European governments, British politicians were seeking a forum for negotiating on the economy with the major interests. The main Council consisted of government ministers and civil servants, representatives of the CBI and TUC, and a few 'independent' members, such as the chairman of the Consumer Council and chairmen of the nationalized industries. In addition there were over twenty councils, or 'little Neddies', for particular sectors of industry, again with representatives of the other interests.

The 1974–9 Labour government established other tripartite forums. In 1974 the Manpower Services Commission (now defunct) was established to deal with job creation, placement, and training. The Health and Safety Commission was set up in 1975 to deal with health and safety at work and the Advisory, Conciliation, and Arbitration Service (ACAS) in 1976 to deal with industrial relations disputes. Each of these bodies had three nominees from the TUC and CBI, three 'independent' members, and a further three members drawn from other interests, all working under a government-appointed chairman. Although ultimately responsible to the Secretary of State for Employment, these bodies enjoyed a good deal of autonomy in shaping public policy in their areas, and were formally linked to producer interests.

By 1979 some commentators claimed that the role of groups in policy-making had made Britain 'corporatist', or 'tripartite' (Middlemas 1979). Although the two terms are often used interchangeably, commentators agreed on the importance of the producer interest groups, that the government–group linkages had grown in number and closeness, and that the contacts bypassed Parliament. In the 1970s governments of both parties sought the cooperation of trade unions and employers in policy covering employment, training, investment, and restraint of prices and incomes, and were prepared to negotiate about other social and economic policies. The producer interests appeared to be 'partners' of the government in managing the economy.

Since 1979 the Conservative ministers have distanced themselves from the tripartite approach. They did not seek cooperation of trade unions for an incomes policy, did not have an 'Industry Strategy' as such, and privatized many government-owned industries. The Conservative government has managed to identify 'corporatism', incomes policies, and 'Social Contracts' with the poor economic performance of the 1970s. NEDC's role was reduced and it was finally abolished in April 1992, soon after the general election.

The Thatcher government vs. the interests

Since 1979 there has been more conflict between interest groups, largely in the public sector, and the Conservative government than at any other period in the post-war era. Three factors may explain this:

1. The Thatcher government made no secret of its wish to change the direction of policy in much of the public sector. Groups which have an interest in the *status quo* are therefore likely to be offended.

2. Ministers wished to constrain the growth of public spending on many services and to reduce state subsidies; again, it is not surprising that interests dependent on such expenditure, notably on health, social welfare, education, and local government, will complain. The Major government has continued with policies designed to encourage market disciplines and competition as spurs to efficiency in the public sector.

3. Ministers, finally, took seriously claims that the authority and autonomy of an elected government should not be compromised by bargains with sectional interests, particularly the trade unions.

A theme running through the government policies has been a general *distrust of the producer interest groups*. Ministers complained that there was too often an 'unholy alliance' between a department and its client interest group. Interest groups and bureaucrats are, allegedly, interested in maximizing their own advantages—in the form of autonomy, salaries, and conditions of work—rather than responding to the consumers of their services. A more elaborate statement of this case is made by Mancur Olson (1982), who has claimed that interest groups ('distributional coalitions') use their power to resist change and slow down innovative policies; their veto power produces an 'institutional sclerosis'. In Germany and Japan, by contrast, the interests were either smashed or severely weakened by the rise of totalitarian governments or defeat in the war, and both countries have enjoyed post-war economic miracles. Olson claims that the collapse of the regimes destroyed many tradition-bound forces. Innovation was helped by a fresh start. In the case of Britain, the continuity of the regime allowed the interests to become entrenched. Much of this analysis has been accepted implicitly by Thatcherites.

The influence and formal powers of two other major interests, local government and trade unions, have been severely curtailed in the past two decades. Some commentators might point to the partisan factor at work here—Labour-supporting trade unions and left-wing local authorities. But the Conservative government has also battled with *middle-class professional groups*. The claims to self-regulation and possession of professional expertise by lawyers, doctors, and teachers have been challenged by the government. These groups have claimed to be the expert judges of what is a 'good' service, be it legal advice, health care, or education. School teachers have found that their pay-bargaining rights have been scrapped, and a core curriculum, national testing of pupils, and a contract of service imposed on them. University teachers have lost tenure and the quality of their teaching and research is regularly assessed by independent bodies. Finally, in 1989 the doctors had new contracts imposed on them by the Ministry of Health. These limit their budgets and link a greater part of their pay to the number of patients they treat. In 1989 the Lord Chancellor proposed changes to the legal profession, notably ending the barristers' exclusive rights of audience in the higher courts and the solicitors' monopoly conveyancing services. The offended barristers recruited Saatchi and Saatchi, the advertising agency, and employed lobbyists to advise MPs and water down the proposals. In much of the public sector there is a new emphasis on audit performance pay and value for money. Not all of the measures have enhanced the power of central government. Some of the changes have given more power to members, for example, ballots for trade union members, schools opting out of local authority control and managing their own budgets, hospitals and GPs controlling their own funds.

The pace of change has continued under the Major government, whether it be the Citizens' Charter for consumers against the deliverers of public services, or the 1993 Trade Union Reform and Employment Rights Act.

Conclusion

The role of pressure groups and the development of the 'group politics' style of decision-making are crucial to an understanding of the development of British politics. They also raise important questions concerning power, accountability, and democracy.

The claim that Britain is a pluralist political system rests on the belief that several autonomous groups are involved in policy-making and that no group dominates the process. The late R. T. McKenzie defended groups as an ancillary form of representation, enabling voters to convey more specific views to the government than can be represented by broad party programmes at general elections every four to five years. Democracy, he claimed,

BOX 8.6. THEORY AND PRACTICE: PRESSURE GROUPS

Theory

Pressure groups remedy the shortcomings of representative democracy:

- they permit a continuous dialogue between government and the governed
- they provide opportunities for political participation
- they provide government with information and expertise
- they articulate and defend minority interests
- they act as a check on abuse of government power
- they compete with one another to influence policy outcomes

Practice

- not all groups have equal access
- resources and leverage vary between groups
- clientelism may arise and be detrimental to national interest

Case-Study

The National Farmers' Union (NFU) and the Ministry of Agriculture, Food and Fisheries (MAFF)

The 1947 Agriculture Act created a legal obligation for MAFF to consult with representatives of producer interests — in effect the NFU — and establish annual price reviews setting guaranteed prices for agricultural products. This initiated a close and continuing relationship. The union's representatives frequent the MAFF and sit on its committees. There is also a degree of sidewards movement between the two — senior officials from the ministry have moved to top positions in the NFU.

In 1988, Edwina Currie, junior Health Minister, announced that almost all egg production in Britain was infected with salmonella poisoning. After two months of argument and counter-argument from health and farming experts, Mrs Currie resigned and the government set out to reassure the public that it was safe to eat eggs again. Many argue that this demonstrated the 'producer' strength and leverage of the 'insider' NFU in contrast to the weakness of, or absence of, groups organized to articulate the 'consumer' interests or national interest in the health of the population and public safety.

includes the 'right to advise, cajole, and warn [the authorities] regarding the policies they should adopt' (1974 280). Others suggest that a kind of free market operates and prevents one set of interests being dominant for too long: the government's fear of public opinion, the existence or likely emergence of rival or counter-groups, and certain general 'rules of the game' combine to produce over time a rough balance of power among interests. Because citizens are members of different groups and have different loyalties they do not become too closely attached to one interest.

Critics, however, make two contrary arguments. One is that interests differ in their organizational strength, resources, leverage over society, and access to decision-makers. *The group system, in other words, is biased in favour of some interests and against others*. The minority who benefit from a policy may be better organized and more articulate than the larger number of non-beneficiaries, for example, the taxpayers. Another matter of concern is the definition and defence of the public interest amid all the sectional pressures. Is the government able to pursue a coherent, long-term set of policies, or does it reflect the balance of pressure-group forces? But in the 1970s the interests with which government shared power (for example, the unions in the Social Contract) were sectional, subject to few legal controls, and not publicly accountable. Critics were impressed by the power of groups to resist government policy which they disliked. Britain, it was often said, suffered from 'pluralistic stagnation' and government was unable to embark on a new policy without the cooperation of powerful groups (Beer 1982, Olson 1982).

In the United States 'interest-group liberalism' has been blamed for producing policies which are justified largely on the grounds of their acceptability to the major interests. Such an approach results in conservative policies designed not to disturb the *status quo*. In Britain, critics have claimed that sensible investment decisions have been sacrificed to consumption and tax cuts made for the short term to win elections. The 'bargaining' with groups and the 'bidding' for the voters' favour hardly encourage coherence in policy-making.

Since 1979 Conservative governments have turned away from both the traditional 'group approach' of bargaining with interests and the use of incomes policies. In part this stems from a belief that the corporatist methods were not working (by 1979 few would disagree) and a belief in the superiority of the market. In part it also derives from a belief in the virtue of strong government, of a government that does not have to share its power with other groups—particularly local government and public-sector trade unions. But it also stemmed from the government's determination to change many existing policies and the policy style, and this resulted in conflict with many groups. By the end of Mrs Thatcher's premiership many commentators thought that British government had moved from being weak and 'overloaded' to being too dominant.

Summary

- Interest groups extend opportunities for participation in the state and provide a channel for continuous communications between government and those affected by government policy.
- They can be seen as a potential threat to democracy—anonymous, unaccountable groups exerting sectional power over government. Alternatively, they can be seen as watch-dogs, checking against government abuse of power and safeguarding the rights of minorities within a majoritarian system.
- Interest groups and political parties have some shared characteristics but some distinctions can be drawn: political parties seek office, general influence, and ultimately responsibility; interest groups primarily seek access to decision-makers and to exercise sectional influence.
- The growth of group activity accompanied the rise of the interventionist state. Governments depend upon the expertise of groups, but at the same time governments impinge more extensively on citizens' lives and this prompts citizens to seek to infuence the nature of such intervention.
- There are three types of groups: promotional groups which advocate a cause; interest groups which are usually based on economic interests; and peak groups which articulate the views of a set of interests.
- Groups in Britain operate in the context of a highly centralized state. Consultation with affected interests forms a part of the 'rules of the game' underpinned by a belief that government should be by consent.
- The effectiveness of a group at any particular time will depend upon such factors as: the compatibility of its aims with the programme of the government of the day; the prevailing climate of opinion; and the degree of support for the group's demands.
- Techniques: groups may exert pressure throughout the cycle of policy-making and may try to gain access to the different institutions involved. Some groups succeed in achieving 'insider' status, giving them access to Whitehall and the crucial drafting stage when detailed policy is drawn up. But insider status brings constraints on a group's freedom of action. As a general rule pubic campaigning can be seen as a measure of a group's weakness. Some groups may use, or threaten to use, sanctions to press their claims—this is usually a weapon of last resort.
- Trade unions are prominent interest groups. From 1945 to 1979 they had a major consultative role in the development of government economic policy, partly because governments needed the expertise and consent of unionists, partly because unions possessed considerable leverage in terms of sanctions. A change in political culture since 1979 has served to weaken trade union influence.

- Whilst the CBI is the major 'peak' organization of business, 'business' is not monolithic. There are distinctions in the scale of businesses and the nature of their activities. This results in some internal incompatibility of businesses' interests. Financial interests seem to have had greater success in influencing policy outcomes than have manufacturers.
- In the 1970s Britain appeared to be becoming a corporatist state: much economic policy was the outcome of tripartite negotiations between the government and the major producer groups.
- Mrs Thatcher rejected the policies of the post-war consensus and the consensual style of policy-making which reduced the government's authority and restricted the government's role to that of being negotiator with, or referee between, competing interests.
- Pressure groups provide an important channel of political communication, but the inequality of resources, leverage, and access between groups requires that governments be aware of, and compensate for, bias in favour of some groups.

CHRONOLOGY

1787 The Abolition Society, one of the earliest promotional pressure groups, was founded by William Wilberforce and Thomas Clarkson, succeeded in abolishing slavery by 1807

1839 Anti-Corn Law League

1903 Women's Social and Political Union, pressing for votes for women

1945 Enormous increase in interest-group activity accompanied the rise of the interventionist, welfare state

1962 National Economic Development Council ('Neddy' or NEDC) created as a forum for tripartite discussions on the economy

1965 Confederation of British Industry (CBI) formed

1974 Heath calls 'Who Governs?' election

1975 Health and Safety Commission created. Labour minority government makes Social Contract with the unions

1976 Advisory, Conciliation and Arbitration Service (ACAS) created

1978 Winter of Discontent — many local services paralysed by unions

1979 Mrs Thatcher introduces a less consensual political culture

1988 Edwina Currie resigns her position as junior Health Minister after offending egg producers and The National Farmers' Union by claiming that almost all egg production was infected by salmonella

1990 Widespread anti-poll tax demonstrations

1992 NEDC abolished

1995 Animal rights demonstrations against the export of live animals

ESSAY/DISCUSSION TOPICS

1. Do interest groups enhance democracy?
2. Interest groups are not an unmixed blessing. Discuss.
3. Why have Conservative governments since 1979 been more successful than their predecessors in altering the legal position of trade unions?
4. Why are some pressure groups more powerful than others?
5. What kind of problems face environmental pressure groups?
6. 'All citizens have an equal opportunity to influence public policy through interest-group activity.' Discuss.
7. Did Mrs Thatcher bring an end to the corporatist state in Britain?

RESEARCH EXERCISES

1. A location near your town has been chosen as a possible site for a nuclear waste processing plant. What kind of action could you take to stop the project?
2. With reference to examples, distinguish between 'insider' and 'outsider' groups. Explain, using examples, which is likely to be the most successful in pressing its cause, and why.

FURTHER READING

Dated but still interesting is S. Finer, *Anonymous Empire* (London: Pall Mall, 1958). On corporatism see K. Middlemas, *Politics in Industrial Society* (London: Andre Deutsch, 1979). On producer groups see W. Grant, *Business and Politics in Britain* 2nd edn. (London: Macmillan, 1993) and D. Marsh, *The New Politics of British Trade Unions and the Thatcher Legacy* (London: Macmillan, 1992).

The following may also be consulted:

Baggot, R., 'Pressure Groups and the British Political System: Change and Decline?', in B. Jones and L. Robins (eds.), *Two Decades in British Politics* (Manchester: Manchester University Press, 1992).

Grant, Wyn, *Pressure Groups, Politics and Democracy in Britain* (London: Philip Allan, 1989).

Norton, P., 'The Changing Face of Parliament: Lobbying and its Consequences', in id. (ed.), *New Directions in British Politics* (Aldershot: Edward Elgar, 1991).

Watts, Duncan, 'Lobbying Europe', *Talking Politics*, 5/2 (1993).

9 THE MASS MEDIA AND POLITICS

Reader's guide

The media in their various forms provide a vital channel of political communication. The relationship between politicians and the media is symbiotic: politicians need the media to get their message across; the media need 'copy'. The prominence of some Downing Street press officers such as Joe Haines (1974–6) Bernard Ingham (1979–90), the fact that in 1995 the government employs 470 information officers, the fact that the Conservative party's erstwhile public relations firm, Saatchi and Saatchi, became a household name, and the fact that the current Labour party frontbench have submitted to 'Folletting' (guidance from image consultant Barbara Follett) are a testament to the importance politicians place on media relations and the need to create a media 'friendly' image and style.

But the media are more than a neutral channel of political communication. They are pervasive and persuasive. They do not merely provide information about the world of politics, they are influential participants in the political system. Some argue that the media are now the fourth estate: they have to some extent usurped the role of the Opposition and Parliament. Question Time on the TV; radio and TV in-depth political interviews; documentaries; polemical articles in the quality press serve as watchdogs to alert the electorate and to keep the government on its toes. Politicians are less fazed by a gruelling session at the Parliamentary dispatch box than by a 'grilling' from 'infamous' political interviewers like Jeremy Paxman and John Humphrys.

This chapter explores the relationship between politics and the media in Britain. It begins by establishing what constitutes the media. It goes on to address the following questions: how influential are the media? do the media set the political agenda? what part do they play in election campaigns? what has been the impact of television on British politics? The chapter concludes by noting that although the media may act as a check on politicians there seem to be few checks on the media themselves.

THE vast majority of the population gain their information about politics from the mass media, from press, radio, and television. The media not only report but provide interpretations of the news. They can help to set an agenda—by highlighting certain issues and neglecting others—and shape popular perceptions and images, for example, 'loony left' Labour councils, a 'strong' Mrs Thatcher, or a 'wimpish' John Major. Indeed, many politicians are so convinced of the influence of the media that they think that, in effect, what the media reports is, virtually, political reality. Politics is largely a *mediated activity,* in which the media mediate between voters and politicians.

The media are more than channels of reporting and influence; they are now a powerful interest group in their own right. For example, in Britain the multinational News International Corporation controls not only BSkyB Television but three national daily newspapers and two national Sunday newspapers, including the best-selling Sunday and daily papers. Critics complain that sections of the press have their own agenda, are unelected political actors, and operate free from regulation. In 1990 a Press Complaints Commission was established (replacing the Press Commission). This is largely a self-regulating body and was set up as an alternative to Parliament's providing statutory regulations.

Britain has eleven national daily newspapers (see Table 9.1). Some four-fifths of British households take a daily paper and three-quarters of the population aged 15 or more claim to read one. These figures are higher than those in many other comparable states. In 1994 two popular daily papers, the mass circulation *Sun* and *Daily Mirror* (with sales of 4.1 and 2.5 million respectively), were read by 10 and 8 million respectively. There is a division between the six popular papers or tabloids (e.g. the *Sun* and *Daily Mirror*), and the five broadsheet papers (e.g. *The Times* and *Guardian*). In 1994 the former had daily sales of around 11 million, the latter of 2.5 million. In 1959 there were two tabloids, with combined sales of 5.7 million and seven broadsheets, with combined sales of 10.4 million.

The tabloids often present a simplified, exaggerated, and personalized view of politics. Pictures and graphics appear to have driven out words, and partisanship is blatant in elections (Seymour-Ure 1995). The broadsheets provide a more extended and serious coverage of politics and current affairs and also have a more middle-class readership; the *Express* and *Mail* draw their readers fairly evenly from across the social spectrum, while the *Sun* and *Mirror* readership is largely working class. The readerships differ in their evaluation of the media. Broadsheet readers are likely to regard the press as their main and

Table 9.1. National newspapers

Title and foundation date	Controlled by	Circulation[a] average February–July 1994
Dailies		
'Populars'		
Daily Mirror (1903)	Mirror Group Newspapers (1986) plc	2,497,076
Daily Star (1978)	United Newpapers	671,373
Sun (1964)	News International plc	4,101,988
'Mid market'		
Daily Mail (1896)	Associated Newspapers Ltd	1,796,795
Daily Express (1900)	United Newspapers	1,358,246
Today (1986)	News International plc	595,468
'Qualities'		
Financial Times (1888)	Pearson	296,634
Daily Telegraph (1855)	The Telegraph plc	1,013,860
Guardian (1821)	Guardian Media Group plc	400,856
Independent (1986)	Mirror Group consortium	275,447
The Times (1785)	News International plc	507,894
Sundays		
'Populars"		
News of the World (1843)	New International plc	4,769,105
Sunday Mirror (1963)	Mirror Group Newspapers (1986) plc	2,560,234
People (1881)	Mirror Group Newspapers (1986) plc	2,006,393
'Mid market'		
Mail on Sunday (1982)	Associated Newspapers Ltd	1,972,012
Sunday Express (1918)	United Newspapers	1,544,404
'Qualities'		
Sunday Telegraph (1961)	The Telegraph plc	633,112
Independent on Sunday (1990)	Mirror Group consortium	327,689
Observer (1791)	Guardian Media Group plc	495,483
Sunday Times (1822)	News International plc	1,205,457

[a] Circulation figures are those of the Audit Bureau of Circulation

most reliable source of news. Readers of the tabloids regard television as the main and most reliable source of news (Negrine 1994: 2–3).

The regions have some important local newspapers, like the *Scotsman* in Edinburgh, the *Telegraph* in Belfast, and the *Western Mail* in Cardiff. But the press has, on balance, been a force for political nationalization and centralization. Virtually every household in Britain can receive a daily paper which has been printed overnight in London or Manchester.

The British Broadcasting Corporation (BBC) was established as a public corporation in 1926, and is financed by fees from the sale of licences to owners of radios and televisions. It held a monopoly in the field until in 1954 the government established an Independent Broadcasting Authority (now Independent

Television Commission). The latter subcontracts programmes to independent companies and is financed out of advertising revenues. Although the two major broadcasting authorities are independent of the government, the latter possesses important levers. It decides on the size of the BBC licence fee, and so determines the BBC's finances, and the Prime Minister appoints the director-general of the two authorities and their boards of governors. Both bodies operate under charters which are subject to review and renewal by the government. Both also accept that they are under an obligation to be impartial in their political coverage and balanced in their treatment of the parties. The satellite and cable channels are free from such requirements.

Television was slow to cover politics. Until 1959 broadcasters ignored the general election campaign apart from carrying party election broadcasts, although news of the general election dominated the national press. Only in 1989 were proceedings in the House of Commons televised.

Political influence

The political influence of the media is often misunderstood. Newspapers and television are certainly important as sources of information about politics. They are, however, only one among several shapers of political attitudes. Attempts at direct pressure on government usually fail. The efforts of the press lords Beaverbrook and Rothermere to shape Conservative party policy in the inter-war years failed spectacularly. Lord Rothermere's demand in 1931 that he have a say in the choice of Stanley Baldwin's Cabinet in return for his press support was contemptuously dismissed by the latter as a demand 'for power without responsibility, the prerogative of the harlot through the ages'. Much of the Conservative-inclined press campaigned for the unseating of John Major in the party's leadership election in 1995—the *Mail*, *Telegraph*, and *Sun* strongly so. Conservative MPs and, according to surveys, Conservative supporters did not accept the advice.

Theories about the political impact of the media are subject to qualification. Because individuals have different predispositions the media have different effects on different people. Readers, listeners, and viewers are not passive. Early academic research suggested that the media operated more to reinforce than change political views. The *selectivity* theory argued that users of radio and press interpreted what they heard and read to *reinforce* existing loyalties. People seemed to be selective in their *exposure* to political communications, selective in their *interpretation*, and selective in their *retention*.

Later studies of the new medium of television adopted what was called a *uses and gratifications theory* (Blumler and McQuail 1968). This emphasized that the voters' values and expectations led them to use the media for, variously, information, reinforcement of values, voting guidance, and entertainment. More recently, both theories have been challenged on the grounds that declining partisanship results in fewer voters having firm party loyalties to reinforce, and the media therefore may have more scope for forming and changing views.

A third, so-called *radical*, model argues that

the media reflect the assumptions and values of dominant groups in society and are biased in particular against parties of the left and trade unions which question those assumptions. The Glasgow University Media Group (1976, 1982), in its coverage of industrial disputes, claims to detect a bias for employers and against trade unions; the latter are often presented, unfavourably, as making 'demands' and 'threatening' to strike. The Glasgow group regard the mass media as part of the ruling order in a capitalist society. This group's research on industrial relations, however, has been subject to damaging reassessment (Harrison 1985).

Politicians frequently express concern about *political bias in the press*. It is understandable that complaints have most often come from the left, for most national newspapers support the Conservative party. There has long been an imbalance in the partisan leanings of the press, and it has increased since 1970. In that year the Conservative party was supported by papers which had 57 per cent of the national daily circulation, Labour by papers which had 43 per cent (Seymour-Ure 1991). By 1992 the figures had shifted to 70 per cent Conservative, 27 per cent Labour. The switch of the popular *Sun* newspaper in 1974 to the Conservatives accentuated the imbalance. As long as 80 per cent of the working class read a tabloid, most Labour voters will be exposed to a Tory-supporting paper. Interestingly, since the 1992 general election, some of the formerly staunchly pro-Conservative papers (including the *Sun*) have become highly critical of John Major's government.

In the first national study of British voters Butler and Stokes (1969) reported that people often chose newspapers to fit in with their existing party loyalties, in line with the *selectivity thesis*. Martin Harrop (1986) doubted the effect that the press could have in the four weeks of a campaign and thought that on balance the support of the press was worth only a 1 per cent advantage (or some ten seats) to the Conservatives. Table 9.2 shows that less than half of the *Sun's* readership in 1992 followed its political line.

Other work, however, allows that over the long term the press can influence attitudes. The continued imbalance in press partisanship and the steady diet of anti-Labour propaganda in the 1970s and 1980s may have helped to make voters more resistant to the party's

Table 9.2. Party supported by daily newspaper readers (%)

	Conservative		Labour		Lib Dem	
Election	1987	1992	1987	1992	1987	1992
Daily Telegraph	80	72	5	11	10	16
Daily Express	70	67	9	15	18	14
Daily Mail	60	65	13	15	19	18
Financial Times	48	65	17	17	29	16
The Times	56	64	12	16	27	19
Sun	41	45	31	36	19	14
Today	43	43	17	32	40	23
Daily Star	28	31	46	54	18	12
Independent	34	25	34	37	27	34
Daily Mirror	20	20	55	64	21	14
Guardian	22	15	54	55	19	24

Source: MORI.

policies, reinforcing negative images of the party and putting party spokesmen on the defensive by raising 'scares'. Interestingly, partisanship among the tabloids has increased at the same time as voters have become less partisan.

The questions of press bias and influence came to a head in 1992. After the election the *Sun* newspaper proclaimed 'ITS THE SUN WOT WON IT'. That paper had campaigned ruthlessly against Labour and Neil Kinnock and on polling day ran a nine-page special, with the front page reading 'NIGHTMARE ON KINNOCK STREET'. The paper's boast was endorsed by Lord McAlpine, a former Treasurer of the Conservative party, and by Neil Kinnock, who attacked sections of the press when he announced his resignation as Labour leader. McAlpine called the Tory tabloid editors 'heroes' because of their strong support for the party.

Never in the past nine elections have they (the Tory press) come out so strongly in favour of the Conservatives. Never has the attack on the Labour party been so comprehensive . . . This was how the election was won, and if the politicians, elected in their hour of victory, are tempted to believe otherwise, they are in real trouble next time. (*Sunday Telegraph*, 12 April 1992)

Although the paper may have slightly increased perceptions of Labour as being divided and extreme, subsequent analysis has cast doubt on the *Sun*'s claims that it produced the alleged late swing to the Conservative party. The pro-Conservative swing occurred across readers of most papers, including those of the Labour-supporting *Mirror*. A careful study of the subject in 1992 concluded that partisanship, however attenuated, still operates, leading most voters to screen out unwelcome or divergent political messages. In other words, the old factor of selectivity still operates (Semetko, Scammell, and Nossiter 1994).

Some part of the increasing polarization in the press may have been a consequence of Mrs Thatcher and her determination to break with consensus politics. The Conservative tabloids were less partisan when Mr Heath was Prime Minister and have been unhappy with John Major since his 1992 election victory. The tabloids strongly supported her 'tough' stand on law and order, vigorous advocacy of the British cause in the EC, and campaigns against the trade unions and left-wing local councils. Nigel Lawson complained that Mrs Thatcher, her press secretary Bernard Ingham, and the partisan *Sun* fed one another. Mr Ingham, according to Lawson (1992), would often provide a story for the *Sun*, the paper would carry it, and then Ingham would highlight it in his digest of the daily press he prepared for Mrs Thatcher. It was not surprising that she thought she had a hot line to the British people!.

Agenda setting

In spite of the many tensions which exist between politicians and the media there is also a mutual dependence between them. The former need publicity for their policies, speeches, and initiatives, while the latter need the cooperation of political actors to write their stories. Political parties, groups, and government departments have media officers who are specialists in liaising with the media, and they try to put a favourable 'spin' on relevant news items. In 1995 central government had 470 information officers and the Ministry of Defence 160, more than work on some national newspapers (*Guardian*, 19 April 1995). A good example of the spinners' work was seen in the Conservative leadership election in 1995. Before the result many commentators, and even supporters of Major, claimed that he would need at least 230 votes to be safe: between 210 and 220 would not be a convincing victory. As soon as the result was known (Major 218, Redwood 89, absentions, spoilt ballots, etc. 22) Major's spokesmen dominated the airwaves and declared it an 'impressive victory', 'the largest margin of victory, by any Tory leader facing a serious challenge', 'a larger share of the vote than Tony Blair gained in Labour's leadership election'. The media duly reported it thus; the 'spinners' had succeeded.

A growing number of activities by politicians, parties, and governments are prepared exclusively for the media, involving the staging of media events and photo-opportunities which would not take place without the media. A particularly famous example was Mrs Thatcher's holding of a calf in a Norfolk field for over fifteen minutes for the benefit of

Bernard Ingham, press secretary to Margaret Thatcher (1979–90), reflects upon the art of news management:

'The media always felt we were up to no good, though. This reflects the normal state of tension between the Government Information Service and journalists; indeed between journalists and press officers, however good, the world over . . . I would deal with only one aspect of the relationship: news management. This is a most heinous offence in journalists' eyes and is the crime — I do not jest — with which press officers, and not least Chief Press Secretaries, are most frequently charged. Journalists see us all as consummate Machiavellis. I plead utterly, completely and wholeheartedly guilty. Of course, I tried to manage the news. I tried — God knows, I tried — to ensure that Ministers spoke with one voice, if necessary by circulating a standard speaking note which I wrote myself. I was hit by all kinds of journalistic avalanche if they spoke out of turn. I tried to ensure that Ministers were aware of what each other was doing and whenever they were likely to cut across each other. Dammit, that was what I was supposed to do. And if I failed, then the media would fall like wolves upon Government and condemn it as useless. I tried — hell's teeth I tried — to make sure that the media had early, embargoed copies of important documents so that they had plenty of time to digest them and prepare their stories before publication. And what did some of these pious, sanctimonious characters do? Occasionally, they made it impossible for me to help them because they broke embargoes or, more often, put opponents of the course of action in a position to issue instant, damning comment. . . .

But news management, in the sense of ensuring that nothing is allowed to get in the way of the story the Government wants to get over, is impossible in the modern world. A chief information officer can plan and plot and generally bust a gut in trying to clear the way for an important announcement. But he is not in charge of events, or journalists. Nor has he any influence over a Minister who suddenly goes ape and commands the front pages. Unfortunately, he is not in command of other events in the global village in this age of instant, telephonic and, more important, televisual communications. An earthquake here; a famine there; an horrendous aircrash elsewhere; or quite simply some appallingly visual event anywhere — and his news management cause is lost. The real news managers today are the media themselves. It is television which predominantly dictates news values for the masses: either there are pictures or there are not, and if there are no pictures there is no news. It is the editors of this world who receive the raw material in the form of reports from the journalist on the spot. Then they get to work on it: developing it, exploiting it, angling it, massaging it and eventually presenting it as polished fact and unvarnished truth. And they dare to accuse Government press officers of news management?'

Bernard Ingham, *Kill the Messenger* (London: Harper Collins, 1991), 187–8

Alan Clark, Parliamentary Under-secretary of State for Employment (1983–6), turns his hand to news management:

'Department of Employment *Thursday, 29 March*

I was in Peter Morrison's room early. Out of the blue he told me that my suggestions for reforming the Lady's Private Office would in all probability be put into effect over Easter. But who was going to be put in charge? None other, or so he claimed, than David Young.

This is appalling. I hardly know the man. But from what I've seen he's simply a rather grand H.R. Owen, the big Rolls-Royce dealers' salesman. I got to know the type well when I was working as a runner in Warren Street just after the war. Very much not one of the 'Club'.

Worse was to come. Peter told me that he was going to have a Red Box, Minister of State rank, and 'operate from the Lords'. It is virtually signed up, as Peter has to find a replacement as head of MSC. The co-ordinating structure at Number 10 will be promulgated during the Easter recess. I said I was not too keen on the idea. The Party never likes outsiders getting high ministerial rank without going through the mill. . . . as the appointment was so imminent there was nothing that could be done to alter it.

But I determine to have one try. There is only one journalist influential enough to make an effective scene about this if he minded to do so, and that is Peter Riddell of the FT. He is well informed about all three Parties and writes with great insight.

But he might approve. Supposing he likes DY? I contemplated leaking the story to him, and made an assignment to speak to him in the Lords' corridor. Then I got cold feet; too easily traced.

Almost immediately afterwards a brilliant idea, my old friend and standby for many a dirty trick, Jonathan Aitken. I told him the problem. He was very understanding, got the point at once, and promised that he would attend to it immediately.

House of Commons *Friday, 30 March*

The fish has taken! A critical account, on the front page of the FT, setting out the Prime Minister's intentions, her decision to appoint David Young, the rank intended for him, etc., plus a beautifully restrained piece of comment about 'reservations' in the Party concerning DY's 'controversial past' in property development, etc.

It is very late, but might just do the trick, partly because it is a leak of the intention, partly because it is couched in such distinguished, though disapproving language.'

Alan Clark, *Diaries* (London: Weidenfeld & Nicolson, 1994), 67–8

photographers in the 1979 election campaign. The statements of leading politicians, press conferences, and annual party conferences are also prepared with media coverage in mind.

Parties in government are not passive in their relationships with media. They may decide not to supply party representatives to a programme if the subject is potentially embarrassing. During election campaigns, party strategists may decide to take media questions and grant interviews only on the party's chosen themes of the day. Politicians may also leak material, to advance their own career or causes, or to denigrate rivals. Over the Westland crisis in 1986 senior ministers and officials were a party to leaking a letter of the Attorney General which was damaging to Michael Heseltine (see p. 262). Nigel Lawson has written of how, as Chancellor of the Exchequer, he briefed friendly newspapers and provided quotes on Labour's tax plans in the 1987 election. This led the *Daily Mail* to lead with 'LABOUR'S LIES OVER TAXATION' and the *Daily Express* the following day led with a story 'EXPOSED: LABOUR'S TAX FIASCO'. 'Friendly' proprietors and editors were frequently in touch with Mrs Thatcher's and John Major's offices during the 1987 and 1992 general elections respectively, and Labour strategists similarly had close contacts with the *Mirror*.

A problem is that the continuous and more intensive coverage of politics speeds up public awareness of crises, scandals, statements, and actions by political rivals—and therefore

BOX 9.1. THEORY AND PRACTICE: A FREE PRESS AND INDEPENDENT BROADCASTING

FREEDOM

Freedom of the press and independence of broadcasting are defining characteristics of liberal democracy. Voters require independent information about politics and government in order to make an informed choice.

CONSTRAINTS

1. Charter

Broadcasters, although not journalists, are required to adhere to the strict impartiality code set out in their charters.

2. Legislation

- Official Secrets Acts 1911 and 1989
- laws on libel, slander, obscenity, race relations, sedition

3. Culture

The habit of secrecy is deeply ingrained in British government. Whitehall and Westminster collude to ensure that the mysteries of government remain just that. This tradition is reinforced by the Official Secrets Acts. John Major's thrust for greater openness, more intensely competitive journalism, and a declining willingness to play by the rules on the part of journalists, politicians, and officials indicate signs of a cultural change taking place.

4. Appointments, funds, and franchises

- The Director General and Board of Governors of the BBC and the Independent Television Commission are government appointments
- The government sets the BBC's licence fee
- The government also oversees the allocation of independent television franchises

All of these offer opportunities for informal pressure.

5. Regulatory bodies

- Broadcasting Complaints Commission
- Advertising Standards Authority
- Broadcasting Standards Council
- Radio Authority
- Press Complaints Commission

All perform watchdog roles over the media. Their terms of reference, composition, and powers are largely a matter for the government.

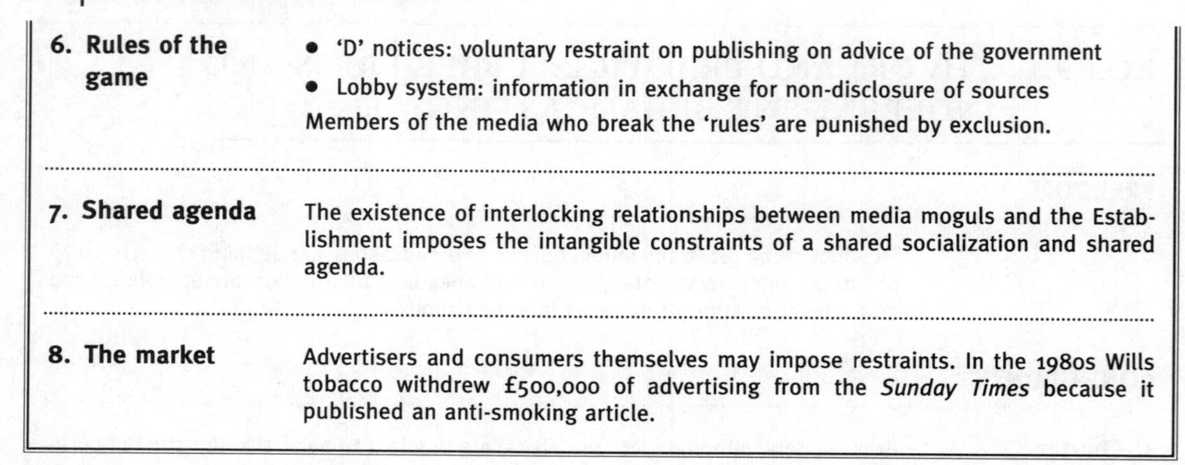

6. Rules of the game	• 'D' notices: voluntary restraint on publishing on advice of the government • Lobby system: information in exchange for non-disclosure of sources Members of the media who break the 'rules' are punished by exclusion.
7. Shared agenda	The existence of interlocking relationships between media moguls and the Establishment imposes the intangible constraints of a shared socialization and shared agenda.
8. The market	Advertisers and consumers themselves may impose restraints. In the 1980s Wills tobacco withdrew £500,000 of advertising from the *Sunday Times* because it published an anti-smoking article.

increases pressures on politicians to speak and to act. Speedy reactions do not necessarily produce the wisest responses. In the past decade ministers have been a casualty in a number of press stories. For example:

- reports of his affair with a former secretary led to Cecil Parkinson's resignation as Secretary of State for Trade and Industry in 1988
- revelations in 1994 by the *Sunday Times* that Conservative MPs accepted payments for asking questions in the House of Commons led to two being censured, fined, and suspended from the House
- press revelations of their adultery, or non-declaration of interests in the MPs' register, or acceptance of hospitality which laid them open to influence led in the three years April 1992 to April 1995 to the resignations of one Conservative Cabinet minister and twelve junior ministers

Politicians often complain about 'biased' television coverage. Tensions arise between politicians and broadcasters simply because they have different agendas. Both Labour and Conservative ministers have long complained that good news is ignored or under-reported by the broadcasters, whereas bad news, particularly about the economy, is prominently reported. BBC coverage of the Falklands war in 1982 offended Conservative leaders. Mrs Thatcher did not think that this was the time to be even-handed between 'our' side and Argentina. There was more trouble over a BBC *Panorama* programme on the SAS killing of IRA suspects in *Death on the Rock* in 1989. In the same year the government imposed a ban on the broadcasting of interviews with Irish extremists. In recent years Conservative pressure on the BBC has taken the form of pre-emptive attacks prior to a general election. In 1986 the Conservative party Chairman Norman Tebbit set up a monitoring unit in Central Office and in 1991 his successor Chris Patten urged Conservative viewers to protest about unfair coverage by the BBC. This sensitivity is not confined to Conservative leaders. Labour's Harold Wilson frequently accused the BBC of 'plotting' to undermine him.

Press and television not only compete, but may feed off each other in setting the agenda—television following up stories carried in the press, or print journalists covering a television programme. In the 1987 general election Neil Kinnock gave a television interview with David Frost early on Sunday morning. When he was asked how Britain would defend itself in the event of a Russian invasion the Labour leader replied that we would not use nuclear weapons but resort to guerrilla tactics. The interview made little impact until it was picked up by Tory tabloids, which then headlined the story for four successive days. Three days later the *Daily Mail* ran a front page 'KINNOCK: THE MAN WITH THE WHITE FLAG' and the *Express* had a virtually identical headline. This was a gift to the Conservative campaign, which wanted to make defence an issue in the campaign, while Labour, committed to an unpopular policy of unilateral disarmament, wanted to ignore the issue.

In the same election the *Sun* carried a front-page story about a private hip operation which the wife of Denis Healey, Labour's Shadow Foreign Secretary, had received years earlier. This was an embarrassment to Labour, which was opposed to private health care. The next day Healey was interviewed on TV AM, and when the issue was raised his discomfort was obvious. He lost his temper with the interviewer and the row was fully reported in the tabloids the following day.

The 1992 general election provided a dramatic case of the interaction of press and television, as both media followed up Labour's election broadcast which contrasted the NHS's delay in treating a little girl's ear complaint with the speedy treatment provided for another girl with a similar complaint but who was able to afford private care. This was the famous 'Jennifer's ear' broadcast (Harrison 1992). The broadcast was designed to support Labour's claims that the underfunding of the NHS had led to long hospital waiting lists. It also warned that a two-tier health service was emerging, with those able to afford private treatment getting a better deal. But these issues rapidly got lost in the furore that erupted over the revelation of the little girl's name, the disagreement between the parents about whether the case should have been broadcast, and the truthfulness and ethics of the film. For the next few days the main story was how certain newspapers had acquired the name of the girl featured in the film. This in turn gave rise to major rows at Labour and Conservative press conferences which were then covered in detail on television and in the press. The media by now were concentrating on the alleged dirty tricks of the two parties over the broadcast and the parties found it difficult to get the media to report anything else. Labour wanted to talk about an NHS under threat and the Conservatives about Labour's unethical broadcast. In a revealing radio interview Michael Heseltine pleaded with an interviewer 'to give us the chance to get on to the issues . . . we depend on you, there is no other way we can get over what we want to say' (*The World This Weekend*, 28 March 1992).

The media and election campaigns

The interdependence between politicians and media is at its closest during elections. The parties recruit technical experts from the public relations, media, and advertising industries to assist with campaign publicity and media presentation in general. Established film producers like John Schlesinger and Hugh Hudson and writers like Ronald Millar and Colin Welland are involved in the filming and writing of party election broadcasts. Much of the leading politicians' campaign day—the morning press conference, walkabout, and even rally—is now largely shaped by the requirements of the media, particularly television. Setting the election agenda for the mass media is the main purpose of a party's communications strategy (Kavanagh 1995).

An election is also a period when tensions between the two groups are particularly intense and fraught with the possibility for rows and misunderstandings. All parties scrutinize broadcast coverage for signs of deliberate or accidental bias or unfairness, and put pressure on the broadcasters. The latter strive to satisfy the parties' demands for balance and fairness while also striving, as professionals, to report items on the basis of their newsworthiness. Usually this means appearances of party splits, gaffes, personal attacks, and shifts in the opinion polls. Differences arise from the different role of newscasters and politicians. For the former, '. . . communications was a tool of public enlightenment, to the latter (politicians)—against rival parties *and* professional journalists' (Blumler Gurevitch, and Nossiter 1989: 159).

Campaign managers try to ensure that the media cover the party's agenda for the day or days, by:

- holding a number of press conferences on the same topic on the same day
- putting forward only one or two major speakers on the hustings, to discourage cameras from reporting somebody who does not voice the chosen issue of the day
- coordinating the party leader's activities during the day so that the same theme is reflected in what he or she says and does
- declining interviews on subjects not on the party's agenda (Blumler, Gurevitch, and Nossiter 1989)

In their turn, journalists as professionals try to create a role for themselves in shaping the campaign agenda. They concentrate on the 'horse race' (using opinion polls to report which party is in the lead) or provide 'state of play' interpretation of the campaign, and perhaps ask the questions of party leaders that the latter would rather not have posed.

Television effects

The politicians are increasingly concerned with agenda setting, and this is another cause of their sensitivity over television. A feature of the more 'permanent' campaign is that each party's communications officers and public relations advisers are regularly developing media strategies for the politicians (Franklin 1994). At times, it may appear that at least as much attention is paid to the presentation as to the substance of politics. The government of the day has many opportunities to shape the news agenda by exploiting its use of office, via ministerial announcements, policy initiatives, the budget, and Prime Minister's Question Time to dominate the airways. Party conferences, like American party conventions, have increasingly become stage-managed for the televised projection of the positive party image and strong leadership. Press coverage of Parliament has steadily declined and it is no surprise that MPs queue up to be interviewed on the *Today* programme or television, rather than speak in Parliament. Indeed the local and regional media provide opportunities for media-oriented backbench MPs to become well-known figures. The broadcasting of debates in the Commons has not lessened the desire for politicians to be interviewed outside Parliament. On major occasions like a leadership election, Cabinet reshuffles, ministerial resignations, or the Budgets, the strip of turf opposite Parliament on College Green resembles a crowded marketplace.

How the media cover politics is increasingly becoming an issue in its own right, part of the political agenda. Political parties now routinely make charges of bias, for example, over the amount and tone of coverage, the running order of stories, and the selection of interviewers. Mrs Thatcher tried to avoid aggressive interviewers like Robin Day and sought less adversarial forums in the popular programmes hosted by Jimmy Young and Michael Aspel (Jones 1992). A favourite target for politicians' criticism is the early morning *Today* radio programme. On one occasion an irate Mrs Thatcher phoned from Downing Street to complain about an item. In March 1995 a Cabinet minister, Jonathan Aitken, complained on air that Conservative ministers were being questioned more aggressively on the programme than were opposition politicians. During elections all parties (as well as the broadcasting channels) conduct a stopwatch analysis of coverage, to ensure that they receive at least their agreed shares of airtime, as well as monitoring the output for 'fairness'.

Yet in spite of the politicians' complaints about television bias, survey evidence about the perceptions of voters consistently fails to support the claims. Some four-fifths of people detect no bias in election coverage and regard television as trustworthy and truthful compared with less than a quarter who think the same of the press (Butler and Kavanagh 1980).

Television has encouraged a presidentialism in politics in so far as it focuses more on the activities of party leaders. Between general elections something like one-third of television coverage of a party's politicians is of the leader and more than half during an election. This means that a party's campaign messages are carried through the leader. The concentration is partly a consequence of television resources. Camera crews and major reporters are assigned to follow each party leader—hence there is more film of the leader. Michael Foley (1992: 121) has commented on the relationship between party managers and media: 'television's inclination to personalising the treatment and presentation of politics has been matched by the willingness of parties to provide their leaders with the prominence and licence to fit the party product to the optimal form of communications.'

Television to date has not influenced the political careers of politicians, although this may be changing. Mrs Thatcher, Michael Foot, and Mr Heath were not particularly skilful at public relations and presentation, certainly not at the time of their election as party leader. In Britain politicians usually get to lead the party by displaying parliamentary skills, being acceptable to all or most strands of the party, and being regarded as an election winner. Yet both Mrs Thatcher and Neil Kinnock worked hard to develop their television skills, and this may increasingly be regarded as an essential skill for a would-be leader. Alternatively, lack of presentational skills (being 'bad' on television) may be a barrier to political advancement. One wonders if Clement Attlee, generally regarded as one of the more successful premiers but also thought to be colourless, could have got to the top today.

Within the political parties people with communication skills have risen in importance, particularly with regard to election campaigns. Peter Mandelson (Labour) and Sir Gordon Reece and Shaun Woodward (both Conservative) all had a background in television before assuming key communications posts with their parties. Parties are also now more likely to employ an advertising agency (the Saatchi and Saatchi agency worked on all Conservative general election campaigns from 1979 to 1992), opinion pollsters, and film directors to help make political broadcasts (Kavanagh 1995).

BOX 9.2. PRESSURES FOR CHANGE: PRESS REGULATION

Issue	Proposed reform
Press intrusion: intensified competition plus changes in technology have resulted in the press becoming increasingly intrusive.	Statutory control in the shape of privacy laws similar to those of some European countries, e.g. France.
Inadequacy of regulation under the auspices of the Press Complaints Commission • insufficient sanctions • insufficiently independent • partly self-policing	Give existing regulatory bodies more teeth/sanctions and greater independence
Concentration of ownership of the press plus broadcasting, e.g. Murdoch's News International empire, raises questions about control and potential influence	Close scrutiny by Monopolies and Mergers Commission
Excessive government secrecy vs. the need for greater openness and freedom of information	Freedom of Information Act proposed by Lib Dems and also Labour (with some reservations). John Major has taken some steps in this direction, pledging greater openness, and 'D' notice system is under review.

Conclusion

Relations between media and politicians certainly seem to have deteriorated in recent years. Many politicians express fears about the agenda-setting influence of the media. The media's role in driving politicians from office, not on grounds of incompetence but because of their private lives, has also provoked widespread criticism. In his resignation speech in 1992 in the House of Commons the ex-Cabinet minister David Mellor spoke of media providing an 'alternative criminal justice system'. A Press Complaints Commission was established in 1990 to receive and adjudicate complaints about press intrusions on privacy. In so far as newspapers are represented on the Commission it is a form of self-regulation. It is, however, widely seen as a toothless body. Some reformers advocate the introduction of a privacy law, but this has proved difficult to frame and most of the press are strongly opposed. In their defence the media say that the politicians have brought some of their troubles on themselves by their private behaviour. They warn of a possible return to the silence of the British media about Edward VII's affair with the divorced Mrs Simpson in the 1930s, in contrast to the coverage by the foreign press.

Summary

- Politics is largely a mediated activity. Media are a channel for information about politics but they are also influential in shaping opinions and an interest group in their own right.
- The media in Britain include numerous daily and weekly, national and local newspapers and periodicals; the BBC and independent national and local radio stations; plus four terrestrial TV channels and a satelite service. They will soon be augmented by a cable TV network.
- The extent and nature of the political influence of the media are keenly debated by both academics and politicians. Some argue that the media serve to reinforce rather than change political views. Others argue that declining partisanship has increased the political influence of the media. The radical model asserts that the significance of media influence is not in persuading voters to support one or another party, but in reinforcing the *status quo*, making radical alternatives seem dangerous, 'un-British'.
- The press is overwhelmingly Conservative. The broadcasting authorities are required to adhere to a strict code of political impartiality. Politicians from both sides, however, frequently accuse broadcasters of bias.
- There is a mutual dependence between politicians and the media. Many political events are staged for, or geared towards facilitating, media coverage. The 'political' leak has become a commonplace of political communication. Inevitably 'bad' news and political indiscretions attract the greatest coverage. The media largely determine 'what's news', and there is a growing trend towards media coverage itself forming part of that news.
- Election campaigns are increasingly media events, a fact reflected by the increased dependence of the parties on media experts. Daily press conferences and campaign events are coordinated with media coverage.
- Television has helped to presidentialize election campaigns and personalize political issues. The political role of television and the question of potential bias have become issues in their own right.
- Relations between politicians and the media have deteriorated. Regulation of the media is itself a controversial issue.

CHRONOLOGY

1912 'D'-notice system begins: voluntary suppression of information on government advice

1924 First party political broadcast on radio

1926 BBC created

1929 First Downing Street press officer appointed — George Steward

1936 BBC began TV broadcasting

1947 Hugh Dalton, Chancellor of the Exchequer, resigns; disclosed budget secrets to press before Parliament

1950 BBC's first transmission of general election results

1951 First party election broadcast on TV

1954 Independent Broadcasting Authority established — commercial TV

1963 Press Council set up

1973 Commercial radio began broadcasting

1981 Rupert Murdoch bought *The Times*; Lonrho bought the *Observer*

1982 Launch of Channel 4

1986 The *Independent*, a new quality daily, launched; also *Today*, a new tabloid. Leon Brittan, Secretary of State for Trade and Industry, resigns, taking responsibility for press leaks of his department's officials

1987 The Government obtains an injunction to prevent BBC showing a programme about the Zircon spy satellite

1988 Broadcasting Standards Authority created

1989 House of Commons televised. Ban on broadcasting interviews with Irish extremists

1990 Press Complaints Commission established

1992 David Mellor, Arts and Heritage Secretary, resigns and complains that the media have become almost another criminal justice system

1993 The Queen sues the *Sun* for leaking her Christmas speech

1994 The *Sunday Times* reveals the 'cash for questions' scandal

ESSAY/DISCUSSION TOPICS

1. To what extent do the media set the political agenda?
2. A free press is a myth. Discuss.
3. The media enhance democracy. Discuss.
4. What kind of problems are associated with attempts to regulate the mass media?
5. Why is control of the media frequently on the agenda?

RESEARCH EXERCISES

1. The government proposes to establish a media ombudsman to police the press. Write a leading article (600–700 words) on this subject for (a) a tabloid (b) a broadsheet.

2. Select a current item of political news. Analyse the way in which it is covered by *The Times*, the *Sun*, and the *Daily Mirror*, indicating any examples of bias in the reporting.

FURTHER READING

For a good introduction see R. Negrine, *Politics and the Mass Media*, 2nd edn. (London: Routledge, 1994). On press and television during a general election see the chapters by Harrop and Scammell and by Harrison in D. Butler and D. Kavanagh, *The British General Election of 1992* (London: Macmillan, 1992). More generally see B. Franklin, *Packaging Politics* (London: Arnold, 1994).

The following may also be consulted:

Curran, J., and Seaton, J., *Power without Responsibility: The Press and Broadcasting in Britain* (London: Routledge, 1990).

Grant, Moyra, 'The Politics of the Media', *Talking Politics*, 6/2 (1994).

Seaton, J., and Pimlot, B. (eds.), *The Media in British Politics* (Aldershot: Avebury, 1987).

10 THE ROLE OF GOVERNMENT

Reader's guide

When Mrs Thatcher expressed the view that government had a limited capacity to do lasting good but a great capacity for doing harm, she was contributing to the perennial debate about the appropriate role of government. It is indeed *the* key debate that divides those on the left from those on the right of the political spectrum. The former believe that government should play a positive role in delivering not only defence, law and order, and market regulation, but also economic and social well-being to citizens. The latter see the role of government as regulatory in a pluralist society in which individual liberty is guaranteed by providing an environment conducive to enterprise but in which economic and social intervention is kept to a minimum. From the perspective of the late twentieth century one could be forgiven for thinking that it is a debate which the left have won. In Britain the twentieth century has been the century for positive government: few areas of social and economic life are beyond the reach of state intervention. It can be argued, however, that the 1970s mark a turning-point The views of those believing that government could solve all society's problems were discredited and those of the right began to enjoy a new vogue.

This chapter defines the concept *government* and examines how it is organized in Britain. It explains how the functions and responsibilities of government have increased during this century. It discusses the changing attitudes towards and policies of state intervention in industry, exploring the debate about nationalization and privatization.

Public expenditure is one indication of the 'size' of government. This chapter examines public expenditure, the controls upon it, and the obstacles in the way of attempts to reduce it, illustrating the intensely political nature of public expenditure questions. The chapter concludes by considering the debate about the appropriate role of government and the critique offered by the New Right.

THE term *government* may refer to a set of office-holders, the executive institutions, or a system of ordered rule. A country may have a government in the first two senses, but the office-holders or institutions may simply lack authority or the ability to gain compliance with their laws, and so the country will lack government in the third sense. We may similarly have different images in mind when we refer to the *British Government*. In everyday language when we talk of, for example, the Conservative government of John Major, we refer to the Prime Minister, Cabinet, and the controlling party. But we may also refer to a more enduring set of institutions, Parliament, the Cabinet, the senior Civil Service, and the Monarch. A foreigner in search of where the British government is located may choose Westminster, Downing Street, Whitehall, or Buckingham Palace.

This chapter explores a number of issues surrounding government. It looks, first, at the central government departments, then at how the responsibilities of government have steadily increased in the twentieth century—notably over the economy and welfare—and how public spending has grown, and the attempts to reverse both trends. Finally, it considers the political debate about the proper role of government, and the critique offered by the New Right.

Departments

The growth of government activities in the twentieth century and the need for coordination have had important consequences for the organization of the central administration. Before 1914 most government departments were small in size and their work so circumscribed that not much coordination was required: when it was, the Cabinet, the Committee on Imperial Defence, and the Treasury provided it. The role of the Treasury was enhanced by a decision in 1919 to make the Permanent Secretary of the Treasury also the head of the Civil Service. New departments have been created to provide new services (e.g. National Heritage or Transport). The proliferation and specialization of functions have added to the pressures for the creation of more ministries. Interest groups are vociferous in pressing for a department to be established to deal with their concerns. According to the Civil Service Year Book for 1994 there are sixty-one government departments, nineteen of them headed by a Cabinet minister.

But this pressure for specialization has been partly offset by the need for coordination. The growing workload and heterogeneity of some departments have led to most of them having a number of ministers. The simple structure of the minister directing the department has given way to the division of responsibilities between ministers in a department, for example, the Foreign Secretary and four Ministers of State in the Foreign Office, or the Secretary of State for the Environment and two Ministers of State and a Minister for Housing. A final consequence is that the ministerial hierarchy has also become more complex and the government has grown in size. In 1900 most departmental ministers could expect to sit in Cabinet; in recent years fewer than half of the ministers have been in Cabinet at any one time.

Both parties in government have tinkered with departments in the search for a more satisfactory performance from central government. The 1970 White Paper, *The Machinery of Government*, defended the creation of larger or 'super' departments, along functional lines. Such departments, it was claimed, would provide the opportunity to achieve more integration and coordination of related policies within a department, encourage ministers to weigh the costs and benefits of alternative schemes, achieve economies of scale in management, and prevent the minister from becoming too 'departmentalized'. The possible disadvantages are that some policy disagreements may be better discussed in Cabinet rather than being confined to a department and that the burden of coordination may prove to be too great for the 'super' minister.

Between 1960 and 1979, thirty-one departments were abolished and twenty-nine new ones created. By the early 1970s five super-departments had emerged: the Foreign and Commonwealth Office (1968); Health and Social Security (1968); Environment (1970); Trade and Industry (1970); and Defence (1964). They amalgamated nineteen ministries which had existed in 1952. Mrs Thatcher was not particularly interested in departmental reorganization, although in 1988 she split the DHSS into separate departments for Health and for Social Security. In theory if the government relinquishes responsibilities departments can be closed down. On forming a new government in 1992 John Major abolished new departments for Energy in 1992 and Employment three years later. The recent trend has been to create quangos and 'executive agencies' to carry out government policies. This may or may not amount to *The New British State* (Dynes and Walker 1995), but there are over 5,000 such bodies and they spend some £46 billion each year.

The growth of government

Before 1914, most ordinary citizens could live their lives with little awareness of central government. Only a small proportion of the population paid income tax and there was

BOX 10.1. QUANGOS

Quango **Quasi-Autonomous Non-Governmental Organization**

They are many, approximately 1,500, and varied in terms of size, composition, function, and powers.

W. Jones (1994) notes that they share the following characteristics:

- appointed at ministerial level
- non-accountable
- secretive–not usually open to the press or public
- members are paid, some generously, e.g. Lord Crickhowell, the Chairman of the National Rivers Authority, receives £51,00 per year for this part-time job
- membership is predominantly middle class, white, and male: 3:1 gender ratio and only 2% of members are from ethnic groups
- they dispose of 20% of all public spending
- they have greatly increased the scope of ministerial patronage

Examples of quangos:

- NHS trusts
- University Funding Council
- National Rivers Authority

When Mrs Thatcher was elected she promised to reduce the number of quangos. The number has indeed halved since 1979, but during the same period the amount of income they dispose of more than doubled from £6,150 million to £13,750 million.

John Redwood, in a polemical article entitled 'It Takes Two to Quango' (*The Times*, 14 August 1995), explains the seductiveness of the quango option:

> 'Governments are easily tempted to set up a quango for every problem. Shouldn't people eat fewer chips and more apples? Let's set up a Health Promotion Authority . . . Wouldn't it be good if government did not have to make difficult decisions about how much money to give to university and colleges? Let's set up a Funding Council to do it.'

He claims that neither Conservative nor Labour governments can resist the temptation to create quangos, and that once they have succumbed, bureaucratization sets in:

> 'Each one needs a chairman, a board, a chief executive and a 'corporate structure'. Before long, Whitehall and the new body are up to their eyes in memorandums, articles, guidance, corporate plans, budgets and accounts. Whitehall breathes a collective sigh of relief that a problem has been tackled . . . a press campaign countered. The quango is at arm's length from the minister and the department. Its very establishment creates the impression of doing something.'

He goes on to reflect that all too soon the quango goes from being the answer to being the problem. Questions are raised about control of expenditure and accountability. Many quangos in effect become government-created pressure groups. He concludes:

> '. . . one cheer for quangos and three for democracy. Let the politicians decide direct and debate. Quangos should exist only when they can be given clear instructions and a clear purpose . . . '.

little support for the government regulating the nation's economy beyond balancing its income and expenditure, what was called a nightwatchman's role. Most employees in the public sector were soldiers, sailors, tax collectors, and postmen. The state's main activities were in the *basic* areas of defence, police and law and order, diplomacy, and finance, although there was some regulation of health and industry. Indeed, it is the oldest departments in British government—the Home Office, Foreign Office, and Treasury—which still deal with these functions. Some would argue that these 'primary' tasks (in the sense of being the earliest assumed by government) of maintaining law and order, defending the state's borders, and controlling the currency remain the essential functions of government; if it does not control these, then it ceases to be an effective government. The same claim cannot be made for health, education, or pensions, for example, which can be provided by the market (Rose 1976).

Since then, and particularly since 1945, the major expansion of state responsibilities has been in the provision of such social services as health, education, and welfare, with the bulk of public expenditure shifting into this last category (Fry 1975). In 1994 the three areas accounted for 57 per cent of public spending. Governments make many far-reaching rules—in the form of laws and regulations—for social conduct. They distribute benefits and goods in the form of transfer payments, services, and capital, and raise revenue in the form of direct taxes on income and indirect taxes on goods purchased. Increasingly, the substance of election campaigns and the promises of political parties in their election manifestos reflect this change in the agenda—reducing poverty, promoting economic growth, boosting employment, improving the quality of life, and intervening to correct market failures.

For the first three decades after 1945 British governments, like those of most other Western states, presided over a set of policies designed to promote the mixed economy and the welfare state. These changed the relationships between government and citizen and government and the private sector. The package of policies in this consensus is familiar enough and included the following:

- full employment budgets
- greater acceptance, even conciliation, of the trade unions, whose bargaining position was enhanced through a larger membership and full employment
- public ownership of basic services or utilities
- state provision of social welfare, notably education, health, housing, and social security. In turn, these programmes required high levels of public expenditure and taxation
- economic planning of a sort via a large public sector and a reduced role for the market

These policies have sometimes been described in shorthand as modern capitalism and social democracy. Many of the policies were already in place in wartime. Indeed, the war experience was a good example of how events can alter the expectations of policy-makers, particularly their perception of what is politically and administratively possible. Middlemas (1979: 272) notes that 'Slowly but inevitably the state came to be seen as something vaster and more beneficent than the political parties.' The salient themes of the policies were as follows:

1. *A positive role for government.* John Maynard Keynes for economics and William Beveridge for welfare provided the most important justification for active government. As employer, taxer, and distributor of benefits the government played a larger role, one that seemed to be popular with the voters. Processing issues through the political arena rather than the free market seemed to produce greater social peace, at least in the 1950s and 1960s.

2. *The provision of the Welfare State.* During the war there was greater acceptance of the claim that citizenship needed to move beyond the achievement of legal and political rights to include a range of social rights. The Welfare State soon acquired an ideological life of its own, incorporating ideas of fairness, a common society, and collectivism (Timmins 1995).

3. *The pursuit of economic growth*, to provide welfare and protect the take-home pay of workers. During elections parties made competing claims about their ability to improve the rate of economic growth and thereby improve public services.

4. *The conscious pursuit by governments of full employment* as a goal of economic policy, acknowledged in the famous 1944 White Paper on Employment. Mass unemployment during the inter-war years discredited free-market (or liberal economic) ideas: the achievement of full employment via Keynesian economics strengthened the collectivist case.

5. *Optimism that these goals* could be achieved and the belief that the relevant knowledge for improving social conditions was available. After all, full employment policies had worked, so why should not education, housing, and regional policies also succeed?

The role of government as extractor and distributor of revenues and as manager of society increased over time. A lesson of the wartime Keynesian revolution in economic management was that government, by regulating aggregate demand in the economy, could end the mass unemployment of the inter-war years. Governments were expected to try and affect the aggregate levels of economic activity through their fiscal and monetary policies. Their policies on investment, manpower training, industrial relations, tariffs, wages, and prices have a major impact on the private sector.

There are several indicators of the increasing scope of government in Western states: the aggregate number of central and local government employees, public expenditure and taxation as a share of Gross Domestic Product, the amount raised by taxation, or the cumulative number of laws and regulations. In 1993, general government expenditure and total taxation accounted for 44 and 38 per cent respectively of the country's GDP. Some 30 per cent of the population derive their primary income from government as recipients of pensions, unemployment or sickness benefits, etc., over 70 per cent of households contain at least one beneficiary of a government programme, and nearly 30 per cent of the workforce are employed by central and local government or the public corporations. The main categories of public employment are the Civil Service, National Health Service, local government, and public corporations. In the post-war period the numbers employed in education and health have increased sharply, while those in defence, nationalized industries, and public transportation have declined. Finally, around 3,000 laws already exist and governments add to them annually (Rose and Davies 1995).

BOX 10.2. KEYNESIAN ECONOMICS

JOHN MAYNARD KEYNES, 1883–1946, was educated at Eton and Cambridge. A polymath, he was gifted in philosophy, economics, finance, and the arts. He is remembered for his economic theories on the causes of, and cures for, prolonged unemployment. Keynes became a civil servant in the India Office, then returned to Cambridge as a don. He served in the Treasury during the First World War, but in 1919 resigned and returned to Cambridge because he strongly disagreed with the reparations arrangements of the Treaty of Versailles.

Keynes wrote several books and numerous polemical articles, but is best remembered for his General Theory of Employment, Interest and Money (1936), in which he repudiated classical economic theory, which maintained that mass unemployment only existed if wages were too high. The mass unemployment and falling wages in the 1930s discredited such *laissez-faire* theories; left to itself, the market seemed unable to solve the problem. Keynes maintained that mass unemployment indicated deficient aggregate demand (the total spending of consumers, investors, and public agencies) in the economy. Governments could solve mass unemployment by *budgeting for a deficit*: increasing public spending and stimulating private consumption and investment by using *fiscal* and *monetary* policies (lower taxes and lower interest rates and easy credit controls, respectively). If inflation existed governments should reverse these policies: budget for a surplus and raise interest rates.

During the 1930s Britain's ruling élite remained unconvinced by Keynes's revolutionary economic theories (although Keynes's influence can clearly be seen in America at the time, in the shape of Franklin Roosevelt's New Deal programme). It was the 1944 White Paper on Employment, in which the government accepted responsibility for maintaining full employment, that marked the start of the Keynesian revolution in Britain. Keynes, it appeared, had provided governments with tools of economic management which would enable them to reconcile government control of the economy with political freedom.

Using Keynesian techniques, governments succeeded in keeping unemployment below 3% between 1945 and 1970. By the 1970s the worsening trade-off between employment and inflation discredited Keynesian theory and led to the emergence of neo-classical economics, emphasizing monetarist rather than fiscal policy, and a revival in faith in the efficacy of the market.

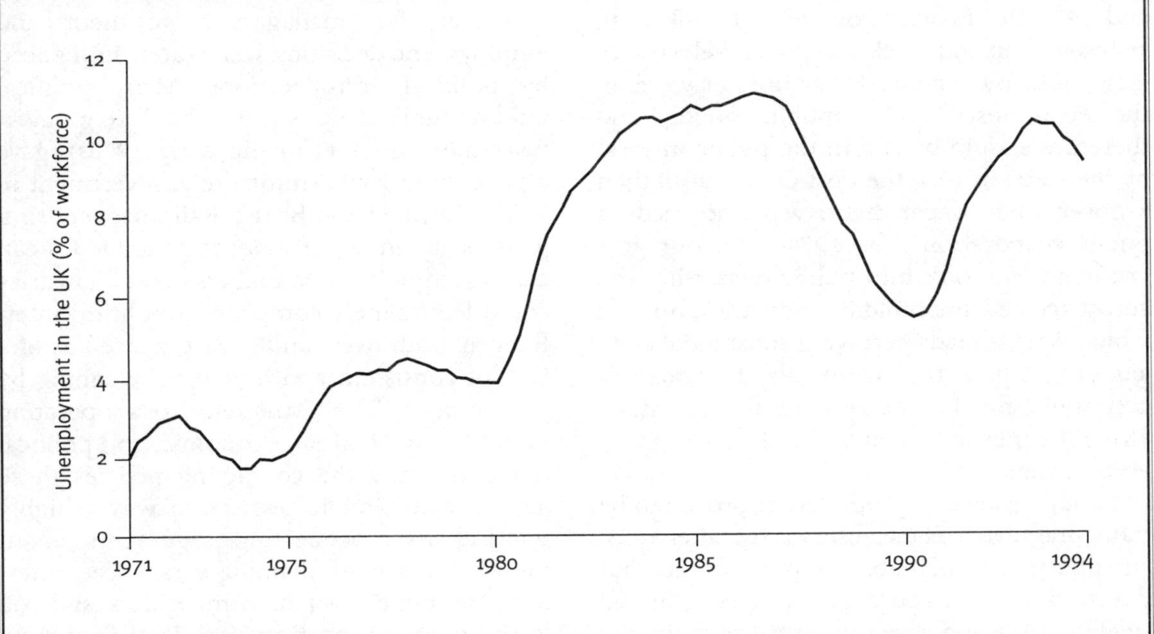

Intervention in industry

State intervention in industry has taken various forms. The most direct is government ownership of an industry or firm through nationalization or public ownership. Before 1939 the state had created the Port of London Authority (1908), the Central Electricity Generating Board (1926), and the London Passenger Transport Board (1933), all of which were managed as public corporations. The main features of public corporations are that each industry has its own management board and chairman, appointed by a minister; the minister is responsible to Parliament for general policy but not for the day-to-day running of the corporation. The idea was that ministers would give general directions and managers would make decisions on commercial criteria, unless precluded from doing so on pressing grounds. The public corporation model was applied to the industries and activities which were nationalized after 1945. Between 1945 and 1951 the Labour government took coal, railways, iron and steel, and gas and electricity into public ownership. These utilities were, in the economists' terms, public goods, and therefore should be run in the public interest by the state. In 1969 the Post Office, until then a government department, was also made a public corporation. The 1974–9 Labour government also took into public ownership the aerospace and shipbuilding industries, both of which were already receiving substantial state aid. By 1981, the nationalized industries accounted for 11 per cent of the country's Gross Domestic Product and 8 per cent of employment.

In office, however, both parties proved to be more pragmatic. The Conservatives, after 1951, accepted much of Labour's programme (but did not add to it, and denationalized iron and steel in 1953), and gave substantial state finance to private industry. Between 1964 and 1970, Labour brought about only a modest increase in public ownership. Many of the industries taken over after 1945 needed large sums of money for investment; if the state provided the funds then, it could be argued, it should decide their policies. It was also argued that public ownership would further greater efficiency, encourage a greater identification of workers with their industries, facilitate economic planning, and promote social objectives which the private market neglected. In the 1970s, however, the purpose of state intervention was increasingly to rescue ailing industries from the threat of bankruptcy and prevent an increase in unemployment, although Labour's left wing continued to regard it as a means of replacing capitalism.

The device of the public corporation has failed to deliver either the hoped-for benefits for the consumers or the intended independence for managers. Investment and employment decisions were often influenced by political considerations. Many political interventions were short-sighted (e.g. over-investment in steel in the early 1970s), gave a poor commercial return (e.g. investment in British Leyland and British Rail), or were simply misguided (e.g. supporting the De Lorean car firm in Belfast). Managers in the industries could legitimately complain of political interference, both overt and covert, as well as of a lack of consistent and long-term planning by government. They suffered from operating under a mix of social, economic, and political objectives, and the conflicting policies these gave rise to. Public ownership was a highly political issue, reflecting arguments about the rival merits of planning versus free enterprise, monopoly versus competition, and collectivism versus individualism. For many party activists it was the crucial division in politics. Not surprisingly, this atmosphere made it difficult for the nationalized industries to be considered on their own merits.

Privatization

Mrs Thatcher's governments sought to reverse the growth of government and the ideas of collectivism. They argued for the merits of a 'smaller state' and sought to reduce the government's responsibilities in a number of areas, or at least to share responsibility with the private sector. The goals included:

- a reduction in the size of the public sector and the Civil Service
- a reduction in public spending as a share of GDP in order to cut the burden of taxation
- encouragement of the private sector, by the sale of state shares in public companies, the contracting out of local and central government services, and the sale of council houses to tenants
- use of market testing in the public sector to encourage the provision of more services by the private sector and achieve better value for money

Since 1979 state intervention in the economy has been sharply reduced. Government has foresworn prices and incomes policies and cut back subsidies for industry. The most ambitious schemes for rolling back the state has involved the selling off, or privatization, of state-owned utilities and firms. The government has claimed that private-sector practices will deliver better quality service to customers and remove the taxpayer's responsibilities for losses. To date, major privatizations include: British Aerospace (1981 and 1985); Britoil (1982 and 1985); Associated British Ports (1983 and 1985); Enterprise Oil (1984); Jaguar (1984); British Telecom (1984); British Gas (1986); British Airways (1987); Rolls-Royce (1987); British Airports Authority (1987); British Steel (1988), Electricity (1990), and Water (1991). By 1995 only the slimmed down coal and railway industries remained of the original nationalized giants, and parts of both were being prepared for privatization. To cope with potential abuses of monopoly or market power a number of regulatory bodies have been created, for example, Ofgas (for gas) and Ofwat (for water). By 1991 some 900,000 jobs, and two-thirds of the state-owned sector of industry, had been transferred from the public to the private sector. The programme, added to contracting out and deregulation, has resulted in a significant redrawing of boundaries between state and private sectors.

Public spending

Public expenditure is the sum total of money spent by central and local government, nationalized industries, and public corporations. This includes all current and capital spending of central and local governments; capital spending of all public corporations and any losses they incur; and the grants, loans, and debt interest carried by any of the above. In 1994–5 the total amounted to £290.2 billion, 42 per cent of GDP. Some of this public money is spent directly by the government on *goods and services* and approximately half of central government expenditure goes on *transfer payments*, such as pensions, grants and loans to industry, and various social security benefits. The former represent the public sector's claims on the resources of the economy, while the latter is a transfer of purchasing power between individuals. The government's main sources of revenue are taxes (direct and indirect), national insurance charges, and trading surpluses of the public corporations.

Over time, the share of public expenditure spent on different services has altered. Before 1914, some 40 per cent of the expenditure went on the 'primary' activities of defence and administration; sixty years later this combined total was down to less than 10 per cent. In the 1990s the break-up of the Soviet Union and the end of the cold war has brought about a decline in the claims of defence on the total budget. But there has been no easing of pressure in the areas of social security, education, health, and personal social services, which account for two-thirds of all government spending. As societies grow richer, people live longer, and technological change

Fig. 10.1. Public money, 1994–1995

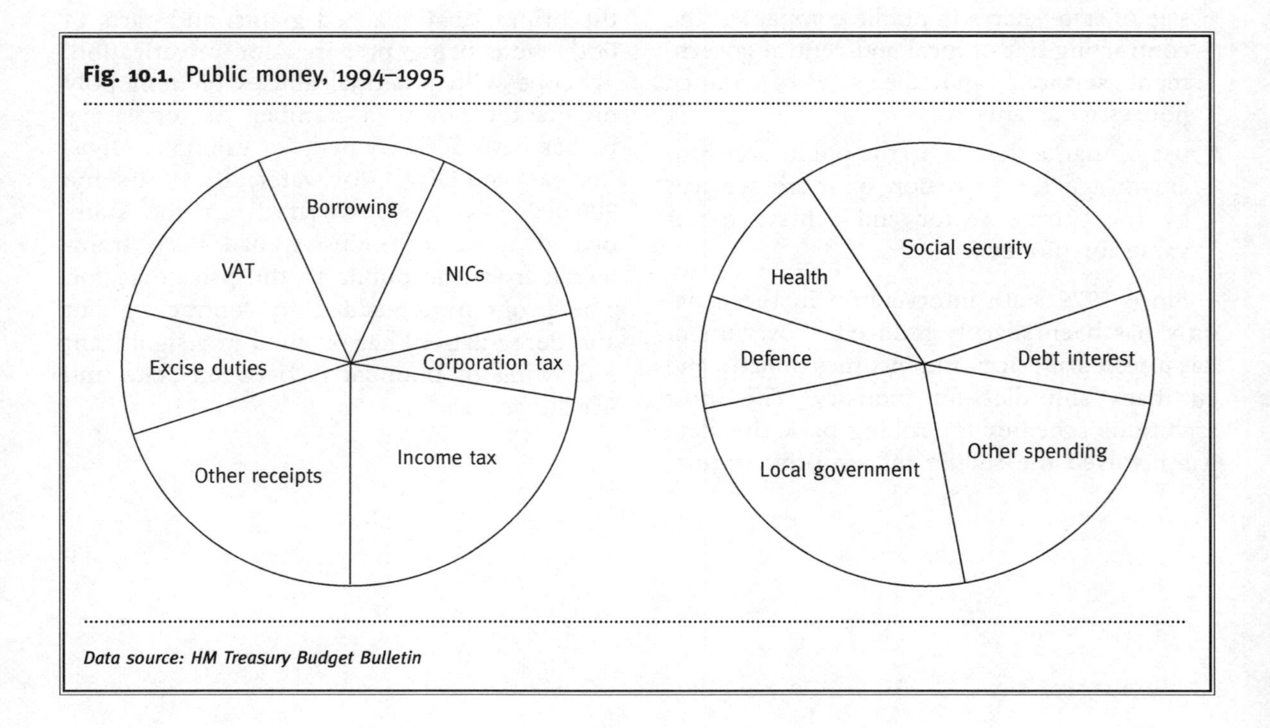

Data source: HM Treasury Budget Bulletin

increases the need for a more skilled and better educated workforce, so there is more demand for these services.

The role of public expenditure—its total distribution across different programmes, and attempts to increase, cut back, or balance the total in relation to government revenues—lies at the heart of policy-making. Academic studies, memoirs of ministers, and media reports of the spending process remind us of the clashes between the conflicting ambitions, values, and purposes of ministers. Departments like Education, Health, Social Security, and Defence, for example, have major 'spending' programmes. The reputations of the ministers for these departments *vis-à-vis* their respective clienteles depend in some measure on how successful they are seen to be in extracting resources for their departments. They have to contend with the Treasury, represented by the Chancellor of the Exchequer and the Chief Secretary, which is more concerned with the overall spending picture and the reactions of financial markets; it therefore, frequently, resists claims for more public expenditure so that taxes or borrowing, or both, do not have to be increased. If the 'gap' between revenue and expenditure is to be reduced, then the Treasury has three options—to borrow money, cut spending, or increase taxes. Borrowing may displease the financial markets and force a rise in interest rates. Increased expenditure on health brings plaudits for the Minister of Health and Social Security, but raising the taxes to pay for it brings criticism on the Chancellor.

Parties in government may also have different priorities. The Conservative government of 1979 promised to increase spending in real terms on defence, law and order, and health, and protect the value of the state old-age pensions against inflation. Conservatives regularly talk of cutting out 'waste' and 'bureaucracy' in the public sector as a means of getting greater value for money, and cutting public expenditure as a proportion of GNP. They are usually vague when it comes to talking of specific cuts in the programmes of departments. Labour in the 1979 general election promised to cut expenditure on defence and increase it on practically every other programme. The same pattern of promises was largely repeated for both parties in the 1983 and 1987 general elections. There are persistent themes in the Conservative and Labour manifestos. Conservative governments have usually pledged to increase efficiency, contain government expenditure, and cut income taxes. Labour has usually come to office armed with a long list of social programmes which would, hopefully, be financed by economic growth, or, if necessary, by higher taxes on the well off.

During the nineteenth century, government expenditure as a proportion of Gross Domestic Product (GDP) actually declined, and by 1890 it was only 9 per cent. It rose to over 50 per cent in the First World War and 75 per cent in the Second World War. The proportion steadily fell with the end of both wars. In 1957 it started to increase again, and by the mid-1970s leaders in both Labour and Conservative parties regarded it as a 'crisis'—raising economic, constitutional, and moral dilemmas. In 1975–6 it reached 49 per cent of GDP—a post-war peak—and by 1980 it was cut back to 43 per cent. By 1994, after a decade and a half of Tory governments that made spending controls a key element of economic policy, the figure was still 43 per cent. For the past thirty years the figure has been 40 per cent or more for all but the years 1986–8. It is worth noting that the British share is not high by the standards of other West European states, and Britain is the only G7 country in which public spending has fallen as a percentage of GDP since 1980.

There are many theories to account for the growth of state spending. Economic growth is one plausible explanation for the rise, as more resources become available for the government to spend. In the post-war period such demographic trends as an increase in the number of old-age pensioners (who have also been living longer) and, until recently, of schoolchildren (who stay in school longer) resulted in higher demands for public services and expenditure; the extra resources became available to meet the rising demands for pensions and education. Education, health, and administration are labour-intensive services, with much of the budget going on salaries and wages. It has been calculated that simply to maintain existing commitments (with no improvement in services) in health and social security requires a real increase of 1 per cent in public spending yearly. Some critics have claimed that the desires of politicians—to please voters who demand new or more generous programmes—and of bureaucrats, who want to expand their departments' programmes as a means of increasing their budgets and boosting their careers, also operate to increase expenditure. Finally, the expansion of the Welfare State and a resulting increase in entitlements to state benefits may also make it difficult to cut programmes or control budgets. Existing programmes also become more expensive as the value of some benefits (such as the old-age pension) is protected against inflation, regardless of economic growth, and the number entitled to the benefit increase. Some of the growth in expenditure is difficult to control because spending is beyond the direct reach of central government, for example, payments to the EC.

Public expenditure control

A major step in the effort to control public expenditure was the report of the Plowden Committee (1961). As a result of the committee's recommendations, a Public Expenditure Survey Committee (PESC) was set up to present each year the government's expenditure plan for the following five financial years, based on the continuance of existing policies. The annual PESC exercise starts in autumn when departments list all items of expenditure and future spending proposals at 'constant' prices. After negotiations with different departments the Treasury draws up a set of public spending plans for the next four years. These are carried in a White Paper, usually published in the following December or January: the plans refer to expenditure starting the next financial year beginning in April. The spending plans are then subject to further revision before being presented in the Budget in March or April.

It is easy to misperceive what is happening when there is talk of spending 'growth' or 'cuts'. In the 1970s, in spite of several rounds of cuts, public spending increased three and a half times, although only by 42 per cent in real terms (i.e. allowing for inflation). The 'cuts' may mean different things—a reduction in the future rate of growth of a programme, in the money costs of delivering a similar volume of services, or in the actual volume of services but without a fall in money costs. The actual 'cuts' might be across the board, affecting all public expenditure programmes, or in one programme or in sectors of programmes (Rose 1984*b*). Many programmes acquire a political and administrative momentum, however, and are difficult to cut, as even the Thatcher government found.

If there are limits to the taxable capacity of a society there is an almost infinite list of demands which groups and individuals make on the state for services. Governments have to make choices, not only between different claims, but also in weighing the likely consequences for their political support and for the state of the economy of alterations in spending or programmes. Nor is it simply a matter of considering the rival claims of taxpayers and beneficiaries of programmes—many people are both taxpayers and beneficiaries.

The 'politics' of public expenditure has been entertainingly described in a book called *The Private Government of Public Money* (Heclo and Wildavsky 1974; also Dunsire and Hood 1989). Every summer each department submits its spending proposals for the next year to the Treasury, which then 'pools' the claims and sets them against what it judges the economy can 'afford'. A useful rule-of-thumb guide to what a department will spend in the next year, *other things being equal*, is to start with the present total and allow a margin for growth or 'improvement'. Another rule for policy-makers is to look at how other departments are faring: is there a 'going rate' of increase? Both figures represent superficially 'hard' guidelines. There follows a series of bilateral negotiations between the departments and the Treasury, with the latter usually intent on cutting the programme and the former defending as much as they can. There is great pressure in Whitehall to settle disagreements lower down the hierarchy, before they reach the Cabinet. Most negotiations are settled by a department's principal finance officer and the Treasury officers. The sensitive issues which civil servants will refer 'upwards' to the minister usually concern cuts which will be politically unpopular or involve large amounts of money. A minister may be tempted to take a disagreement to Cabinet when he feels confident of the support of colleagues, or when an issue is highly 'political'. Totals which are still the subject of disagreement go to the full Cabinet for arbitration.

In practice, only a few issues can be referred, but the Chancellor, backed by a Prime Minister, usually gets his way. In 1976, the Labour government debated at length the IMF's conditions for a loan, particularly the proposed reductions in the amount of government borrowing and public expenditure, both favoured by the Chancellor. A Cabinet majority was opposed to cuts on the scale demanded by the Treasury, but, in the end, came round when Mr Callaghan supported the Chancellor.

The present system, introduced by the then Chancellor Norman Lamont in 1992, is for the Cabinet to agree a total in advance. Ministers who are unhappy with their budgets may appeal to a Cabinet committee, EDX, which contains the Chancellor and other ministers, most of whom have already agreed their budgets. The new system helps the Chancellor by involving more ministers in the process.

Spending cuts

The Thatcher government came to office pledged to reduce the role of government in employment, expenditure, and taxation. This was part of the Thatcherite strategy of reducing the size of the public sector, to make way for tax cuts and encourage an enterprise economy. Within Whitehall the 'bidding' strategies of departments were no longer based on expected economic growth rates. As the economy failed to grow in line with expectations so public expenditure plans, which had been based on a higher growth rate, were cut back, but the total as a share of GDP still grew. The first three years of Mrs Thatcher's 1979 government were dominated by periodic spending reviews. The public spending/GNP ratio actually increased, largely because the effects of economic recession restricted economic growth, and higher unemployment increased demands on the social security budget. With the economy growing again after 1983, the proportion came down, only to rise again after 1990, as the recession took hold—increasing the unemployment totals and social security spending and hitting revenues.

Cutting exercises are usually made on various grounds:

- *Government policy.* A government may be pledged to make certain cuts, perhaps on the ideological grounds of disapproving particular items (for example, Labour's abolition of tax relief for mortgages on second homes, or Conservative cutbacks on subsidies to industry or help for local authority housing)
- *External pressures*, usually lack of confidence in the currency. Foreign holders of sterling have welcomed 'sound' economic policies and reacted negatively, as in 1967, 1976, and 1992, to large public-sector deficits and an overvalued pound
- *Slower than expected economic growth*, which means that resources are not available for planned spending programmes or that some programmes have to be cut if other areas are to be protected or expanded
- *Fall in demand for a service*, such as a reduction in the number of school-age children

In an ideal world, governments make cuts after deciding on priorities. There are different criteria, however, for establishing what constitutes a 'priority' area. Use of 'objective' approaches are helpful up to a point in achieving greater efficiency and economy in the use of resources. But at some point ministers have to make political judgements and acknowledge the potential contradictions between objectives. A programme's importance may be based on its electoral popularity, the demands of some party supporters or a powerful pressure group, the political influence of a minister, or its 'usefulness' for the economy. Frequently, the 'political priorities' that emerge reflect the relative political standing of particular ministers and the interest of the Prime Minister. Most ministries have a vested interest in protecting their budgets. The language of priorities—except where it is a rationalization of what has already been done—is often avoided. An economy-minded Treasury therefore cuts where it can. Spending reviews may often end up with 'formula' cuts or across-the-board reductions of x per cent, with departments left to decide how the cuts are effected in their programmes.

Reviews to cut expenditure are messy because they are made through negotiations, compromises, and adjustments. A Chief Secretary to the Treasury commented on his experience in the 1974–9 Labour government (Barnett 1982):

> **Expenditure priorities were generally decided on often outdated and ill-considered plans made in opposition, barely thought through as to their real value, and never as to their relative priority in social, socialist, industrial or economic terms.**
>
> **More often they were decided on the strength of a particular spending minister. In later years, when Jim Callaghan was prime minister, I have seen him snap the head off Shirley Williams, then Education Secretary, although he was fond of her and would usually apologise almost immediately after. But it might well be enough for her to lose her bid.**

The debate about government

For at least the last century, politicians and commentators have disagreed about the desirable size and scope of government: what tasks should the government assume and what should be left to the market and to individuals and families? are there limits to the responsibilities the state should assume, or share of national wealth it should control? what would be the consequences for democracy of an ever-expanding role for the state? The debate was emerging before 1914 between the Liberal and Conservative parties, but became sharper with the rise of Labour, with its commitment to state ownership of utilities and an enhanced role for government in economic management and welfare provision. There were Conservatives who also believed that government had an important role in these areas.

For much of the post-war period the role of government was decided on pragmatic grounds. But in recent years debate has flourished. Defenders and critics offer value judgements about what is desirable. Advocates of a more interventionist government argue that such government is functionally more effective, that is, it is better able to solve problems, respond to popular demands, and to coordinate different policies. They also claim that, compared with the results of the free market, decisions by government are likely to be made more openly and more purposefully, since the government is politically accountable to the electorate. They may also claim that some problems, requiring coordination and planning, are best dealt with collectively, and that, in the interest of social justice and common equity, there should be uniform rules for welfare benefits, school-leaving age, and road safety, etc., and common policies for education, health, and so on. Above all, government intervention has been justified by a perception of *market* failure—for example, poverty, homelessness, or large-scale unemployment.

Governments, in contrast to most private organizations, are primarily interested in providing public goods (i.e. those that concern all citizens) and in taking decisions for society as a whole. Foreign policy, safety standards at work, policing, and national defence, for example, are almost invariably controlled by government, and citizens help to finance these services through taxation. If this were not done, some people would be tempted to be what economists call 'free riders', that is, while not themselves contributing they would nevertheless continue to enjoy the facilities financed by others. By tradition, the provision of education and health care, and protection against a fall in living standards during illness, unemployment, and old age, were left to charity or perhaps initiative. Only in this century has the British government come to play the major role in these areas, with a small area left for the private sector.

Critics of government action are usually found on the political right, though also among some members of the political centre. Many of the 'New Right' ideas were largely imported from the United States and their influence was seen in the Reagan presidency (1980–8). In both countries critics have complained that government is inefficient and wasteful in its use of money, manpower, and services because it is not subject to the commercial discipline of the marketplace. In the United States there was disillusion with the results of more active federal government programmes, as in the New Deal in the 1930s and President Johnson's 'Great Society' programme in the 1960s. They also complain that although politicians pass laws and vote funds for programmes, the relevant information and techniques for solving many social problems are often not available. In some areas, where such problem-solving knowledge is available, the political means may be regarded either as 'unacceptable' to many people or conflict with other desirable objectives.

Some would also claim that there is a risk to liberty if the state spends 'too much' money (say, over 40 per cent of GDP) or regulates and legislates in 'too many' areas. The more the state spends and taxes the less discretionary income citizens have in take-home pay (or what remains after deductions for tax and national insurance). Government rules usually involve an element of coercion or, at least, a restriction on the citizen's degree of choice. An incomes policy, for example, usually sets limits to free collective bargaining between trade unions and employers, and central government's concern to have a uniform national policy for the school curriculum conflicts with the autonomy of local education authorities and schools.

Disagreements about the desired role and size of government are also related to different conceptions of democracy. Advocates of limited government emphasize the importance of pluralism, checks and balances between institutions and individuals, and groups having as much freedom from restraints as is compatible with maintaining order.

Finally, it is held that the more a government tries to decide issues, the more areas of life become 'politicized'. As they become the subject of disagreement between political parties, there may be sharp discontinuities in policies each time there is a change of government. The extension of government responsibilities also has a constitutional aspect. If a government assumes responsibilities which it proves unable to carry out, then the resulting disappointment may lead to the authority of the government being called into question. J. A. Schumpeter (1942) had this in mind when, in considering the conditions favourable for a liberal democracy, he concluded that the effective range of the political decisions should be limited. The answer to the claim of *market failure* now was *government* failure. Its interventions were actually undermining the economy, as government 'overloaded' with responsibilities and popular expectations and weakened as a result (King 1975, Rose 1980*b*). The result was seen in soaring inflation rates, public spending out of control, and public sector strikes. The answer was that government should limit what it does to its core responsibilities—defence, order, and stable prices.

Carried to extremes, the proponents of limited government would return to something like a *laissez-faire* society, while advocates of big government might find that the realization of their schemes allowed no formal scope for activities which were not approved or directed by the state. There are, however, few advocates of either extreme. It is more a question of adjusting the balance of rights and obligations between citizens, groups, and the government. In fact, of course, in the nineteenth century many proponents of *laissez-faire* justified a limited government role on the pragmatic grounds that it would promote desirable *social* ends, and did not deny the need for the government to make occasional interventions and maintain general rules.

The Keynesian and social democratic ideas, which were dominant between 1945 and the mid-1970s, came under challenge in the 1970s and 1980s. Within the Labour party they were challenged by the left wing—which called for more state control and regulation of the market—and in the Conservative party by the ideas of what is loosely called the 'New Right'. At this time, the thesis of 'decline', referring both to economic performance and the political system, took hold. When the Heath government was undone by the miners' strike in 1974, the Labour government by strikes in 1979, and the IMF was called in to help rescue the economy in 1976, the critics had plenty of material to hand. In the 1970s the slowdown of international economic growth and the minute improvement in living standards in that decade increased taxpayers' resistance to funding large government programmes (Rose and Peters 1978).

On the left, the Marxist economist James O'Connor (1973) warned that the demands on state spending exceeded available taxes and borrowing. Firms faced a crisis of profits and government's ability to invest and safeguard profits was under pressure. Moreover, the rise in unemployment added to the demands for welfare spending. This was a theory of *the crisis of capitalism*. Mrs Thatcher complained that the state, by high levels of taxation and public spending, was limiting the individual's room for choice, and by managing and subsidizing firms and industries was undermining efficiency and competitiveness. There was a *crisis of social democracy*. Mrs Thatcher's policy goals included lower rates of income taxes and public spending, a smaller role for central government in the economy and social services, and the revival of free enterprise and the market economy. But realizing these goals called for government intervention, particularly in education, local government, industrial relations, and the civil service.

The record by the time she left office in 1990 was mixed: taxation as a proportion of GNP had increased compared with 1979, and not until 1988 was public spending as a proportion of GNP below the 1979 level. However, the rate of growth in both was slowed down. Under John Major both have increased again. Progress was made in the privatization of state assets and transfer to the market of state services, particularly in housing, and this has been extended throughout the public services under John Major.

Labour has traditionally portrayed government as a potential protector of the family and the weak through the welfare services, and as the creator and defender of jobs through public services and government spending. Labour's commitment to public ownership in 1918 placed the issue of the state's role at the centre of political debate. It was reinforced by Keynesian economics and expressed in the acceptance in the post-war years of the mixed economy. Under Mrs Thatcher a clearly opposite set of values was articulated. The success of the Thatcher project (Gamble 1994) is seen in the changed policies of Labour. Under Tony Blair Labour has accepted more openly a limited role for the state, the place of markets and private ownership, and abandoned the old Clause Four (see ch. 7). The role of the government in the economy—a subject of bitter controversy during the 1980s—is now largely agreed between the leaderships of the main parties.

Since the early 1990s the New Right has resolutely tried to suggest that government, both local and central, should be like a business, and civil servants should regard themselves as providers of services, competing in the marketplace as entrepreneurs. Many services and utilities, once provided largely or solely by the government, have been contracted out to agencies, or privatized. The role of markets or market mechanisms has been increased in such services as health and education and the Civil Service, and much of the public sector is much more commercially minded and fragmented than was the case in 1979. It is not only the market, however, which has grown at the expense of government. The greater influence of the European Union has also limited the discretion of central government. These trends amount to what Rod Rhodes calls 'a hollowing out of the State' (1994) or David Marquand 'the downward trajectory of the post-war British State' (1995: 5).

Conclusion

The more modest role of government raises questions about democratic accountability. Are the conventional mechanisms of ministerial responsibility to the elected House of Commons or elected local government still adequate? In particular, how useful are they for making accountable decisions which are largely made elsewhere, either through markets, by quangos, or in Brussels? The New Right argues that accountability is rendered to consumers through competition between providers of public services, regulatory agencies for the privatized utilities, and independent audit of agencies which spend public money. Although there is a persistent theme to these developments—towards the market and bodies which are separate from government, and away from the state—they have occurred ad hoc, without any serious debate about what tasks the government should be performing and what it should shed, and on the constitutional issues which they raise.

Public opinion, however, has not been so clear-cut. In spite of four successive Conservative election victories since 1979, surveys suggest that more voters still favour increasing taxes to pay for greater expenditure on such services as health, education, and welfare than tax-cutting and programme-cutting (Crewe 1989). The role of the state in the provision of welfare remains dominant; even after a decade and a half of Conservative rule, less than 10 per cent of school children are privately educated and less than 10 per cent of people are in private health schemes. Health and education remain overwhelmingly tax-financed services. Surveys also show that most voters agree on many issues and on the goals of government. There is wide support for the mixed economy—a private enterprise economy subject to government controls—as well as for state provision of welfare services. The role of the government and of the state is viewed neither in the negative terms of the Conservative right nor in the positive terms of Labour's left wing.

Summary

- Government is a complex term used to refer simultaneously or separately to office-holders, institutions, and a system of ordered rule.
- British government is organized in a system of departments. The expansion of the role of government and the changing nature of its responsibilities is manifest in the number of departments and their changing functions.
- Increased government responsibilities leading to an increase in the number of departments has led to problems of coordination and to the development of multi-minister departments.

- The oldest departments — Home Office, Foreign Office, Treasury — deal with what some argue are the state's primary tasks: maintaining law and order, defending the territory, and currency and exchequer functions. Governments have added to these functions throughout this century, but particularly when, after 1945, they assumed responsibility for delivering a wide range of services associated with the welfare state.

- 1945 ushered in a period of optimism that governments, using the tools of Keynesian economic management, would be able to ensure the social and economic well-being of citizens. 1945 also ushered in the age of 'big' government, measured in terms of government's share of GDP, taxation levels, the number of central and local government employees, and the number of laws and regulations coming into force.

- The nationalization programme can be largely explained in terms of economic rationale, but for Labour governments it also had an underlying ideological rationale.

- Privatization likewise has economic and ideological explanations. Mrs Thatcher adopted a policy of privatizing former state industries and services, partly in her bid to create a property-owning democracy and partly in an attempt to create a leaner and more efficient state.

- Determining overall levels of and distribution of public expenditure lies at the heart of policy-making. Attitudes to public expenditure have traditionally distinguished the parties from each other, but can also be a source of conflict within governments as departmental ministers are partly judged according to the share they win for their own department's sector of work.

- Governments have resorted to a variety of different techniques in their attempts to control levels of public expenditure and to establish priorities within the levels set. Attempts to reduce public spending are usually frustrated by the proportion of previously committed expenditure. Spending cuts are usually restricted to reducing levels of growth and levels of future commitments.

- The expansion of the role of the government during this century has been accompanied by a debate about what the appropriate role of the government should be. Has big government led to more or less freedom? Has disappointment with the performance of government and resentment with the levels of taxation undermined the authority of government? Between 1945 and approximately 1970 optimism about the government's ability to improve the lot of its citizens prevailed. Persistent economic problems brough disillusionment and the ideas of the New Right gained wider acceptance. Scepticism about the role of the government has begun to permeate the political culture and has been manifested in the attempt to roll back the state by privatizing many functions previously performed by the government.

CHRONOLOGY

1944 White Paper on Employment: the government accepted responsibility for maintaining full employment

1946 Nationalization: Bank of England, Coal Mining, Cable and Wireless, Civil Aviation

1947 Creation of NHS. Nationalization of Electricity, Gas and Transport (railways, canals, road haulage)

1949 Nationalization of Iron and Steel. Housing Act encourages council house building

1953 Denationalization of Iron and Steel

1961 Plowden Report on Public Expenditure. Incomes policies begin

1968 Labour government creates giant Department of Health and Social Security

1970 Heath government produces White Paper on the Machinery of Government: 'super' departments of Environment, Trade and Industry, to improve coordination of government functions

1971 Rolls-Royce taken into public ownership

1976 Government expenditure reaches a post-war high of 49% of GDP. IMF loan. Keynesian economics officially abandoned

1979 Privatization programme begins: British Petroleum and council houses sold off

1981 Privatization: Aerospace, Cable and Wireless

1982 Privatization: Amersham International, National Freight Consortium, Britoil

1983 Privatization: Association of British Ports

1984 Privatization: Enterprise Oil, Jaguar, British Telecom

1986 Privatization: Gas

1987 Privatization: British Airways, Royal Ordinance, Rolls-Royce, Airports Authority

1988 Privatization: Rover Group, British Steel, Water. DHSS divided into Department of Health and Department of Social Security

1990 Privatization: Electricity

1993 Privatization: Railtrack

1994 Post Office privatization fails

1995 Department of Employment absorbed into Department of Education

ESSAY/DISCUSSION TOPICS

1. Rolling back the state was more rhetoric than reality. Discuss.
2. To what extent is Britain becoming a 'quangocracy'?
3. In what respect and why have attitudes towards the role of government changed since 1945?
4. Why has privatization been a key government policy since 1979?
5. What criteria determines which industries and services should be in the public or private sector?
6. Discuss the view that 'Treasury control' is the key to understanding public expenditure in Britain.

RESEARCH EXERCISES

1. The Prime Minister has just announced that the government intends to cut public expenditure by £x billion in the coming year. You are the Secretary of State for Education. What arguments and tactics could you use to ensure that the education budget is increased?

2. In 1994 privatization of the Post Office was on the agenda but failed to gain approval in Parliament. Examine the arguments of those proposing and those opposing this privatization. Identify which, if any, of the points raised are ideological.

FURTHER READING

See R. Rose, *Ministers and Ministries* (Oxford: Oxford University Press, 1987) and C. Pollitt, *Managerialism and the Public Services*, 2nd edn. (Oxford: Blackwell, 1993).

The following may also be consulted:

Dobson, Alan, 'Approaches to the Economy and the State: Post-War Concepts of the Political Economy', *Talking Politics*, 5/3 (1993).

Jones, W., '"The Unknown Government": The Conservative Quangocracy', *Talking Politics*, 6/2 (1994).

Moran, Michael, 'Reshaping the British State', *Talking Politics*, 7/3 (1995).

Thomas, Graham, 'British Politics 1945 to Date: The Postwar Consensus', *Talking Politics*, 7/2 (1994/5).

11 CABINET AND PRIME MINISTER

Reader's guide

In 1986, Norman St John-Stevas, a former member of Mrs Thatcher's Cabinet, said: 'There is no doubt that as regards the Cabinet the most commanding Prime Minister of modern times has been the present incumbent, Mrs Thatcher. Convinced of both her own rectitude and ability, she has tended to reduce the Cabinet to subservience.' During, and since, Mrs Thatcher's eleven-year premiership, politicians, journalists, and academics focused attention not only on her policy agenda but also upon her conduct of cabinet government. The question of whether or not cabinet government has been superseded by prime ministerial government has been an enduring theme of modern British politics. From time to time the debate subsides, only to be rejoined with renewed vigour when a conspiciously dominant tenant, like Mrs Thatcher, occupies No. 10. Rules about, and relationships within, the executive branch form part of Britain's unwritten constitution; much is, therefore, left unsettled, and this enables each new group of office-holders to develop their own working relationships. The

Prime Minister and the Cabinet exercise formal powers but the ratio of power between these institutions will depend upon informal relationships and changing variables such as personalities, circumstances, and issues.

This chapter examines the Cabinet. What is its role? How well does it perform its functions? What is the role of ministers? It describes how the Cabinet is organized and discusses the part played by Cabinet committees and the Cabinet Office. It goes on to examine the formal powers of the Prime Minister, but also explores their practical and political limits. The chapter concludes with contrasting case studies of Mrs Thatcher and John Major, ending on a word of caution that the legacy of Mrs Thatcher remains inconclusive: a British Prime Minister still operates within the constraints of a party and parliamentary system.

ALTHOUGH British government is often termed 'cabinet government', the Cabinet is actually a committee of the government. The close association of the Cabinet and government reflects the fact that for at least the last century and a half the sovereignty of the Crown in Parliament has been vested in the Cabinet. Reference to the British executive may include the government departments or ministries, the senior ranks of the Civil Service, the Prime Minister and Cabinet, the Prime Minister's private office, or ministers. Collectively these are sometimes called 'the core executive'.

This chapter discusses the political and constitutional role of the Cabinet today and how it and the committee system work. It also considers the theoretical and practical powers and constraints on a Prime Minister.

Role of the Cabinet

The fact that the Cabinet is a collective body and most of its members are ministers with departmental responsibilities gives it a dual focus. The Cabinet and its members fuse political and executive functions (Hennessy 1986, James 1992). Its combination of political and executive work may be considered under four headings.

1. *It is the body in which many of the most important political decisions are taken or to which they are reported*. This is true of a decision to go to war, approve a policy line on education or something else, and settle the total figure for public spending. Some decisions—such as whether the President of the USA should address a joint gathering of the two Houses

BOX 11.1. KEY CONCEPTS: CABINET AND PRIME MINISTER

Cabinet government	The Cabinet is a collective executive: members share in decisions and are held to be collectively responsible. The Cabinet is the peak institution in a hierarchical and centralized system of government. It • determines policy • oversees and broadly controls government • reviews all major decisions The PM is head of the Cabinet, a little more than first among equals but not a singular executive taking all decisions and being held individually responsible for the government, as, for instance, an American president.
Prime Ministerial government	The PM takes all major decisions in consultation with whichever advisers are picked for the occasion, possibly, but not necessarily, the Cabinet or members of it. Secondary level decisions are taken by Cabinet committees and committees of officials.
Collective responsibility	A convention of the constitution relating to operations of cabinet government. Cabinet ministers share collective responsibility for decisions of Cabinet and its committees and must publicly support them or resign.
'Kitchen Cabinet'	A group of close advisers surrounding the PM, some of whom may be preferred members of the Cabinet but other officials or personal friends and cronies. Controversy usually arises about the 'kitchen cabinet' if it is believed that PMs are taking decisions based on the advice of members of such a group instead of the advice of the Cabinet. Harold Wilson was accused of having a 'kitchen cabinet'; some argue that Mrs Thatcher had a 'kitchen cabinet' which included Charles Powell and Bernard Ingham.
Bilaterals	Informal meetings between the PM and his/her advisers and the relevant minister(s) with their advisers in order to thrash out policy decisions.

of Parliament—may seem trivial, but they will be considered at Cabinet level because of their potential political sensitivity. According to Lord Wakeham, who served in Cabinet under Mrs Thatcher and John Major, Cabinet is '. . . a reporting and reviewing body rather than a decision-taker' (1994).

2. *It plans the business of Parliament*, approving the details and timing of legislation which is to be laid before Parliament. Such planning is subject to the opposition's rights to Supply Days and to propose motions of censure, and to the rights of private members. Most Cabinet ministers sit in the Commons and ministers determine much of the work of Parliament by preparing major bills, establishing its agenda, and organizing opinion and votes to get the legislation passed.

3. *It is the arbiter of policy differences between departments*, for example, for disagreements which cannot be resolved in bilateral negotiations or in Cabinet committee. Ministers

would also expect to be a party to sensitive decisions which affect the responsibility of the government as a whole. John Major's Cabinet has deliberated over such issues as:

- preparing for the Maastricht treaty negotiations (1991)
- deciding on proposals to extend to the qualified majority voting in the EU (1994)
- replacing the poll tax with the council tax (1991)

4. It is the body which provides a *general oversight and coordination of the government's policies.*

It is often argued that this last role is not well performed, largely because the pressures of departmental work prevent Cabinet ministers from considering policies of other departments and overall strategy. Some of the institutional changes in recent years have been an acknowledgement of this weakness and an attempt to rectify it.

Most Cabinet ministers head major departments whose interests they represent in Cabinet. *They wear a departmental as well as a Cabinet hat.* Some ministers, like the Prime Minister, or such non-departmental ministers (as Leader of the House or Lord President of the Council) have the opportunity to *take a view which transcends departments.* Their own responsibilities or their membership of many interdepartmental committees bring them into contact with other departments. But a contrary pressure works on many other ministers. The reputation of the typical minister with his civil servants, constituency of pressure groups, specialist media reporters, and subject committee of back-bench MPs depends on his doing a good job for his particular department, which often turns on increasing its budget.

The Cabinet's political authority derives from having the support of a majority of members of the House of Commons. It is not independent of the Commons but represents a fusion of the executive and legislature, a contrast to the United States where the separation of powers precludes Cabinet officers from being members of Congress. The rise of one-party majority government has weakened the ability of the Commons as a body to check the Cabinet. Membership of the same political party also helps to unify Cabinet members; their short-term survival is bound up with that of the party. As elected politicians they are well qualified to make political assessments of the impact of policies on the House of Commons, party, and electorate. In the post-war period only a handful of Cabinet ministers have not had prior experience in the House of Commons, and, apart from the trade union leader, Ernest Bevin, they have not been very successful. In contrast, the President of the United States may appoint to the Cabinet people from other parties or from none. He may even draw on people who have never stood for election or who are hardly known to him. At the end of the day an American Cabinet is responsible to the President, not to Congress.

Size of the Cabinet

The size of the government has doubled in the course of the twentieth century. The number of paid government appointments has grown from less than fifty in 1914 to more than a hundred today, with most of the growth in the ranks of junior ministers

BOX 11.2. THE CABINET SYSTEM

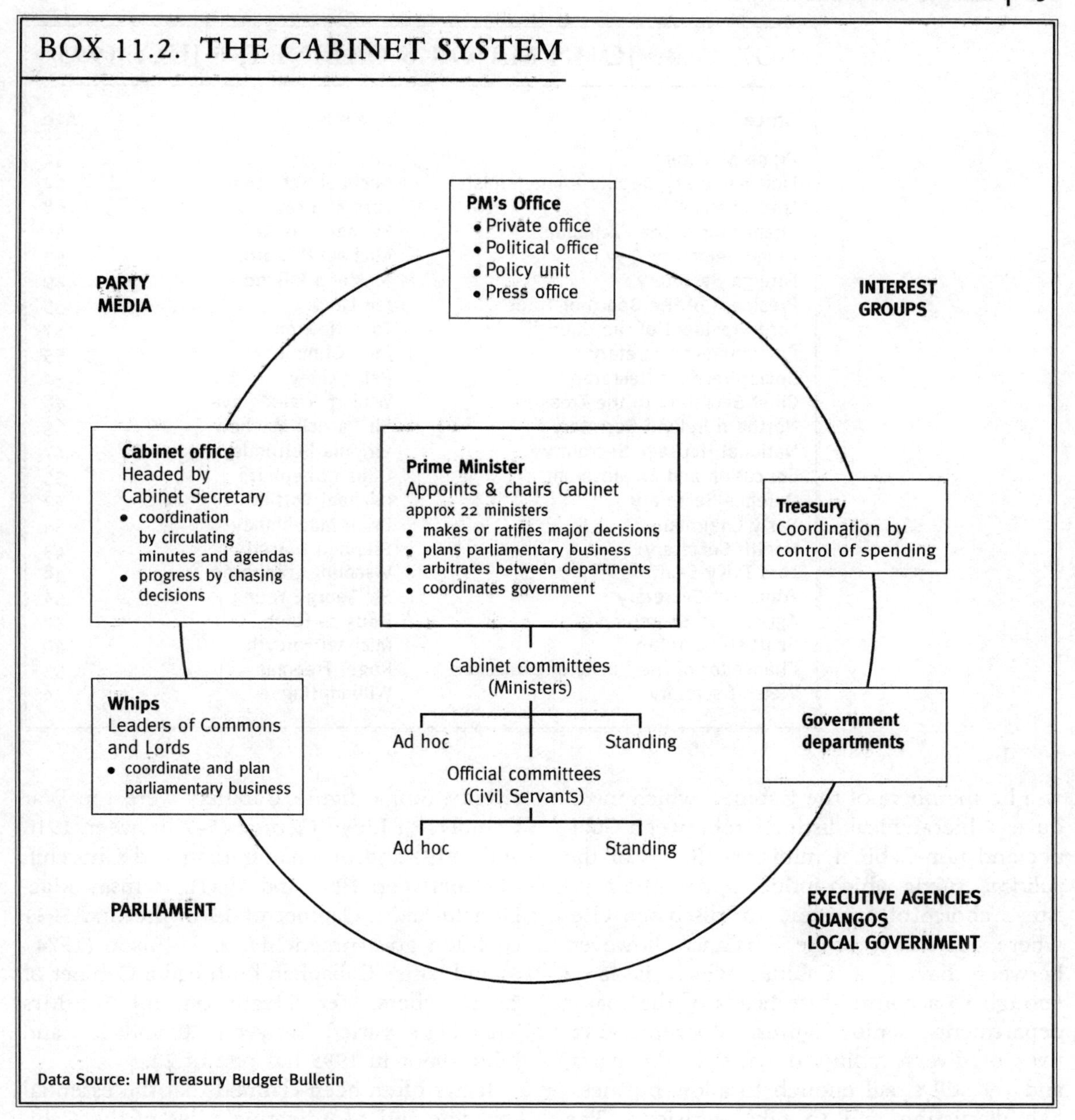

Data Source: HM Treasury Budget Bulletin

and Parliamentary Private Secretaries. Before 1914 most ministers in charge of a department could expect to be members of the Cabinet. An increase in the number of departments and ministers means that the Prime Minister now has to choose which ministers

BOX 11.3. JOHN MAJOR'S CABINET, 6 JULY 1995

Office	Minister	Age
Prime Minister	John Major	52
First Secretary/Deputy Prime Minister	Michael Heseltine	62
Lord Chancellor	Lord Mackay	68
Chancellor of the Exchequer	Kenneth Clarke	55
Home Secretary	Michael Howard	53
Foreign Secretary	Malcolm Rifkind	49
President of the Board of Trade	Ian Lang	55
Lord President of the Council	Tony Newton	57
Environment Secretary	John Gummer	55
Social Services Secretary	Peter Lilley	51
Chief Secretary to the Treasury	William Waldegrave	48
Northern Ireland Secretary	Sir Patrick Mayhew	65
National Heritage Secretary	Virginia Bottomley	47
Education and Employment	Gillian Shephard	55
Defence Secretary	Michael Portillo	42
Party Chairman	Brian Mawhinney	54
Health Secretary	Stephen Dorrell	43
Lord Privy Seal	Viscount Cranborne	48
Transport Secretary	Sir George Young	54
Agriculture Secretary	Douglas Hogg	50
Scottish Secretary	Michael Forsyth	40
Chancellor of the Duchy of Lancaster	Roger Freeman	53
Welsh Secretary	William Hague	34

will be members of the Cabinet, which introduces a hierarchical distinction between Cabinet and non-Cabinet ministers. Some of the *political criteria* which influence a Prime Minister's choice of ministers are discussed elsewhere. He has to strike a balance, however, between having a Cabinet which is large enough to accommodate heads of the major departments, senior figures, and representatives of diverse points of view in the party, and yet still small enough to allow business-like discussion and to take decisions. The major lobbies, such as those on education and health, as well as the nations of Wales, Scotland, and Northern Ireland, also expect 'their' minister to be in the Cabinet. Indeed the role of the Cabinet as a representative body depends in part on its membership reflecting the major interests.

The usual size of the Cabinet is just over twenty. Single figure Cabinets were the War Cabinets of Lloyd George (5–7 between 1916 and 1919) and of Chamberlain and Churchill (8–9 between 1939 and 1945). Ramsay MacDonald had a Cabinet of 10 in the first 1931 coalition government. Harold Wilson (1974–6) and James Callaghan both had a Cabinet of 22 members, Ted Heath one of 18, Mrs Thatcher's varied between 20 and 22, and John Major in 1995 has one of 22.

It has often been claimed that the essential strategic and coordinating roles of the Cabinet would be better performed by having a so-called 'super-Cabinet' of half a dozen ministers. They would be free of day-to-day departmental responsibilities and could concentrate on the general direction of government policy. When Churchill became Prime Minister for a second time in 1951 he experimented with a system of 'overlords', that is,

BOX 11.4. DEPUTY PRIME MINISTER: A TITLE NOT A JOB?

On 6 July 1995 Michael Heseltine was appointed First Secretary of State and given the title of Deputy Prime Minister.

According to a No. 10 spokesperson, Mr Heseltine will 'assist the Prime Minister generally and will have specific responsibility for the competitiveness agenda and the working of government and its presentation of policies. He will chair a number of Cabinet committees and will be an ex-officio member of Cabinet committees generally'.

Mr Heseltine's 'promotion' was widely portrayed in the media as the price Mr Major had paid for Mr Heseltine's decision not to stand in the party leadership contest, and for encouraging his supporters to vote for John Major.

The title of Deputy Prime Minister has been used rarely, and then usually to 'buy off' a powerful rival:

1942–5 Clement Attlee was deputy PM in Churchill's wartime coalition

1964–8 George Brown, Harold Wilson's rival for the Labour leadership in 1963

1979–88 William Whitelaw, one of Mrs Thatcher's rivals in 1975, was informally regarded as deputy PM

1989–90 Sir Geoffrey Howe acquired the title from a grudging Mrs Thatcher, along with his appointment as Leader of the House and Lord President of the Council, but he received no substantial recognition of additional status as a result

Mr Heseltine seemed intent upon transforming the title into a job. He quickly acquired spacious office accommodation and chaired several Cabinet committees.

ministers sitting in the House of Lords, free from departmental (and constituency) duties and charged with coordinating the work of groups of departments. The system was not a success and was abandoned in 1953. Most Prime Ministers operate an informal version of this approach through an 'inner' Cabinet or holding ad hoc meetings with trusted and/or senior ministers to prepare the ground for sessions of the full Cabinet.

The arguments against having a formal super-Cabinet are threefold. One is *political*: the pressures of political management, maintaining political balance, and satisfying the ambitions of colleagues would be increased in a smaller Cabinet. A second is *constitutional*: what would be the relations between departmental ministers, who are formally answerable to Parliament for the work of departments, and the super-Cabinet? A third is *administrative*: ministers with actual experience of a department's work are more likely to have an effective grasp of issues than ministers who do not have this hands-on experience.

Another device for reducing the size of the Cabinet is to amalgamate some departments. For example, the present Environment, Trade and Industry, and Foreign and Commonwealth Office departments have all resulted from earlier mergers of departments. The change, it is claimed, allows for more consideration of strategic matters within departments and eases the burden on the Cabinet agenda. On the other hand, Mrs Thatcher in 1988 felt that the sensitive issues were not

BOX 11.5. CABINET OVERLOAD

[Factors that contribute to the problem of Cabinet overload:]

- lack of time
- complexity of the issues
- volume of work
- departmentalism

Cabinets are often forced to abandon effective coordination of government and strategic planning of policy and settle for governing for the day and short-term crisis management.

In the post-war period prime ministers have tried a variety of remedies—none so far have succeeded in solving the problem, although some have brought temporary improvements.

1945–51	Attlee continued and expanded the Cabinet committee system that had developed during the war. Numbers of committees have fluctuated since. Whilst solving some problems they create additional problems of coordination.
1951–3	Churchill introduced his overlord experiment.
1964	Wilson allowed ministers to appoint a limited number of special advisers.
1970	Heath created (1) super-ministries to reduce the size of Cabinet and thereby improved coordination, (2) the Central Policy Review Staff (CPRS) to aid strategic thinking and provide ministers with government, as opposed to departmental, briefs.
1974	Wilson continued the special advisers experiment and established the Policy Unit to provide alternative (non-civil service) briefing service to the PM to aid his overall coordination of government and act as an early-warning system to alert him to potential problems.
1990–5	Ministers without Portfolio—Duchy of Lancaster, Lord Privy Seal, Lord President of the Council—have frequently been given the task of aiding with the overall coordination of government. John Major appointed David Hunt to the Duchy of Lancaster in July 1994, giving him coordination responsibilities and appointing him chairman of several major Cabinet committees. In the wake of the party leadership election in 1995 Michael Heseltine was appointed Deputy Prime Minister and given a wide-ranging brief to coordinate government strategy.

Other suggested remedies:

1. A Prime Minister's Department to enable the PM effectively to oversee government strategy and coordination. But there are constitutional and political objections to this.
2. Recreation of a think-tank something on the lines of the CPRS to challenge departmental orthodoxies and broaden ministerial horizons.
3. Ministerial Cabinets: small executive committees in each department composed of civil servants and political advisers to improving briefing of ministers and ease routine workload.

properly handled in one Health and Social Security department and divided the work into separate departments.

A reader of the memoirs and diaries of recent Cabinet minsters may be struck by the following recurring complaints:

- the lack of both a central directing mechanism and a sense of strategy in the government
- the alleged influence of the Prime Minister's 'kitchen' cabinet and advisers
- the alleged bypassing or downgrading of the Cabinet, as more decisions are made in committees, within departments, or in bilaterals between the PM and a minister

It is worth adding that the first charge was rarely made by critics of Mrs Thatcher's governments, although the third (by the resignation of Heseltine and Howe) and the second (by the resigning Lawson) were. Cabinet ministers who resign under protest often claim that the Prime Minister has been acting in some way unconstitutionally and neglecting Cabinet (e.g. George Brown in 1969, Michael Heseltine in 1986, Sir Geoffrey Howe in 1990). This may be a cover for hurt pride or disagreement over policy. It was to meet the first complaint that Mr Heath established a Central Policy Review staff in 1970. Mrs Thatcher had little time for it and scrapped it in 1983.

Ministers' roles

A head of department is constitutionally responsible for the work of the department. No matter how idle a minister or how subservient to senior civil servants, this responsibility cannot be delegated. The minister is expected to administer the department, defend it in Parliament and public, negotiate with its network of interest groups, and represent it in Cabinet and its committees. If the minister is a Cabinet member he will also have the opportunity to contribute to the formation of general policy. Ministers differ in their perceptions of their roles. On the basis of interviews with Cabinet ministers, Bruce Headey (1974) concluded that ministers had predominantly departmental rather than Cabinet perceptions of their roles. Most saw themselves as *representatives* or spokesmen for the department, particularly *vis-à-vis* Parliament, Cabinet, and the public. As few as one in six regarded themselves as *initiators* of policies, in the sense of defining policy options for the department. Mrs Thatcher made clear that she preferred ministers who were *initiators*.

Various forces work against ministers being policy initiators. In many departments a minister may not be committed to promoting a change in policy or not have thoroughly worked out a strategy suitable for achieving a promised change. Such circumstances make it likely that a minister will rely heavily on the advice of civil servants. A minister may be overburdened, not least with other, non-

BOX 11.6. 'A USEFUL EXERCISE WITH INTEREST': THE CENTRAL POLICY REVIEW STAFF

Colloquially	Think-tank
Created	1971 by Heath as part of his reorganization of central government
Composition	hybrid: a body of approximately 20, half of whom were 'fast stream' civil servants, remainder seconded from industry, the universities, the City, local authorities, and public corporations
Directors	1971–4 Lord Rothschild, scientist and chairman of Shell Research 1974–80 Sir Kenneth Berrill, former Chief Economic Adviser to Treasury 1980–2 Sir Robin Ibbs, seconded from ICI 1982–2 Sir John Sparrow, merchant banker from Morgan Grenfell
Location	Cabinet Office
Supervision	PM and Cabinet Secretary
Brief	Established to work for ministers collectively • as an apolitical think-tank to support the flagging cabinet system • to remedy overload and departmentalism
Form of Work	• regular strategy reviews of government's overall performance • research projects on subjects which cross departmental responsibilities, e.g. energy, race relations • constant flow of economic advice • collective briefs to tackle a minister's lack of any brief if his department had no stake in the topic on the Cabinet agenda
Evaluation	• strategy reports soon stopped, as emphasis shifted to short-term economic crisis management • departments resisted a central unit which might challenge their specialist monopolies • one or two 'own goals'—report on the Diplomatic Service 1976; future of the Welfare State 1982—which discredited the think-tank and embarrassed the government of the day • became 'leaky' and therefore was distrusted • its apolitical nature suited Heath but not Wilson or Thatcher • gradually became a creature of the PM on the one hand and colonized by the Civil Service on the other

Fate	Abolished by Mrs Thatcher 1983 on grounds that • the PM now had a Policy Unit • the Cabinet Secretariat now provided improved briefing for the Cabinet • the departments had their own research/planning sections
Retrospective	• dismissed by Mrs Thatcher as a freelance 'Ministry for Bright Ideas' out of touch with the government's philosophy • described by Peter Hennessy as 'A firework that fizzled' • described by Lord Rothschild as 'a useful exercise with interest'

departmental, duties. Finally, the rapid turnover of ministers (just over a two-year spell in a department, on average) may also weaken the 'initiator' minister. It usually takes at least two years to see a major policy through from its inception to implementation. At the end of the Wilson–Callaghan period (1974–9) only four Cabinet ministers were in their original posts and after thirty months less than half (9) of John Major's twenty-one Cabinet colleagues appointed in April 1992 were still in the same posts.

Different departments may acquire their own administrative styles because of their particular mixes of demands, responsibilities, and traditions of work and policy. Some will involve the minister in heavy parliamentary duties, a highly public role, or attendance at several Cabinet committees. The Foreign Office, because of its involvement with other EU states and the need to present a British view of issues to other countries, requires the minister to travel a great deal, attend overseas conferences, coordinate foreign policy aspects of other departments' policies, and cope with short-term crises. The Education Department is usually engaged in a battle for resources for its varied responsibilities, which range from nursery education to university postgraduate training. The Scottish Office imposes diverse demands on the minister because of its various responsibilities for policies in Scotland covering agriculture, education, health, development, and the economy.

The Treasury's responsibility for economic policy means that the Chancellor has an interest in many departments, particularly Environment, Trade and Industry, and, until its abolition in 1995, Employment, and Agriculture. It also negotiates the spending levels for each department in the annual review of public expenditure. Pay policy, particularly for the public sector, will also involve negotiation between the Treasury and other departments as a figure (or 'factor') is allowed for pay in the departmental budget. The Chancellor is also usually a member of more Cabinet committees than any other minister.

The Duchy of Lancaster, Lord Privy Seal, and Lord President of the Council are non-departmental posts which usually leave their incumbents free for other, often coordinating duties. Such ministers are particularly useful in chairing Cabinet committees.

The Cabinet at work

In normal circumstances the Cabinet meets for about two hours each Thursday when Parliament is sitting, although it can be summoned at any time to deal with pressing issues.

The proceedings of Cabinet are businesslike and even formal. The agenda and papers are circulated in advance and minutes are kept. Remarks are addressed to the Prime Minister and ministers are referred to by their titles. The seating of ministers is arranged hierarchically, with the Cabinet Secretary on the Prime Minister's immediate right and holders of the senior posts seated on the Prime Minister's left or across the table. Ministers have frequently paid tribute to the serious atmosphere of Cabinet deliberations. Until the Prime Minister calls the meeting to order the mood is usually informal, as ministers gossip and use Christian names, but when the meeting begins, 'At once the whole atmosphere changes. There is no longer a set of individual men but the collective sovereign power of the State' (Gordon Walker, 1970: 60). There are informal rules governing the intervention of speakers, with precedence granted to a senior minister or ministers with departmental interests in the item under discussion. Votes are rare. They advertise Cabinet divisions, fail to register the political weight and strength of feeling of individual ministers, and limit the Prime Minister's ability to formulate the 'sense of the meeting'. Voting, particularly if it is along factional lines, may also weaken the Cabinet's collective sense. But in some circumstances it may be unavoidable. In January 1968, Mr Wilson's Cabinet took a series of votes on proposals to cut spending on domestic and overseas programmes.

The Cabinet's proceedings are supposed to be secret; all members take a Privy Councillor's oath of secrecy and sign the Official Secrets Act. The agenda and the subsequent minutes, though circulated, also remain secret. When there is a change of government, only a record of the decisions remains on the departmental files and other material, including discussion papers, is withdrawn. Defenders of confidentiality argue that it facilitates freer discussion and is essential for collective responsibility. When the Cabinet discusses party political matters, like preparations for a general election or political strategy, the civil servants withdraw.

A Cabinet is also a political body, and therefore beset by divisions and rivalries. Some divisions are fairly predictable, in terms of spending ministers versus the Treasury, or spending ministers against each other in the battle for fixed resources, or political differences between left and right in a Labour Cabinet. In Mrs Thatcher's 1979 Cabinet, many so-called 'wet' ministers opposed the Chancellor's anti-inflation strategy but disagreed among themselves about alternatives. Ministers also differ in their influence. A Prime Minister will give special consideration to some ministers' views because of their personal standing with the party or the public or because of their office. Mr Foot's influence in the 1974 Labour government and Mr Whitelaw's in the 1979 Conservative government depended more on their standing than on the offices they occupied. Some ministers may contribute decisively on matters outside their departmental business, and on matters of general strategy. A Chancellor of the Exchequer, in particular, is well placed to intervene on most items. Most ministers, however, usually keep to their departmental business.

The Prime Minister has to judge the suitable moment to intervene, either before or after different views have been considered and when a view is emerging. On many issues Mrs Thatcher declared her preferences at the outset and was as much a committed participant as a neutral chairman, concerned to reflect the view of colleagues. John Major, by contrast, has been more concerned to hear the views of colleagues and reflect the collective view. Cabinet ministers are not of equal political weight—the views of a minority which are carefully considered or held strongly may be more important than those of a majority not so considered or strongly held—and summing up gives the Prime Minister the initiative in stating the view of the meeting. Attlee's view of Cabinet management (cited in King 1969: 71) was:

BOX 11.7. THEORY AND PRACTICE: CABINET GOVERNMENT

Constitutional Theory

Cabinet government is government by a collective executive resting on the doctrine of collective responsibility: ministers are collectively responsible for, and publicly bound to defend, Cabinet decisions. The underlying assumption is that such decisions are based on Civil Service advice and agreed after full and confidential discussion in Cabinet.

Political Practice

The volume of government work has vastly increased over the century since this theory evolved, with the result that:

- Cabinets are too large for adequate deliberation
- ministers struggle to master their own sector of work and lack the time and information to make a useful contribution to the general directing of government

In practice:

The Cabinet is informed of major decisions taken in its name elsewhere in the Cabinet system, either in Cabinet committees or in bilaterals between the Prime Minister and the ministers involved in a particular policy area.

Cabinet ministers still bear collective responsibility for decisions in which they did not participate and of which they may have been only vaguely, if at all, aware.

Some ministers seem to accept the inevitability of these developments:

David Howell, Secretary of State for Energy 1979–81, for example, when asked by Peter Hennessy if he resented being excluded from economic policy-making, replied:

> 'I don't think one would have expected, not being in the Treasury, to be involved . . . Cabinet government is only a layer of government and there is a kind of inner Cabinet government whether it's called that or not, under different Prime Ministers, it always tends to develop.' (Hennessy 1990: 315)

Others may balk at the idea of being presented with a *fait accompli*, but can only resort to leaks or resignation.

Examples George Brown 1968; Michael Heseltine 1986; Nigel Lawson 1989; Sir Geoffrey Howe 1990

The job of a Prime Minister is to get the general feeling—collect the voices. And then, when everything reasonable has been said, to get on with the job and say 'Well, I think the decision of the Cabinet is this, that or the other. Any objections?' Usually there aren't.

A Cabinet in not a debating society. The object of discussion is to take a decision, to continue, modify, or replace a policy, or to defer consideration for further study. A Prime Minister's summary of the discussion is rarely challenged. Once the summary is entered in the minutes, it becomes a decision of the Cabinet and the machinery of government proceeds accordingly. Most ministers are extremely busy. Their hours of work, the travel, the public scrutiny of their conduct, and the nervous strain they endure amount to a workload which is far greater than that in most professions. Some of the strain comes from coping with crises under intense public and often critical scrutiny. Coordinating the work of different departments and receiving reports from other committees is time-consuming and a minister may be a member of as many as half a dozen Cabinet committees which will be meeting regularly. Most ministers, for example, also have to attend to constituency, parliamentary, and party affairs. Not surprisingly, they may be so preoccupied with the business of their own departments that they may have too little time to spend on other matters coming before Cabinet (e.g. Castle 1980, Crossman 1975–7, Hennessy 1986). Memoirs suggest that on many Cabinet items only two or three ministers speak.

The regular items before the Cabinet include parliamentary business (i.e. for the following week) and home, overseas, and European Community affairs. Many matters will have already been discussed in committees and the Cabinet will usually endorse these recommendations. An entire Cabinet session or series of sessions may be devoted to a politically sensitive issue. In 1967, Mr Wilson devoted half a dozen Cabinets to the single issue of the Common Market, Mr Callaghan nine sessions to the conditions attached to the IMF loan in 1976, and John Major several sessions to prepare for the Maastricht negotiations in December 1990.

The conduct of Cabinet to some degree reflects the style of the chairman. Attlee was succinct and expected others to follow his example. Churchill, after 1951, tended to ramble self-indulgently. Macmillan was generally admired by colleagues for his ability to listen to different views and to give a lead. Heath appears to have been a particularly dominant leader of his colleagues. Members looked to him for a lead and it is striking that the policy reversals over prices and incomes and intervention in industry were accomplished without a single Cabinet resignation. In the early years Mrs Thatcher's Cabinet was argumentative and discordant, a consequence not only of disagreements about economic strategy but also of her argumentative and directive style.

Some ministers have taken a very sceptical view of the use of Cabinet. Nigel Lawson (1992) offered the following observations on Mrs Thatcher's Cabinet:

- The weekly Cabinet meetings were 'the least important aspect of Cabinet membership' and 'the most restful and relaxing event of the week' (125)
- Cabinet meetings were not important and Cabinet is now a dignified part of the constitution. Key decisions were made in small groups or committees. Cabinet was to keep colleagues informed and allow general political discussion (127)
- Ministers acquiesce in the downgrading of Cabinet as long as they can do bilateral deals with the Prime Minister (129)
- If the Prime Minister and the relevant minister are agreed then it is very difficult for another minister to overturn their agreed line. Most Cabinet ministers see themselves firstly as departmental chiefs. Cabinet, for

BOX 11.8. THE PRIME MINISTER AS POLITICIAN AND ADMINISTRATOR

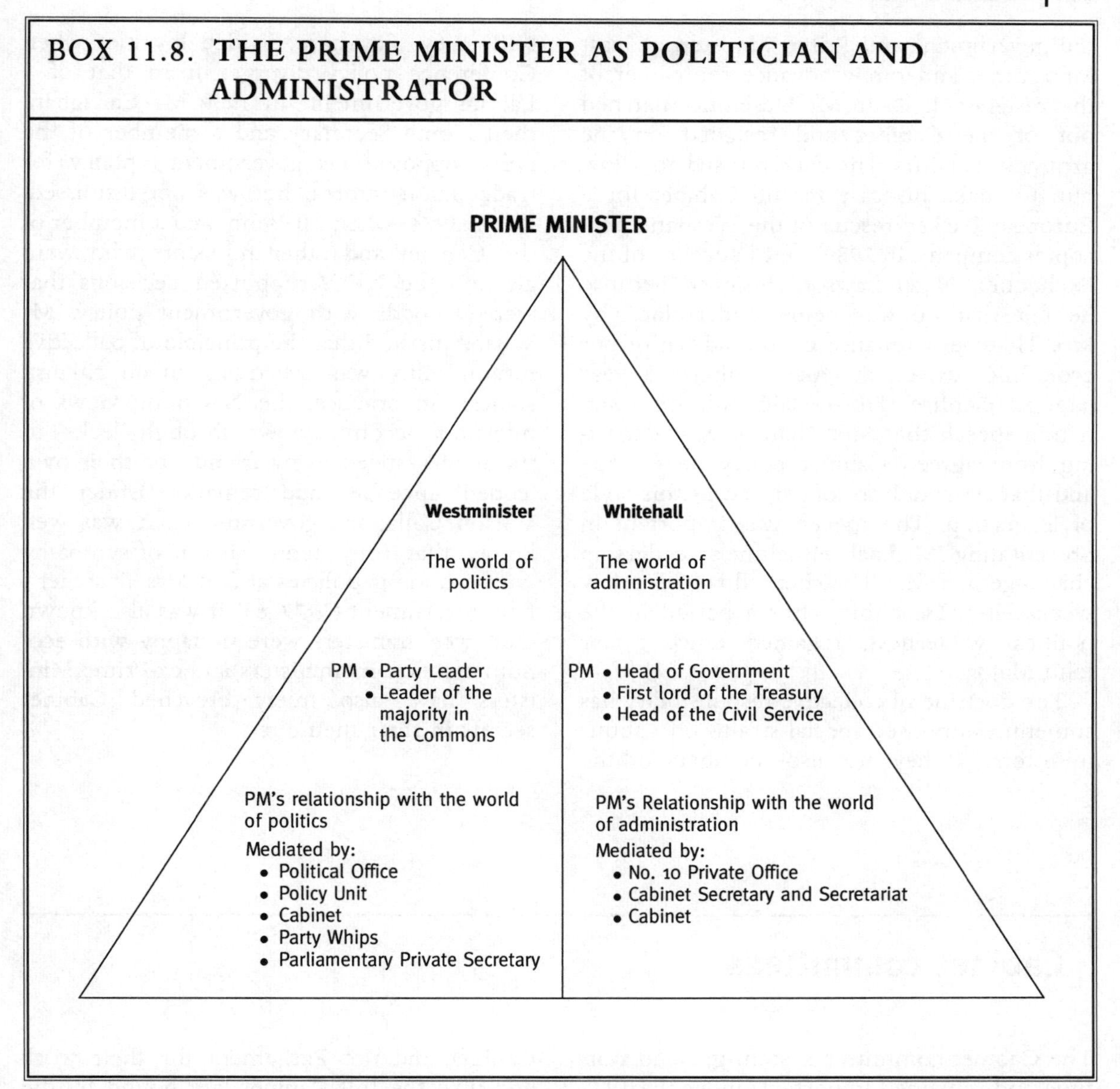

example, did not get involved in the poll tax (562–3)

- A committee of twenty-two members is simply too large to be a really effective forum for discussion and an Inner Cabinet is preferable (127–8)
- Most people around the Cabinet table know little of what is going on.

Under the convention of collective responsibility ministers are bound to support Cabinet decisions in public. Although the convention has been relaxed on special occasions, it remains the case that if a minister is unable to support the Cabinet decision publicly he is liable to be dismissed or is expected to resign. Such resignations are rare, but when they do occur they usually damage the government and Prime Minister, at least for a time, and rarely advance the career of the resigner. In 1986 Mr Heseltine marched out of the Cabinet and resigned on the grounds that Mrs Thatcher refused to allow him to make his case in full Cabinet for a European-backed rescue of the Westland helicopter company. In 1989 the Chancellor of the Exchequer, Nigel Lawson, resigned because he felt that he was being undermined by Mrs Thatcher's reliance on the advice of her economic adviser, Sir Alan Walters. A year later Sir Geoffrey Howe claimed in his resignation speech that Mrs Thatcher was departing from agreed Cabinet policy on Europe and that he could no longer accept this style of leadership. The speech was important in precipitating Michael Heseltine's leadership challenge and Mrs Thatcher's downfall a few weeks later. Heseltine, after a period in the political wilderness, regained office under John Major.

The doctrine of collective responsibility has sometimes imposed special strains on Labour ministers. If they are also members of the NEC they may have divided loyalties when Conference policy diverges from that of a Labour government. In 1969 Mr Callaghan, then Home Secretary and a member of the NEC, opposed his government's policy for trade union reform, but was not dismissed. But in 1974, when Mr Benn, and a member of the Cabinet and other ministers who were also on the NEC, supported decisions that were at odds with government policy, Mr Wilson insisted that the principle of collective responsibility was overriding in all circumstances. In practice, the dissenting views of ministers become known through 'leaks' to the media, speeches by friends, or their own 'coded' speeches and remarks. Under the Wilson/Callaghan governments, it was well known that Tony Benn was out of sympathy with economic policies and in Mrs Thatcher's first government (1979–83) it was also known that 'wet' ministers were unhappy with economic policy. Ex-ministers and ex-Prime Ministers have also freely breached Cabinet secrets in their memoirs.

Cabinet committees

The Cabinet committee system grew in wartime and remained important under the 1945 Labour government. It is now a fixture. The committees are composed of Cabinet (and sometimes non-Cabinet) ministers, usually from departments which have an interest in the issue, and chaired by a senior Cabinet minister. The most important committees will be chaired by the Prime Minister or another senior minister. Until 1992 details of the committees were not made public, on the grounds that members are responsible to Cabinet and not Parliament for their work and, like the full Cabinet, are bound by the doctrine of collective responsibility. In a spirit of open government, John Major decided in May 1992 to reveal the committees' terms of reference and membership.

Reasons for the growth of the committee system are not hard to find. If the Cabinet did not delegate most of its business it would soon become overloaded and have little time to consider major issues. Committees help to iron out details of policy that have already been agreed in principle, clarify areas of disagreement, coordinate the work of different departments, and, occasionally, take an issue out of a divided Cabinet. Like the development of the Cabinet Office, the growth of the committee system is a mark of the adaptation of the Cabinet to the problems of modern government. The right to choose a committee's terms of reference, chairman, and membership may be exploited by a Prime Minister to push a particular policy. For much of her first administration, many Cabinet ministers were not wholehearted supporters of Mrs Thatcher's economic strategy. She ensured, however, that her supporters were in a clear majority on the influential Economic Affairs Committee. Once a policy is approved by the committee it virtually commits the members of the committee and may acquire such momentum that it is difficult to overturn it, even in the full Cabinet. Committee members are expected to support the recommendations that go before full Cabinet and, as a rule, there is no appeal by a committee member against the decision unless the chairman agrees.

There are five main types of committee: separate standing committees, of ministers and of officials; separate ad hoc committees, of ministers and of officials; and mixed committees of officials and ministers. The important standing committees consist largely of Cabinet ministers. They are named and deal with major areas of activity; they include the Public Expenditure Survey Committee (PESC), the Overseas Policy and Defence Committee, and the Social Services, Home Affairs Committee, and Future Legislation Committee. Ad hoc groups are numbered and known as MISC (for miscellaneous) and are set up to deal with particular problems; they revert to GEN (for general) with a change of government and then back to MISC again at the next change. Mrs Thatcher created a good number of ad hoc committees, whereas John Major has relied more on the standing committees.

The Cabinet Office

Before 1916 no minutes of Cabinet meetings or records of decisions were kept. A Cabinet Secretariat or office was established in 1916 at the instigation of Lloyd George, not least to cope with the pressures of war but also to help with the efficient dispatch of business. Initially the Cabinet Office or Secretariat became identified with Lloyd George's almost presidential style of leadership and there were strong pressures for its disbandment after his departure from No. 10 in 1992. It has survived and become an essential part of the government machine. As the work of the Cabinet and of its committees has grown so has the staff and remit of the office. Its work in relation to the Cabinet and its committees is essentially threefold: (1) to prepare the agenda of the Cabinet and its committees, (2) to record the proceedings and decisions, and (3) to follow up and coordinate the decisions (Seldon 1990). It does the first by circulating relevant papers to departments beforehand. The second is done by the Cabinet Secretary, in 1995 Sir Robin Butler, and other staff. The minutes do not report the debate in Cabinet but record the decisions based on the Prime Minister's summing-up. The Cabinet Office coordinates by informing departments of decisions and checking that appropriate action has been taken in relevant departments. Each Cabinet committee is similarly serviced and has its own secretariat, agenda, and minutes.

The close relations that almost invariably develop between the Cabinet Secretary and the PM (notably between Sir William Armstrong and Mr Heath and Sir Robert Armstrong and Mrs Thatcher) can make the former, in effect, the Secretary for the PM.

The Prime Minister's support system

The Prime Minister does not have a department to administer and this freedom gives him the opportunity to take a broad view. The freedom is only useful, however, if he has the resources to know what is going on. 10 Downing Street is not an office or a powerhouse to be compared with the President's White House.

The Prime Minister has a *private office*, headed by his principal private secretary and four assistant private secretaries, all civil servants. Each assistant private secretary is responsible for a different area of policy: home affairs, foreign affairs, economic affairs, and Parliament. The office helps the Prime Minister with preparations for parliamentary business and parliamentary questions, with correspondence, and briefs him on the work of other departments. There is also a *press office*, which briefs the media about the work of No. 10 and a political office which deals with party matters.

In 1974 Mr Wilson created a small *Policy Unit* to service himself and this has been retained by his successors. Its members are political appointees and they help the Prime Minister with speech-writing, liaison with the party, and advice on policy and proposals from other departments. They are free to see most Cabinet papers and examine issues which cut across departmental boundaries, particularly those which have short- or medium-term political significance. They are particularly concerned to look for the party and political consequences of events, trends, and policies (Donoughue, 1987). The staff support at No. 10 is a reminder that the Prime Minister is a political as well as an executive leader (Rose 1980*a*). The two roles are not easily separated, but it is only in recent years that the political role has been formally recognized.

By 1982 Mrs Thatcher had appointed personal political advisers on foreign affairs, defence, and economic policy to provide an independent evaluation of the policies of these departments. Some observers thought that she was trying to create a Prime Minister's Department and intended to downgrade the Cabinet. Ministers, it was claimed, feared that their positions in their own departments might be undermined by experts who were appointed by and reported to the Prime Minister, who in turn might become less accessible to them. In fact no permanent institutional reform came from the moves. Prime Ministers have long been surrounded by advisers, courtiers, and cronies, people whom they trust or find congenial (G. Jones, 1973, 1987). The creation of a Prime Minister's department seems difficult to graft on to a Cabinet system of government.

The main argument for the Prime Minister having a stronger 'support system' is that he or she so often has to speak for the government—handling questions in Parliament, giving media interviews, and attending conferences abroad—and therefore requires more help, if not a department of his own. The total prime-ministerial staff amounts to more than thirty people, with about half being concerned primarily with policy. The British Prime Minister appears to be less well provided with this type of political support than leaders in most other Western states.

Prime Ministerial power

A number of commentators have claimed that the Prime Minister's power has grown to such an extent that we should now talk of 'Prime Ministerial government' rather than Cabinet government. Borrowing Bagehot's language, they suggest that the 'efficient' power in the political system now lies with the Prime Minister and a few colleagues, while the Cabinet is on its way to joining Parliament and the monarchy as a 'dignified' part of the constitution.

John Mackintosh, in his classic work *The British Cabinet* (1962), claimed that the Prime Minister's power had grown over time relative to that of the Cabinet and that a Prime Minister, in good health and with the support of his Cabinet colleagues, was virtually irremovable between general elections. He stated that British government was now more accurately described as Prime Ministerial rather than Cabinet government. Richard Crossman (1972) and Lord Hailsham (1978) have similarly argued that in the post-war period Cabinet government has been transformed into Prime Ministerial government. For Crossman the Prime Minister was able to monitor the work of departments; the rise of disciplined parties made the House of Commons more pliant to the executive; and the unification of the Civil Service gave the Prime Minister the final say over senior appointments in all departments. The Prime Minister, therefore, was the apex of centralized, powerful, political and administrative machines. Tony Benn (1980) has complained about the emergence of an 'Absolute Premiership' and argues that the removal of the Prime Minister's powers of dissolution and of hiring and firing Cabinet ministers are essential in restoring a 'Constitutional Premiership'.

This view is advanced regardless of which party is in government, of how unpopular or unsuccessful it is, and of the Prime Minister's skills or lack of them. The trend is unilinear, towards more centralization in No. 10.

There are, however, other views. An alternative is that the Prime Minister's power is *cyclical*, that strong leaders are often followed by *consolidators* because there is (a) a reaction among Cabinet ministers and backbenchers against strong personal leadership, and (b) if the strong leader comes to a sticky end, his/her style is seen as contributing to this. Examples would be the three most dominant leaders this century, Lloyd George 1922 (followed by Bonar Law), Churchill 1945 (succeeded by Attlee), and Thatcher 1990 (succeeded by Major).

Another view denies that there is any trend, either cyclical or unilinear. Instead it is held that each Prime Minister's power is a function of (a) the incumbent's unique *personality*, including skills, experience, and drive, (b) *opportunities* or circumstances, and (c) *role requirements*, such as making appointments, chairing the Cabinet, selecting the agenda and choosing the date for dissolution, etc. The role requirements are fixed but the first two are variable. According to this analysis each incumbent writes a distinctive page of Prime Ministerial history, depending on the mix of the above three factors. Mrs Thatcher, for example, was more effective in exploiting powers in the office than many of her predecessors and her successors; personality and circumstances clearly account for a good part of her success.

The idea of the dominant Prime Minister is certainly consistent with a public image of British politics. Coverage of general elections and parliamentary politics by the news media is highly personalized and focused on the party leaders (Foley 1993). The Prime Minister so dominates public coverage of politics that we often talk of 'Mrs Thatcher's' or 'Mr Major's' government. For the past quarter-century the Prime Minister alone has regularly received more attention in *The Times* than the three leading Cabinet ministers—the Foreign Secretary, the Chancellor of the Exchequer, and the Leader of the House of Commons—together (Rose 1980*a*: 19–21). The interest of television and the popular press in personalities makes them even more disposed to accentuate the patterns of *The Times* reporting. The involvement of Prime Ministers in more newsworthy activities than their colleagues and their position as leader of the government makes them newsworthy. They have many opportunities to personify the government or, strictly speaking, to be its spokesman. In the mass media, on foreign visits, and in Parliament, the Prime Minister is regarded as the authoritative spokesman for the government. Mrs Thatcher was prone, early in her premiership, to 'make policy' in television interviews and impromptu utterances, almost committing her Cabinet in advance.

Cabinet ministers do not need to be reminded that they are appointed by, and hold office at, the Prime Minister's pleasure. The latter's resignation entails the resignation of the whole government. The morale of some ministers and of aspirants for office may fluctuate according to signs of the Prime Minister's pleasure or annoyance. MPs and Cabinet ministers may be dissatisfied with the leader but, apart from the political damage involved in removing somebody who is unwilling to go, there is rarely unanimity about the successor.

The Prime Minister is also a party leader. In the British system, leadership of a major political party is the first requirement for being a Prime Minister. A politician, however eminent or popular, who lacks that base will not reach or survive at the top, as illustrated in the careers of Lloyd George (post-1922) and Winston Churchill (pre-1940). A leader who hopes to reach or remain in 10 Downing Street has to devote much time to party management, making appointments and policies with an eye on the reaction of party factions. MPs in marginal seats assess their own chances for

Table 11.1. Prime Ministers since the Second World War

Prime Minister	Party	In office
Neville Chamberlain	Conservative	1937–40
Winston Churchill	Conservative	1940–5
Clement Attlee	Labour	1945–51
Sir Winston Churchill	Conservative	1951–5
Sir Anthony Eden	Conservative	1955–7
Harold Macmillan	Conservative	1957–63
Sir Alec Douglas-Home	Conservative	1963–4
Harold Wilson	Labour	1964–70
Edward Heath	Conservative	1970–4
Harold Wilson	Labour	1974–6
James Callaghan	Labour	1976–9
Margaret Thatcher	Conservative	1979–90
John Major	Conservative	1990–

re-election on the basis of the government's popularity and the Prime Minister's standing. Of Conservative leaders in the twentieth century, A. J. Balfour (1911), Austen Chamberlain (1922), Neville Chamberlain (1940), Sir Alec Douglas-Home (1965), Mr Heath (1975), and Mrs Thatcher (1990) were all eventually forced out of the leadership because of the lack of party (and Cabinet in Thatcher's case) support in Parliament. The perceived electoral unpopularity of the last three was a factor in ending their leaderships. The new system of annual leadership election, even when the leader is Prime Minister, is a constant reminder of the power of MPs to make and unmake a Tory leader.

If the office of Prime Minister has become stronger over time, variations in the incumbent's personality and political skill and circumstances determine how the opportunities are exploited. The Prime Minister has more patronage and the resources combined in the Political Office, Press Office Policy Unit, and Cabinet Office have all grown in size and are oriented to serve the Prime Minister, and the work of the Cabinet has increasingly been delegated to its committees. Harold Wilson's verdict was that in peacetime each Prime Minister has been able to secure more power than his predecessor 'because, over the whole period, governments have exercised more power and influence' (1977: 21). He also refers to two related factors which work to increase the Prime Minister's power: a crisis, which leads to demands for Prime Ministerial action, and the tendency of the news media to focus on him. There is no doubt that Mrs Thatcher's tenure convinced many observers that the role of the Cabinet had declined and that of the Prime Minister had reached almost presidential status.

None of this, however, is conclusive. Wilson's alleged sources of power refer as much to an increase in responsibilities and public expectations as to power and resources. By the end of her premiership Mrs Thatcher was the most unpopular PM the opinion polls had ever reported, until John Major plummeted to new depths three years later. What is more important—and variable—is the political context in which power is exercised and what it is that Prime Ministers are supposed to achieve with their increased power. For example, it is clear that perceptions among the public and colleagues of how successful the Prime Minister is are important. Fresh from a decisive general election or by-election victory, or an appreciative party conference, a Prime Minister has more authority with which to bargain with colleagues. Close association with policy success (such as the recapture of the Falklands for Mrs Thatcher in 1982) or failure (the poll tax, or the Major government's forced exit from the ERM) also affects influence. Success feeds on itself, just as failure does. There are several cases of a Prime Minister, including Mrs Thatcher, being overruled by the Cabinet, or not pressing a policy because of the lack of support. As Crossman, formerly the academic theorist (before he was a minister) of Prime Ministerial government, witnessed Harold Wilson's authority in the Cabinet crumbling after 1967—in the face of economic failure and political unpopularity—he revised his thesis.

The bypassing of Cabinet does not strengthen only the Prime Minister. It also provides more scope for the departments. A Prime Minister may find it difficult to battle simultaneously against different members of departments. In the late 1980s Mrs Thatcher was in conflict with the Treasury and Foreign Office on Europe. Departments, as Martin Smith (1994) observes, are the initiators of most policies, for they possess the time, legitimacy, and expertise to do this. Similarly, a PM is dependent on the political support and cooperation of Cabinet ministers—as Mrs Thatcher discovered on the night of her resignation.

Prime Ministers differ in their interests,

BOX 11.9. A PRIME MINISTER'S JOB DESCRIPTION

The Prime Minister is expected to:

- **chair the Cabinet and some Cabinet committees**
- **oversee the security services**
- **choose the general election date**
- **answer Prime Minister's Questions in the House of Commons**
- **appoint to senior Civil Service posts and to Cabinet (in the case of the Labour party this last is subject to a rule adopted in 1981 that the Cabinet is drawn from members of the Shadow Cabinet or parliamentary committee)**
- **manage the party**
- **attend international summits**

personalities, and working methods, and these may change over time in response to political circumstances and the pressure of events. Mr Wilson, before 1970, had a particular interest in foreign affairs, the economy, and industrial relations. His own influence in the Cabinet declined after the devaluation of the pound in 1967 and again after the failure of *In Place of Strife* (1969). Mr Heath took many initiatives and managed to keep his Cabinet united. He was particularly involved in the application to join the EEC and the decisions to impose direct rule in Ulster and introduce a statutory prices and incomes policy. Mrs Thatcher's style of chairmanship was more interventionist and abrasive than that of her predecessors. Her tendency to state her own views on an issue at the outset of a discussion could heighten tension. In her first term of office the Cabinet contained opponents of the government's economic policy and the Prime Minister at times acted like a faction leader rather than as a chairman concerned to arbitrate and seek debate.

Yet there are certain continuities in the Prime Minister's role in government. These include chairing the Cabinet, answering parliamentary questions, appointing ministers, overseeing the security services, and choosing the time for a dissolution of Parliament (see Box 11.9). In the post-war period, foreign affairs and economic management have tended to dominate his concerns. Richard Rose shows that in the course of a year a Prime Minister will, on average, participate in six debates and make six parliamentary statements on policy, probably on the economy, foreign affairs, and government business. Involvement in these areas, together with the need to attend to party management, means that he sees the Chief Whip, Foreign Secretary, and Chancellor on political business more frequently than he does other Cabinet colleagues (1980*a*: 16–17).

Although the loss of Empire has entailed a reduction in Britain's international influence and responsibilities, the growth of jet-age diplomacy and international summitry has resulted in the Prime Minister attending a growing number of meetings between heads of state—as at the biennial Commonwealth conference or the quarterly meetings of heads of EU states—more than when Britain was a world power. The sudden development of a foreign crisis usually requires heavy Prime Ministerial involvement, for example, the Falklands war in 1982, or the Desert War with Iraq in 1991, and John Major has taken a close interest in the peace process in Northern Ireland.

The Prime Minister–Chancellor relationship is crucial for the conduct of government. The close involvement between the two is such that the failure of a Chancellor's economic strategy inevitably reflects on the Prime Minister also. Cabinet rarely has the time or information to debate general economic policy, it has no say over interest rates or monetary policy, and the annual Budget is presented to it as a *fait accompli*. A Prime Minister is expected to back his Chancellor in Cabinet; when Mr Macmillan refused to support the full cut in public expenditure demanded by his Chancellor, Mr Thorneycroft, in 1958, the latter resigned along with other Treasury ministers. Mr Macmillan dismissed his Chancellor, Selwyn Lloyd, in 1962, in the search for a more expansionist economic policy. By contrast, Chancellors under Mrs Thatcher could rely on her support when they resisted spending pressures. Mr Heath worked doggedly to arrive at an agreement with the TUC in 1973 on an anti-inflation policy. Mr Callaghan, following the sterling crisis and IMF intervention in 1976, took a close interest in negotiations to develop a 'safety net' for sterling and participated, along with the Chancellor, Mr Healey, in trying to work out an agreement on wages with the TUC. Mr Callaghan also chaired a secret 'Economic Seminar', to review sensitive economic issues like exchange-rate policy. This was attended by a few senior ministers, civil servants, the Governor of the Bank of England, and political advisers. Mrs Thatcher was closely involved in the shaping of the Budgets of Howe and Lawson between 1979 and 1989. In 1988 and 1989 Mrs Thatcher and her Chancellor, Nigel Lawson, were publicly at odds over whether Britain should join the European Monetary System (EMS) and over exchange-rate policy. The division was damaging to the standing of the government and ended with Mr Lawson's resignation. When Britain was forced from the ERM in September 1992 the failure reflected on the Chancellor, Norman Lamont, and John Major because they had been so strongly identified with the policy.

Formal powers

The Prime Minister's formal duties are not laid down in any document. The decision to dissolve Parliament or appoint Cabinet ministers is formally exercised by the monarch, who acts on the Prime Minister's advice. The Prime Minister's distinctive powers and responsibilities distinguish the office-holder from other Cabinet ministers. Ministers may be a Prime Minister's colleagues, all collectively responsible for what the government does, but they are not equals. In Winston Churchill's words: 'There can be no comparison between the positions of number one and numbers two, three or four' (1949: 14). But the powers are also responsibilities and a Prime Minister's reputation, for good or ill, depends in large measure on how they are exercised. Three of these powers are worth considering.

Dissolution

It has often been claimed that the power to dissolve Parliament strengthens the position of a Prime Minister. Faced by rebellious colleagues, he can force them into line by threatening to call an election. Alternatively, he can manipulate the economy and, aided by opinion polls, choose the most favourable time to go to the country. It is not, however, credible to threaten rebellious backbenchers, for a divided government is likely to lose an election. The opinion polls may also prove fallible. A Prime Minister with a working majority needs a good reason to dissolve before his term of office is near completion. Of the thirteen elections since 1945, the incumbent Prime Minister has lost five, and the 1950 and 1974 victories produced a majority too small to last for a full Parliament. As party loyalties become more volatile and the economy more difficult to manage, governments may find it less easy to manipulate credible election-year economic 'booms'.

Although most Prime Ministers are now careful to consult with colleagues before deciding on the election date, there was no gainsaying the personal responsibility of Mr Heath in February 1974 and Mr Callaghan in 1979, both of whom 'mistimed' the dissolutions. Most experts argued that they could have done better had they dissolved earlier. The election defeats subsequently weakened their positions in the party. Calling an election is a two-edged weapon and nobody has more to lose from defeat than the Prime Minister.

Appointments

Patronage, particularly the appointment and dismissal of Cabinet ministers, is perhaps the clearest illustration of the Prime Minister's superior standing to colleagues. A Prime Minister appoints over a hundred ministerial posts covering Cabinet, non-Cabinet, and junior ministers, and Parliamentary Private Secretaries. If the governing party has around 330 MPs and some of them are eliminated on the grounds of youth or old age, lack of parliamentary experience, and actual or suspected personal, political, or administrative shortcomings, then about a third of the party's MPs can be offered some office. There are several limitations on the Prime Minister's freedom of choice over Cabinet appointments for he also has to take account of the criteria of political representativeness, expectations of senior colleagues, and personal loyalty. A Prime Minister may reflect on the lack of competition for places and on how restrictive it is to staff a government from a fixed pool of a few hundred MPs.

Within the civil service appointments of permanent secretaries also require the Prime Minister's consent. When a vacancy arises a short list of candidates is compiled by a group of senior permanent secretaries, including the Secretary to the Treasury, who has responsibility for appointments. The minister in the appropriate department will be consulted, but the final choice is that of the Prime Minister. A Prime Minister who has a clear policy agenda and has several opportunities to make appointments, like Mrs Thatcher did in her eleven years in office, can certainly influence the outlook of the highest echelons of the Civil Service. A Prime Minister also has the opportunity to appoint peers to the House of Lords. Mrs Thatcher had ample opportunities to exercise this patronage. On her maiden speech in the Lords, she observed that she had appointed nearly a fifth of its members.

Cabinet reshuffles—entailing voluntary retirements and dismissals, and new appointments—usually occur at least once in the lifetime of a government. A new Prime Minister's first Cabinet may be heavily influenced by original Shadow Cabinet appointments and political debts. Indeed, a future Labour Prime Minister is required, under party rules, passed in 1980, to appoint shadow ministers to their posts in the new government. Mrs Thatcher's Cabinet at the end of the 1979 Parliament was more to her political liking than the one she started out with. Her abrupt dismissal of Francis Pym as Foreign Secretary after her election victory in 1983 and the transfer of a reluctant Sir Geoffrey Howe from the same position in 1989 were brutal illustrations of the power to hire and fire.

Reshuffles, however, also have political costs, including the resentments of those dismissed and the disappointments of those passed over (Alderman and Cross 1979). Over time, the accumulated resentments and disappointments may strain loyalty. Ministers summoned for dismissal have testified to the Prime Minister's unease and discomfort on these occasions. Mr Callaghan reputedly refused suggestions that he reshuffle his Cabinet in early 1979 on the grounds that 'You make an enemy of the man sacked and disappoint a dozen MPs who hoped for promotion'. Because some ministers are new to their posts and learning a new set of problems the frequent reshuffles may increase the influence of the permanent civil servants in forming policy. Moreover, if done too often, reshuffles may appear as acts of panic or scapegoating by the Prime Minister. Mr Macmillan's dismissal of a third of his Cabinet in 1962 did not increase his support in the party. Although a dramatic example of his 'power', it is doubtful that he could have repeated it on such a scale and survived. After Mrs Thatcher had suffered the resignation of Nigel Lawson in 1989 her position was bound to be further damaged by another resignation—as happened with the departure of Sir Geoffrey Howe.

Control of the agenda

The agenda for the Cabinet, and for all committees chaired by the Prime Minister, is approved by the Prime Minister in consultation with the Cabinet Secretary. A skilful Prime Minister can certainly exploit the right to schedule items for Cabinet discussion, call speakers, sum up the views of the meeting (which is then a decision), and decide whether an issue should be referred to a Cabinet committee and the membership and the terms of reference of that committee. But this is done within constraints. Much of the Cabinet agenda is fairly predictable, covering such items as reports from the Foreign Secretary, statements about the following week's parliamentary business, and recommendations from the subcommittees. There is also the sheer pressure of events and deadlines against which the Cabinet considers issues. Prime Ministerial control of the agenda was seen in Mr Wilson's refusal to allow Cabinet to consider devaluation for nearly three years until 1967, and Mrs Thatcher's exclusion of discussion of Britain's entry to the ERM for a number of years in spite of many Cabinet ministers' wishes. Neither case benefited the Prime Minister, and in each case, devaluation and ERM entry were seen as defeats for the leader.

This does not mean that the Prime Minister is usually at odds with the Cabinet. If some Cabinet ministers are political rivals of the Prime Minister, they are also colleagues. Most are there because, among other reasons, the Prime Minister thinks they have something useful to contribute. On many issues, particularly short-term crises, a Prime Minister will want the Cabinet's views: a more informed and considered decision may emerge, potential critics may be mollified by having the opportunity to present their views, and the fact that the Cabinet has deliberated and arrived at an agreed view may encourage the spirit as well as the reality of collective responsibility and reassure supporters in Parliament and the country.

Limitations

Shortage of time is an obvious limitation on a Prime Minister's power. His timetable is already partly taken up with duties which only he can perform. Many policies and decisions are decided in the departments, where they may develop their own momentum. Most Prime Ministers have to limit their active involvement to a few areas, usually the persisting problems of the economy and foreign affairs. Intervention in one area means that there is less time for dealing with something else. The sheer pressure of work and the need to take decisions, combined with entrenched departmental interests, limit the opportunities for a Prime Minister to intervene directly. His staff may also lack the expertise to enable him to stand up to the other departments on all but a few issues. The growth of the committee system makes it difficult for the Prime Minister to oversee most of the policies.

It is a limitation as well as a strength of the Prime Minister's position that he combines so many roles in his office: leader of the party, chairman of the Cabinet, political head of the State, and is vested with a special responsibility for such matters as intelligence. There is the possibility for a conflict of interest in these roles. The Labour Prime Ministers Ramsay MacDonald (public spending cuts in 1931), Harold Wilson (trade union reform in 1969), and James Callaghan (pay policy in 1979) all experienced such conflicts when their perceptions of the 'national interest' conflicted with pressures from the party. Moreover, it is likely that Britain's post-war economic failures and gradual loss of world influence, the long-term decline in popular support for the two main parties, and the electoral unpopularity of most governments have all combined to sap the influence of Prime Ministers. The opinion polls show that post-war British governments are not generally considered to have been successful, nor, over time, have Prime Ministers been (see above, Table 5.3).

No Prime Minister is a 'superman', and his own skills and popularity will not compensate indefinitely for Cabinet ministers who are inadequate; their performances reflect on the leader. British government is a collective enterprise and a weak Chancellor, Foreign Secretary, or Cabinet harm the Prime Minister. For all that a Prime Minister may have his own agenda, he has to keep the Cabinet together. He may sometimes get his own way against strong Cabinet opposition, though hardly against a clear majority. But to push for his own view all the time, regardless of the strength and persistence of opposition, entails political 'costs'. The downfall of Mrs Thatcher is an object lesson. Not only had key Cabinet ministers resigned in protest at her style of leadership, but a growing number of MPs regarded her as a barrier to party unity, and another general election victory. The power of the Prime Minister is, as Harold Wilson has observed, 'effectively the power he has *in* and *with* the Cabinet' (1977).

Within as well as outside the Cabinet, the exercise of political power is in large part the exercise of political management. Mr Macmillan was, according to colleagues, prepared to give way with good grace when he could not carry the Cabinet. Mr Wilson found himself in a minority in Cabinet and PLP over proposals to reform the trade unions in 1969 and had to give way. And Mr Callaghan misjudged the mood of the party in fixing a 5 per cent 'norm' for pay policy in 1978–9. Mrs Thatcher had views on most issues and was quick to voice them: she found herself in a minority on occasions. John Major has been careful to position himself with the majority view in Cabinet.

The doctrine of collective responsibility applies to the Prime Minister as well as ministers and means that they have to make compromises and settle for a policy that is broadly acceptable. Cabinet ministers, after all, have to defend the policy publicly.

Unhappy ministers can 'leak', make 'coded' speeches of dissent, or ultimately threaten to resign. A resignation advertises discord in the government, weakens the Prime Minister, unsettles morale, and encourages the opposition to think they have the government on the run.

A Prime Minister has many resources to employ in managing the Cabinet. The most tried political technique is 'squaring', or agreeing a Cabinet line with the appropriate departmental minister, or with key ministers before it reaches the Cabinet, or to make concessions to dissident ministers beforehand. The Prime Minister may promote an agreed Cabinet line simply through strength of personality and force of argument; after all, he has the opportunity to take a broad view, while many ministers are preoccupied with departmental duties. He may choose a point in the discussion, if it looks favourable, to force a decision, or if not favourable, then to postpone it. He may appeal to personal and political loyalty, since colleagues realize that to overrule the Prime Minister on a crucial issue weakens the authority of the government. In the 1976 negotiations with the IMF, already referred to, there was a potential majority in the Labour Cabinet for refusing the IMF terms for a loan. The opposition collapsed at the last moment because of fears of the political and economic consequences which might follow the Cabinet's open repudiation of Mr Callaghan and the Chancellor, Mr Healey.

All this is particularly true when the Prime Minister is trying to move the Cabinet in a new direction. In 1972–3, the Heath government reversed itself on three major policies: it embraced a statutory prices and incomes policy, provided state funds to rescue industries, and abolished the Stormont Parliament in Northern Ireland. These so-called 'U-turns' were accomplished without any Cabinet resignations in protest at the policy changes. There are many reasons why the Cabinet sticks together in such circumstances, including calculations about personal and party political survival. But other factors were also operating in the Heath Cabinet. Over a time a group of people who work closely together may develop a common way of looking at problems—a form of 'group-think'—and a loyalty to the group. Mr Heath's Cabinet appears to have been very loyal to him personally and to have been more politically homogeneous than many other Cabinets. It was striking that when he was defeated for re-election as leader in 1975, all but two of his former Cabinet colleagues are reported to have voted for him, whereas his support on the backbenches was much thinner.

The case of Mrs Thatcher

Mrs Thatcher was such a distinctive and long-serving Prime Minister that she has helped to colour contemporary assessments of the role of the Prime Minister. The decline of the Cabinet certainly seemed to gather momentum under her premiership. Alone amongst twentieth-century Prime Ministers she gave her name to a political doctrine and an undoubted political style. The view that Britain had moved to Prime Ministerial government was encouraged by her record: the ruthless dismissal of Cabinet dissenters in 1981, sacking of the Foreign Secretaries in 1983 and 1989; abolition of the CPRS in 1983; close involvement in the promotion of permanent secretaries; and her tendency, by publicly expressing her views on controversial issues, to depart from an agreed line or to pre-empt Cabinet discussions. There was a reduction in the number of Cabinet and standing Cabinet committee meetings, and papers distributed to the Cabinet—the very stuff of collective decision-making.

Mrs Thatcher seems to have relied heavily on her Policy Unit and ad hoc groups for policy advice and was more interventionist in the work of departments. Her press office presented her as a dominant figure. 'Maggie Acts', 'Maggie Steps In' were familiar tabloid headlines as Mrs Thatcher convened seminars of experts at Downing Street to cope with pressing issues, for example, football violence, the future of broadcasting, environmental issues, schools, and the universities.

One tactic Mrs Thatcher used to retain the initiative was to decide matters outside the formal Cabinet, either in ad hoc committees or in informal groups. A Prime Minister has more opportunity to shape policy by appointing a committee, its members, and the chairman, and she exploited this to the full. She ensured that her supporters dominated the Cabinet committee dealing with economic strategy. Once the Prime Minister and a few senior ministers make up their mind, a decision then acquires what civil servants call 'momentum'. The decision to ban trade union membership for workers at the intelligence gathering centre GCHQ in 1984 was also taken outside of Cabinet and its committees. According to *The Times* (7 February 1984) 'The first most Cabinet ministers knew about it was when Sir Geoffrey Howe announced the decision in the Commons on January 25'. Many other policies, such as the poll tax, the abolition of the GLC, many of the educational and health reforms, all had a strong Prime Ministerial push and were largely drawn up by study groups convened by Mrs Thatcher. Within her private office she relied on her foreign policy adviser, Charles Powell, to challenge the Foreign Office line on the European Community, and used Professor Alan Walters to buttress her suspicion of the European Monetary System and her own Chancellor's exchange-rate policy in 1988 and 1989.

But her power *vis-à-vis* Cabinet varied over time. In her first two years of office she and Sir Geoffrey Howe had to battle hard to persuade Cabinet colleagues to go along with their economic strategy. Mrs Thatcher also had to give way to Cabinet on a number of issues, including public spending, a compromise on the EEC budget, and trade union legislation. In September 1982, faced with opposition, she had to abandon Cabinet discussion of a paper from the CPRS about the implications of the rising trend of public expenditure and its consequences for Welfare State spending. In 1986 following the Westland crisis she supported the sale of Land

BOX 11.10. MARGARET THATCHER: CULTURAL REVOLUTIONARY

Margaret Thatcher, Britain's first woman Prime Minister and the longest serving Prime Minister this century.

Leader of the Conservative Party 1975–1990
Prime Minister 1979–1990

Born Margaret Hilda Roberts, 13 October 1925, she came from a modest family background. Her father was a grocer, local alderman, and Methodist lay preacher in Grantham, Lincolnshire. She was educated at the local girl's grammar school and then Somerville, Oxford where she read Chemistry.

After failing to win Dartmouth for the Conservatives in 1950, she qualified at the bar and married Denis Thatcher. She gave birth to twins. In 1959 she was elected MP for Finchley.

Prior to winning the Conservative party leadership in 1975, Mrs Thatcher held the junior post of Parliamentary Secretary at the Ministry of Pensions (1961–4) in Macmillan's government. She achieved cabinet rank in Heath's 1970–4 government as Secretary of State for Education, a post in which she fought successfully for large increases in her department's budget but for which she earned the title 'Thatcher, Thatcher, Milk Snatcher' for withdrawing free school milk. Mrs Thatcher gained the Conservative party leadership partly by default: none of Mr Heath's senior colleagues was willing to challenge him.

She is one of the few PMs to achieve the status of having an -ism named after her. **Thatcherism** is a doctrine and that evolved and gathered momentum during her period in office. It is a complex mixture of

economic liberalism

- belief in the efficacy of markets
- sound money
- private rather than public provision and ownership

social authoritarianism

- restoration of respect for the authority of state, schools, parents, and police
- individual responsibility
- revival of the virtues of citizenship

She claimed that her mission was to 'kill socialism' and that she came to office with one deliberate intent: 'to change Britain from a dependent to a self-reliant society–from a give-it-to-me, to a do-it-yourself nation'.

Rover to General Motors but was defeated in Cabinet. In 1990 she finally agreed to Britain's membership of the ERM.

By pushing to the outer limits of her authority, making decisions with small groups of ministers and advisers, and closely involving herself in Whitehall promotions and the policies of departments Mrs Thatcher presented a distinctive model of leadership. Measured in terms of winning general elections or carrying through policies, she was undeniably successful. In large part it derived from her determination to achieve key policy changes and, as has often been noted, her own agenda for radical change was rather different from that of many Cabinet colleagues. But her downfall is also a warning of what can happen when a dominant PM is regarded as an electoral liability and neglects Cabinet support. She had built up much ill will over the years and the arrangements for annual election of a Conservative leader—even if a Prime Minister—made her vulnerable to disaffected colleagues.

The case of John Major

John Major became Prime Minister in November 1990 and by the end of 1995 he had served for longer than all post-war Prime Ministers except Attlee, Wilson, and Thatcher. He has already achieved some fame as being the most unpopular Prime Minister registered in the opinion polls. He is usually regarded as the antithesis to Mrs Thatcher in that he:

- allegedly lacks convictions and has no agenda of his own
- is colourless or grey, as portrayed in the satirical TV programme *Spitting Image*
- is more of a team player than an individualist, and provides a conciliator rather than a warrior style of leadership

The first two judgements are criticisms, the third is not. Some part of Major's style comes from his own personality, some from the narrow majority of twenty-one with which he started the 1992 Parliament, which has been steadily reduced. (Mrs Thatcher had a majority of over 100 for most of the time), and some from party divisions, notably on Europe. Since 1992 he has been an embattled leader of an embattled government. If he has not had an agenda on the scale of his predecessor, he has certainly had his policy concerns, including the peace process in Northern Ireland, low inflation, and the Citizen's Charter to improve public services.

The Cabinet under Major has become more collegial. It has effectively decided important issues, for example, the council tax, and Britain's stance on the Maastricht treaty and qualified majority voting. The divisions in Cabinet are less between 'wets' and 'drys' on economic policy, as in the 1980s, and more on Europe between 'sceptics' (Portillo, Lilley, and Redwood, until his resignation in 1995) and supporters (Heseltine, Clarke, Hurd). Major takes a middle stand and the media report the clash in part in terms of who will succeed a weak PM. Given the Tory divisions over Europe, it is likely that a strong pro or anti thrust would split the party. After the Thatcher experience, some might argue that government only works well with a strong No. 10 staff and a strong Prime Minister.

Conclusion

A Prime Minister still operates with constraints imposed by party, Parliament, and Cabinet colleagues; the British Prime Minister cannot be presidential in a Cabinet system. It is too soon to say whether Mrs Thatcher, because of her robust style and commitment to change in so many policy areas, will represent a trend or is a unique case, Yet it is also the case that the Cabinet hardly exists in its classic sense as the decision-making centre. It coordinates the decisions of the inner cabinet and committees, in which the groundwork for the full Cabinet is done. The Cabinet system embodies other important values like collective responsibility, the identification and continuity of decision-makers, and coherence in policy.

Summary

- **The Cabinet combines political and executive functions. It oversees important political decisions, it plans the business of Parliament, it arbitrates between government departments, and it coordinates government policy.**
- **Coordination is possibly the role that the Cabinet performs least successfully. This is partly explained by constraints of time and the sheer volume and complexity of government responsibilities, but also by the centrifugal force of departmentalism.**
- **The fusion of the executive and legislature in the institution of the Cabinet bestows political authority on the Cabinet. The fact that members of the Cabinet are usually drawn from one party fosters Cabinet unity.**
- **There are no formal rules specifying the size of Cabinets but constraints do exist. There is a need to represent particular sectors in the Cabinet, and this sets the lower limits. The need for adequate deliberation sets upper limits.**
- **PMs have resorted to various experiments (overlords, inner cabinets, super-ministries) in an attempt to improve the Cabinet's performance of its coordination and strategic role. Unwelcome political side effects have tended to make these experiments short-lived.**
- **Departmental ministers are constitutionally responsible for their department: they administer it, and they represent it in Parliament, in the Cabinet, and to the public.**
- **Cabinet meets for approximately two hours per week. Its proceedings are formal and secret. Votes are rare but ministers are collectively bound by Cabinet decisions. A hierarchy exists within the Cabinet.**

- There is a network of Cabinet committees, standing and ad hoc. It is the PM who appoints the chairmen and members of committees and who determines their powers and terms of reference.

- The Cabinet Office is staffed by civil servants who organize the work of the Cabinet and its committees, preparing agendas, recording and circulating minutes, progress, verifying and coordinating decisions.

- In addition to the services provided by the Cabinet Secretary and the Cabinet Office, the PM also has a support system in No. 10: a Private Office staffed by civil servants; a Political Office staffed by political appointees; a Policy Unit which includes civil servants and political appointees; a Press Office; plus any individual advisers the PM wishes to consult.

- The question of the PM's power has been keenly debated since the 1960s. Some commentators assert that there is a cyclical pattern, others argue that a linear growth in power is discernable; some reject either thesis, claiming instead that the PM/Cabinet power ratio depends on variables such as personalities, oppotunities, etc.

- The PM exercises a formidable range of formal powers, and media attention and the imperatives of international diplomacy have helped to create a general impression of prime ministerial government. But the British party and parliamentary system act as a check. Mrs Thatcher may have dominated her Cabinet but her party could still depose her without reference to the electorate. Whilst there are few constitutional limits on a PM, there are numerous political constraints. There are also physical limits to the PM's power in terms of information, conflicts of interest, personal skills, energy, and time.

- Margaret Thatcher's long tenure and dominant style provided plenty of evidence for those arguing that Britain has a prime ministerial government. She presented a distinctive model of leadership, and in terms of winning elections and carrying out policies she was successful. But she was not invulnerable to the power of the party.

- John Major's style has been more collegial and conciliatory, but this has made him appear weak and indecisive.

- The party and parliamentary system in Britain serve to ensure the survival of the Cabinet as a powerful institution.

CHRONOLOGY

1721 Robert Walpole takes office, he is generally recognized by historians to be the first Prime Minister

1782 Resignation of Lord North's administration: links fate of PM and Cabinet

1878 Disraeli signs the Treaty of Berlin — first official recorded use of the title Prime Minister

1916 Lloyd George creates a Cabinet Secretariat

1937 Ministers of the Crown Act: officially recognizes office of the PM and sets the number entitled to a Cabinet minister's salary. First No. 10 Press Officer appointed

1940 Churchill creates PM's Statistical Section; Cabinet committee system develops

1951 Churchill introduces short-lived 'overlord' experiment

1962 Macmillan dismisses a third of his Cabinet

1964 Wilson creates a Political Office in No. 10; ministers are allowed to appoint a limited number of special advisers

1968 George Brown resigns, citing Wilson's presidential conduct of the Cabinet as reason

1970 Heath creates the CPRS to aid Cabinet strategic thinking and super-ministries to aid coordination

1974 Wilson creates the PM's Policy Unit

1976 Callaghan's discussion of the IMF loan Cabinets appear to restore classic Cabinet government

1982 It is mooted that Mrs Thatcher is about to create a PM's Department

1986 Heseltine resigns, citing misconduct of Cabinet government

1989 Lawson resigns, raising questions about PM's advisers

1990 Mrs Thatcher loses the party leadership and premiership

1992 Major introduces a policy of greater openness regarding Cabinet committees

1995 Heseltine appointed Deputy PM with wide-ranging coordination brief

ESSAY/DISCUSSION TOPICS

1. Is the Cabinet the major decision-making body in British government?
2. Are the Prime Minister's powers unlimited?
3. What are the shortcomings of the Cabinet system and how might they be addressed?
4. To what extent do you agree with Richard Crossman's view that 'The post-war epoch has seen the final transformation of Cabinet Government into Prime Ministerial Government'?
5. Margaret Thatcher's premiership provides ample evidence to support the thesis that British government has become Prime Ministerial government. Discuss.
6. Discuss the view that 'Prime ministers rise and fall but the Cabinet abides' (Norman St John-Stevas).

RESEARCH EXERCISES

1. What reasons did Michael Heseltine in 1986, and Sir Geoffrey Howe in 1990, give for their respective resignations? What implications did these resignations hold for the thesis of Prime Ministerial power?

2. Explain to an American visitor why Mrs Thatcher lost her job as Prime Minister even though she had not lost an election.

FURTHER READING

On the Prime Minister see A. King (ed.), *The British Prime Minister*, 2nd edn. (London: Macmillan, 1985) and B. Donoughue, *Prime Minister*, (London: Cape, 1987). On the Cabinet see Peter Hennessy, *Cabinet*, (Oxford: Blackwell, 1986), *Contemporary Record* (1994), and S. James, *British Cabinet Government* (London: Routledge, 1992).

The following may also be consulted:

Barber, James, *British Prime Ministers since 1945* (Oxford: Blackwell, 1991).

Burch, Martin, 'The Prime Minister and the Cabinet from Thatcher to Major', *Talking Politics*, 7/1 (1994).

Dorey, Peter, 'Widened, yet Weakened: The Changing Character of Collective Responsibility', *Talking Politics*, 7/2 (1994/5).

Foley, Michael, 'Presidential Politics in Britain', *Talking Politics*, 6/3 (1994).

12 PARLIAMENT

Reader's guide

Parliament symbolizes the continuity of the state and the constitution and it is the arena in which the political drama of government versus opposition is acted out. In addition to its symbolic and ritualistic roles, Parliament has an important legitamizing function. Britain has a parliamentary system of government. The political executive, the Cabinet, is recruited from and accountable to Parliament, and it is the consent of the majority in Parliament that legitimizes the government's authority and converts its policy into law. In order to understand Parliament, however, it is necessary to take account of the dominant part played by the political parties in the procedures and work of Westminster.

What are the functions and role of Parliament? Is there a sense in which Parliament can be said to have declined? How does a bill become a statute? This chapter addresses these questions. it goes on to describe Parliament's committee system and then to examine some of the ways in which the parties have transformed Parliamentary control of the executive into executive control of Parliament. It explores the role of Her Majesty's Opposition and describes the formal and informal opportunities for the Opposition to perform its constitutionally recognized functions. It notes the increasing modern tendency towards rebellion — a greater willingness on the part of backbenchers to disobey the party whips. It discusses the role, composition, and powers of the House of Lords and examines some of the proposals for reforming this semi-feudal institution. The chapter concludes by reflecting on the continuing importance of Parliament.

LEGALLY speaking, Parliament consists of the monarchy, the House of Lords, and the House of Commons. The consent of each is required for a statute to be enacted. Each statute contains the clause:

Be it enacted by the Queen's most excellent Majesty, by and with the advice and consent of the Lords Spiritual and Temporal, and Commons . . .

When we talk of the power of Parliament today, we mean, effectively, the power of the Commons, which emerged in the nineteenth century. This chapter discusses the development of Parliament, the legislative process, the Commons' relationships with the executive, and the role of the House of Lords.

Development

Parliament today does not make policy—that is a matter for the executive—but it discusses policy. It also plays a debating role, criticizes and scrutinizes the policies and legislation of the government, is a channel of recruitment to the government, and the Law Lords, sitting in the House of Lords under the Lord Chief Justice, constitute the highest court in the land. The development of Parliament in its present form occurred in two stages. In the first it asserted its independence *vis-à-vis* the Crown; in the second, the House of Commons became representative of the public at large through election by universal suffrage (see Ch. 1).

During the seventeenth century the supremacy of the House of Commons over the Lords in financial matters was asserted. The Commons carried resolutions stating that the House of Lords could not amend finance bills. Practice conformed to the resolutions, although the principle was not formalized until the 1911 Parliament Act laid down that all money bills certified as such by the Speaker of the Commons should be presented for the Royal Assent, regardless of the views of the House of Lords.

The conventions of the accountability of the government to the Commons and its dependence on majority support there—rather than on the support of the monarchy—were established in the nineteenth century. The convention also grew that defeat on a vote of confidence, or the inability of the government to command support in the House, entailed the government's resignation.

The varying relationships between Parliament and the executive have been described in the *Liberal* and *Whitehall interpretations of the Constitution* (Birch 1964). The first, also called a Westminster model, notes how formal political power has been transferred from the monarch to the elected House of Commons. The government is collectively responsible to the Commons and may be dismissed by it; important bills and announcements are first introduced in the Commons; and the work of the executive is accountable to the Commons through the doctrine of ministerial responsibility.

The Whitehall model notes how the rise of one-party majority government and the use of procedural devices have given the government a high degree of control over the Commons. Many elements of the monarch's prerogative power (or powers vested in the

Crown)—such as dissolving Parliament, appointing ministers, declaring war, and making treaties with other countries—have now passed to the Cabinet. Because civil servants and ministers are servants of the Crown, British government has a substantial degree of autonomy from Parliament. In this model, Parliament may debate and criticize, but any realistic analysis of how Britain is governed acknowledges that initiative on legislation and policy lies with the Cabinet.

'Decline'

It has been fashionable for some time to mourn 'the decline of Parliament'. Some critics look back to a golden age when the Commons made and unmade governments and when many MPs were relatively independent of party whips. But this is to look back on a short period of the mid-nineteenth century, and one that has little relevance to the very different political, administrative, and economic circumstances of today. The liberal model of Parliament's role has been undermined by three developments in the twentieth century:

1. *The growth of disciplined party voting* in the Commons from the late nineteenth century, with the vast majority of MPs voting in the division lobbies as directed by the party whips. Effective pressure by backbenchers took place not in debates on the floor of the House but in private party meetings.

2. *The emergence of the dominant two-party system*, largely after 1945, and the control of government by one party with an assured parliamentary majority. This enabled the government to gain control of the business of the House and get its way in the division lobbies.

3. *The emergence of more interventionist government* has meant that Parliament has struggled to cope with an enormous amount of legislation and this has reduced its ability to deliberate and scrutinize.

During the twentieth century there has been little change in the number of government bills which are introduced each session. What has increased has been the length (and probably the complexity) of bills, the amount of delegated legislation, and the number and length of parliamentary sittings. The Select Committee on Sittings of the House in 1992 reported that the annual volumes of public statutes had grown from two to three on average in the 1970s to four to five in the 1990s. Much legislation is now 'delegated', which means that Parliament establishes the broad principles of a measure and permits civil servants and ministers to fill in the details. The House of Commons sits for more days and more hours than any other legislature in the large democracies. Pressures have also come from the growing demands of constituents and lobbies, the growth of legislation from the EC, and the increase in the activity of the departmental select committees.

These changes are not confined to Britain. The emergence of big government, in which a large bureaucracy spends large sums of money and copes with complex social and economic problems, has transformed the role of legislatures in many countries. In Britain, the historic role of Parliament as the exclusive 'consenting' body has also been challenged by the emergence of three other institutions:

1. The development of the *programmatic, disciplined political parties* has led some observers to talk of 'party' rather than 'parliamentary' government. The government's majority in the Commons enables it to push through the 'shopping-list' of proposals it proposed in its manifesto.

2. *Pressure groups*, organizing and representing the many interests in society, negotiate directly with ministers on many issues. Often the policy details are worked out between groups, the minister, and his civil servants and then presented to Parliament.

3. Membership of the *European Community* means that a growing body of legislation is made in Brussels.

Members

MPs may differ in their personalities but the combination of their intense political interest, ambition, and activity makes them distinctive (Searing 1994). To be elected as an MP immediately makes one unrepresentative. The pressures and satisfactions of living in the public gaze appeal to only a minority of people. Some backbenchers find the life in Parliament frustrating, largely because they feel that their opportunities to influence the government are limited. Some try to be all-rounders, seeking to speak on a wide variety of issues. Others specialize in a particular area—defence, the disabled, even the procedure of Parliament. Some are content to be party loyalists. Some calculate carefully how to ascend the ministerial hierarchy, others seem to be natural 'mavericks' and spurn the opportunities to please their leaders—Labour's Denis Skinner or Conservative's Teresa Gorman and Anthony Marlow are notable examples. But for most MPs increasingly it is relations with the constituency, local party, and local organizations, and the problems of constituents, that take up most time (Norton 1994).

The idea that MPs should be paid a salary and, therefore, that they did a full-time job was slow to gain acceptance. Payment was introduced in 1911, largely at the behest of Labour MPs who lacked private incomes. The salary has been raised since on a number of occasions, but still compares unfavourably with the pay of successful middle-aged people in the professions, or members of legislatures in other Western democracies. Salaries are reviewed each Parliament (or at four-year intervals) by the Review Body on Top Salaries and additional allowances are made for travel, secretarial, and research expenses. Many MPs, particularly on the Conservative side, receive additional remuneration from outside 'interests' or draw salaries from other employment. Defenders of these outside activities point to the advantages which the member and House gain from this type of experience. However, the range of such activities is rather narrow, being largely confined to business directorships, legal practice, and journalism or communications. At a time when greater demands are made on MPs, a larger parliamentary burden inevitably falls on the full-time members.

BOX 12.1. MEMBERS OF PARLIAMENT: PAY AND ALLOWANCES

Since 1911 Members of Parliament have received salaries.

Salaries per annum in £ sterling since 1911

1911	400	1980	11,750
1931	360	1981	13,950
1934	380	1982	14,510
1935	400	1983	15,308
1937	600	1984	16,106
1946	1,000	1985	16,904
1954	1,250	1986	17,702
1957	1,750	1987	18,500
1964	3,250	1988	22,548
1972	4,500	1989	24,107
1975	5,750	1990	26,701
1976	6,062	1991	28,970
1977	6,270	1992	30,854
1978	6,897	1994	31,687
1979	9,450	1995	33,000

Source: adapted from *Whitaker's Almanac 1994*.

1924	Free travel was introduced
1969	Office Costs Allowance introduced for secretarial and research expenses; in April 1993 this allowance was set at £40,380
1972	Overnight away from home non-taxable expenses introduced. £10,958 a year was the amount set in 1984

The legislative process

It is ironic that legislation in Britain is largely a matter for the executive and that the most significant stages in the genesis and shaping of a bill have often already occurred before Parliament is consulted. Although there are private members' bills, and MPs have opportunities to press amendments to bills, most legislation arises from government departments. Even before a new government assumes office, civil servants have drafted proposals to try and give effect to the party's manifesto promises. Other stimuli for legislation may come from pressure groups, individual MPs, reports of Royal Commissions, and recommendations from committees, working parties, or other expert bodies. Outside pressures or events may help to give momentum to a framework for legislation. Filling in the substance and detail of a bill, however, usually rests with the department. Parliament remains supreme in making statute law—except for EC law—and once a bill becomes law it is entered in the Statute Book. The operation of an Act, however, depends in part on the interpretation of those who carry it out, in part on the views of the courts which may be involved, and in part on the consent of those who are subject to it. As remarked in a thorough study of law-making, 'Legislation merely provides the legal wrapping paper for discrete packages of policy or of administrative adjustment' (Burton and Drewry 1981: 256).

In the nineteenth century much legislation was private in scope. Today, however, the government and opposition front benches dominate the parliamentary timetable and legislative process. Some 15 per cent of the time is now available for backbench initiatives; this includes twelve days for private members' bills on Fridays, and eleven other days (mostly Fridays) for motions. Friday is not a good day for government business—the House rises at 3 p.m. and many MPs want to go away to their constituencies for the weekend. Most private members' bills are talked out before reaching a vote, and on average only eight such bills have been passed in post-war Parliaments. Under Standing Orders, any proposal affecting government expenditure or taxation may be moved only on the recommendation of a Minister of the Crown. If a private member's bill involves even a minor amount of public expenditure and the government agrees with the proposal, then a minister will move a resolution.

From a bill to a law

The Queen's Speech, which is delivered at the beginning of each session, promises various 'bills', 'legislation', 'measures', 'discussion', and 'proposals'. Some proposals eventually introduced by ministers may not have been mentioned in the Speech, because they were not anticipated when it was written, were not considered important at the time, or are perennial items. To become law a government bill must navigate what has been called the Whitehall 'obstacle race'. Where a department has recognized an issue as suitable for legislation and decides to make out a case for a bill, it will then circulate a draft proposal to other departments which are likely to be affected, and to the Treasury. If the Cabinet agrees the proposal goes to its Future Legislation Committee, then to Parliamentary Counsel, who will draft a bill. Finally, the proposal will be included in the Queen's Speech, along with statements of the government's other plans for the forthcoming session. Once the Cabinet has granted the time, the bill is then forwarded to its Legislation Committee, which is responsible for piloting the bill through Parliament. The Cabinet's decision (or its committee's) is therefore important at three stages: on the substance of legislation in the policy committees, the wording and form in the Legislation Committee, and the parliamentary timing in the Future Legislation Committee.

BOX 12.2. THE LEGISLATIVE PROCESS: GOVERNMENT BILLS

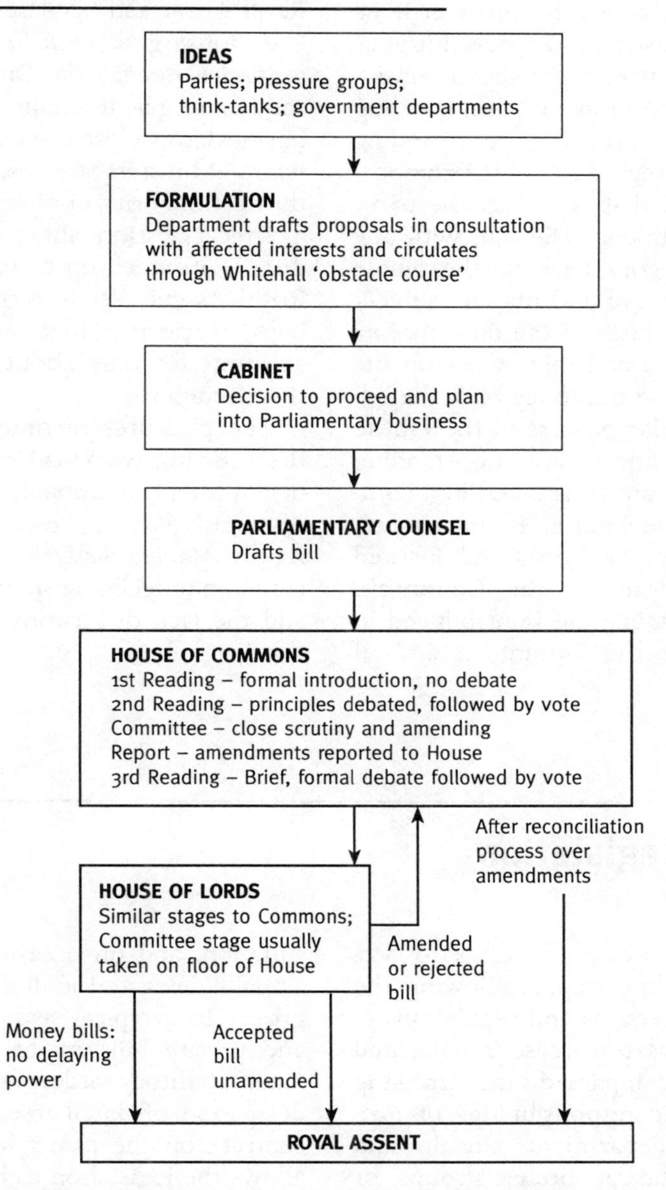

Notes:
1. Some non-controversial bills begin life in the Lords and go through the process in reverse.
2. If the Lords propose amendments, Commons can accept, reject, or propose further amendments and return the bill to the Lords. So on until Houses are agreed.
3. If no agreement can be reached between the two Houses the Commons can insist (except in case of a bill extending the life of Parliament) and the bill can then be passed again through the Commons and then for Royal Assent under the terms of the 1949 Parliament Act.

Within the Commons, the *first reading* of the bill is formal, being a statement of the bill's purpose. On the *second reading* the bill's contents are subject to a fuller debate. If its general principles have been approved it goes to a standing committee for detailed consideration. The composition of the standing committees must, according to Standing Order No. 65, 'have regard to . . . the composition of the House', that is, reflect the party balance of the Commons. The bill, with any amendments, then goes back to the whole House for a *report* stage and may be subject to further amendments. On the *third reading* there is a final debate and then a *vote* on the bill. Except in the case of money bills, the bill goes through a similar process in the House of Lords, but the Lords have no standing committees: all bills are considered in a committee of the Whole House. If the amendments are made by the Lords and insisted on, against the wishes of the Commons, then the bill falls. It may be reintroduced in the same form by the Commons and, if passed again, will become a law, notwithstanding the objection of the Lords. Finally, the bill goes to the monarch to receive the *Royal Assent* and becomes law.

A British government is well placed to get its legislation through the Commons. The rules of the House give it ample opportunity to get its business considered when and in the form it wants. Ministers are responsible for drafting the vast majority of bills and amendments. In a typical session about 150 bills are introduced, of which up to two-thirds receive the Royal Assent. Yet in terms of time it is perhaps surprising that government business accounts for only about half of the time of the Commons.

The pressures of time on the House and the growing workload of a substantial number of MPs has probably impaired the quality of legislation. A recent Hansard Society report, *Making the Law* (1993), expressed concern about declining standards in drafting bills and the lack of scrutiny of delegated legislation.

Delegated legislation

The term *delegated legislation* refers to acts which contain enabling clauses allowing the minister to issue directives and regulations at a later stage. The sharp increase in delegated legislation has accompanied the growing social and economic responsibilities of government. Often a department consults and negotiates with particular interest groups, or with one of its advisory committees on which the groups may be represented. Some statutes indeed specify that there should be such consultation, and on occasion the minister may actually delegate the drafting of an order to a group. In a typical session a hundred public and private bills might be passed but over 2,000 statutory orders and regulations. The delegation of detail frees Parliament to concentrate on the principles of legislation and allows the legislation to be adapted to special circumstances. The Joint (of the Houses, the Commons and the Lords) Statutory Instruments Committee reviews such legislation.

Committees

Most of the work of the House of Commons is now done through committees. There are two main types of official committee for the scrutiny of legislation and policy. *Standing Committees* look at the details of a bill once it has passed its second reading. They meet in the mornings and members are appointed by the party whips in proportion to party strength in the Commons. These committees provide the opportunity to examine and tidy up each clause of the proposed legislation and take account of views expressed in groups outside the House. But the great majority of the amendments accepted are those from the minister who is piloting the bill through the House. And if a government is defeated on an important point in committee it can usually have it reversed on the floor of the House. The standing committees, largely because they deal with legislation (mostly government bills), are subject to close management by the party whips, and party solidarity is the norm. The committee stage of some legislation is taken in the Committee of the Whole House (all MPs minus the Speaker) rather than in a standing committee. These may occur with one-clause bills which can be passed quickly, some financial bills which involve an increase in public expenditure, and bills of a constitutional nature. If the government meets what it considers unreasonable delay or obstructions then it can try to impose a timetable to get a bill through.

There are also a number of *Select Committees*, again set up by the House. These may be ad hoc, set up to inquire into a particular issue, or *sessional* ones, set up at the beginning of each session. The membership of these committees also reflects party strength in the House of Commons. They are *investigatory*, may summon ministers and civil servants to give evidence, commission papers and issue reports, and are concerned more with the application than the formulation of policy.

The Public Accounts Committee (PAC), set up in 1861, is chaired by a member of the opposition, and is supported by a staff of auditors, under the Comptroller-General. Its task is to check the accounts of departments to ascertain if spending has been duly authorized. Other long-standing committees include those for Expenditure, to ensure that government got value for money, Statutory Instruments, Privileges, Members' interests, European Legislation, and Parliamentary Commissioner for Administration. Compared with the standing committees, which reflect the government–opposition adversarial side of Commons life, MPs on the select committees are expected to adopt a 'non-partisan' approach. By and large they do. In 1995, however, there was a party split over the report from the Employment Select Committee on increases in executive pay in the privatized utilities.

Scrutiny of European Community legislation presents different problems. Since 1 January 1973, Britain's membership of the European Community has meant that all EC legislation is automatically enforceable in Britain, as in all member states. The relations of the House of Commons with the Council of Ministers and the European Commission are indirect, as is Parliament's scrutiny of European legislation. With few exceptions the House has no powers to withhold consent. The Select Committee on EC legislation scrutinizes directives and draft legislation from the Commission, but not when these are being prepared, and its role is limited. Moreover, in expressing views MPs compete with British members of the European Parliament, which gained authority as a result of the Single European Act and the Maastricht Treaty.

A major reform of the committees followed a report of the House of Commons Select Committee on Procedure (1978). The report was approved and led to the setting up in 1979 of fourteen departmental committees which would examine expenditure, administration, and policy within a department or two or more related departments, as well as in the nationalized industries and other organizations operating within a department. At the beginning of the 1992 Parliament sixteen were operating (see Box 12.2). Two features have helped the committees to achieve a measure of independence. The committees are established by standing orders for the lifetime of a Parliament rather than on an annual basis. Selection of MPs (usually eleven) to serve on committees is not in the hands of whips but a Committee of Selection. Whips have, however, interfered, notably on the Conservative side to remove potentially troublesome figures like Nicholas Winterton from chairing a committee.

The new system has the merit of coherence: the work of each committee is directed to a specific government department. It has attracted the support of many backbenchers, encouraged a degree of specialism among MPs, and is a force for improving the quality of the debate. It also provides an opportunity for pressure groups to state their case in public. The rate of attendance among MPs is high and a number of useful reports have been published, although pressure of time means that few are debated on the floor of the House (Norton 1994, Drewry 1989*b*). In 1984 two powerful reports from the PAC detailed 'very serious frauds' and 'deliberate swindling' by the Property Services Agency and the misuse of public funds by the De Lorean car company in Northern Ireland. At times a department may anticipate a committee's criticism by changing policy or issuing a consultative document. Yet the frequent complaints about the clarity of an Act reflect shortcomings in Parliament's scrutiny of bills. Too often it is a rubber stamp and shortage of time means that the select committees inevitably give little attention to some measures.

There are also party committees, which are unofficial because they are not set up by the House. The most famous is the Conservative *1922 Committee*. This consists of all backbench MPs when the party is in government and all, except the leader, when in opposition. It meets weekly, elects an executive of eighteen MPs which regularly confers with the party leader and whips, and arranges ballots for the

BOX 12.3. DEPARTMENTAL SELECT COMMITTEES (1992 PARLIAMENT)

- Agriculture
- Defence
- Education
- Employment
- Environment
- Foreign Affairs
- Home Affairs
- Industry and Trade
- National Heritage
- Northern Ireland (from 1994)
- Science and Technology
- Scottish Affairs
- Social Services
- Transport
- Treasury and Civil Service
- Welsh Affairs

party leadership election. The Conservative and Labour parties have a wide range of regional and subject committees and a minister or shadow minister will want to be sure of the support of members of his subject group. The poor reception given by group members to Leon Brittan, the DTI minister involved in the Westland affair, proved fatal for his chances of survival. Election to the chairmanship of a subject group can also be bitterly fought between the party's factional groups.

Control of the executive

The Commons is undoubtedly weaker *vis-à-vis* the executive than it was in the middle of the nineteenth century. The phenomenon of one-party majority government has helped to make a government fairly secure for the lifetime of a Parliament and the passage of its legislation predictable. In the twentieth century, governments have been dismissed by Parliament on only two occasions, in 1924 and 1979, and both were minority Labour governments.

Another factor has been the government's growing control of the proceedings of Parliament. This was largely achieved in the late nineteenth century as a result of agreement between the Liberal and Conservative front benches—a response to the obstructionist tactics of the eighty-plus Irish Nationalist members who were paralysing the work of the House of Commons. The government assumed the powers of 'closure', to limit the time for debate on a bill and force a vote, and 'guillotine', to decide how much time would be allowed to each stage of a bill. Other reforms, revealingly called 'Balfour's Railway Timetable' (after the then Leader of the House of Commons), were introduced to speed the passage of business. Henceforth, the Government had control of the time of the House, subject to the Opposition party's rights of Supply Days and private members' time. The Opposition acquiesced in the redistribution of power because it anticipated that it would soon have a turn in office. *But the House of Commons has become more of a front-bench system, particularly of the government front bench.*

Third, MPs today are less able to extract information from the executive. In the nineteenth century they regularly secured the release of government papers and sat on Royal Commissions. The ability to extract information is more limited today, although the select committee system has helped.

Finally, Parliament has effectively lost control of Supply, that is, the authorization of expenditure. Although the annual Estimates which are presented to the House of Commons show in detail how much is to be spent on each area of activity, they are invariably passed and scrutiny takes place afterwards. Particularly important in this respect is the work of the Comptroller and Auditor-General and his staff of over five hundred. They audit the work of each department, ascertaining whether money has been spent as authorized and whether it has been spent with proper economy. The accounts and reforms of the Comptroller and Auditor-General form the starting-point of the work of the Public Accounts Committee. The Committee has a good record of identifying extravagant items of expenditure by departments.

BOX 12.4. THEORY AND PRACTICE: QUESTION TIME

Theory

The government is recruited from, but also accountable to, Parliament. Question Time is perhaps the best known of several procedures incorporated into the parliamentary timetable to provide opportunities for backbenchers to scrutinize the government's activities.

- Question Time takes place for approximately one hour on Monday to Thursday
- Ministers rotate
- PM answers government-wide questions for 15 minutes on Tuesdays and Thursdays
- 48 hours' notice is required of a question, but one unscripted supplementary question is allowed
- questions may be for written or oral answer

Political Practice

- Ministers are thoroughly briefed, and civil servants are adept at predicting supplementaries
- 'friendly' questions are planted from a minister's own side
- general rowdiness
- lengthy evasive replies
- the procedure can be subverted, serving to reduce the time available for answering and enabling ministers to avoid awkward questions lower down on the list

Question Time has become something of an elaborate parliamentary charade

- it may occasionally lead to the revelation of, and redress of, an injustice, but the dominance of the parties results in the Opposition using QT as an opportunity to embarrass the government and boost morale in its own ranks, and the government defensively closing ranks to outwit the Opposition
- reputations are occasionally on the line and inexperienced ministers do seem to be apprehensive about taking their turn at the Despatch Box

Alan Clark, as a newly appointed Parliamentary Under-Secretary for Employment in 1983, reflected thus on the approach of his turn to answer PQs:

> 'I dread my own Questions, set for 19 July; it must be absolutely terrifying. Once or twice in the last couple of weeks I have sidled into the Chamber in the mornings and held the Despatch Box and looked round. A very odd feeling. . . . This coming week is my test and crisis. First for Questions on Tuesday . . . I wax and wane between confidence and inspiration; sheer terror and fatigue.'
>
> *Diaries*, 20 and 27–8

Richard Crossman, however, reflecting on his first three months as Minister of Housing and Local Government in 1964, had this to say:

> '. . . I can't remember a Question Time . . . when I had any anxiety. And that's not good. Life is too easy for Ministers in our Parliament. Take Question Time. Now that Questions have to be speeded up, the last anxiety has been removed. At the beginning, naturally, I was nervous and took the most tremendous trouble on Tuesdays when I was due to answer Questions. I used to go through them with

my officials and make sure that there were proper draft answers to Supplementaries. But gradually as I grew confident I have done less and less of this . . . in the last resort a Minister can always refuse to answer.'

Diaries, i. 628

The opposition

The legitimacy of the parliamentary activity of political opposition to the government of the day dates back to the eighteenth century and is reflected in the statutory duty levied on the Speaker to designate the leader of the largest party not in government as the *Leader of Her Majesty's Opposition*. Since 1937 the Leader has also been paid a salary out of state funds, making Britain unique among the European states in having such a state-funded office. The working and character of the House of Commons, particularly parliamentary debate and division, are coloured by party. The existence of the opposition reminds the public that there is an alternative government with an alternative 'Shadow Cabinet' and policies.

Hugh Berrington (1968) has demonstrated how a clear government versus opposition dichotomy in the House of Commons emerged during the late nineteenth century. Before 1886, government leaders could often rely on some degree of cross-bench support for their measures. It was Gladstone's adoption of Home Rule for Ireland in that year which sharply divided his Liberal party from the Conservatives and made the two parties more cohesive voting blocs in the Commons. In recent years the simple government versus opposition portrait of the House of Commons has had to be modified, because of the fragmentation of the party system and the rise of a larger, though diverse, third-party force in the Commons. For example, in the October 1974 Parliament, Labour started out with a lead of forty-two seats over the official Conservative opposition. But five other parties had thirty-nine between them, leaving Labour with a precarious overall majority of only three seats. This disappeared by 1977, and Labour depended on the support of the Liberals. The government was able to survive as long as it did because it was not until March 1979 that it was simultaneously in the interests of virtually all non-Labour MPs to combine on an issue of no confidence.

After the 1992 election, John Major's Conservative government had a comfortable sixty-five majority over Labour but only twenty-one overall. The loss of by-elections, dissent among Conservative backbenchers, and the withdrawal of the whips from nine Conservative Euro-sceptics played havoc with the Major government's ability to pass its legislation.

The reactions of the opposition are shaped in part as a reaction to what the government does, in part by pressure in the party, and in part by events. A large part of the parliamentary timetable is taken up with predictable matters, such as six days of debate on the Queen's Speech, days allocated to the Finance Bill. Here the opposition can make both negative and constructive criticisms. On opposition days or no-confidence motions, it can choose its own line of attack. But like a Prime Minister, a Leader of the Opposition has to be concerned with the unity of the party and of the Shadow Cabinet. Disraeli said, 'The duty of the opposition is to oppose'; that remains, but the parliamentary opposition is also expected to be constructive and conduct itself in a way that displays its fitness for office. A principle of collective responsibility operates, by which Shadow Ministers loyally defend Shadow Cabinet policies and take care not to pronounce controversial views on the policy area of another colleague. It was on the last grounds that Mr Heath sacked Mr Powell, the party's spokesman on defence, from the Shadow Cabinet in 1968 when the latter made his 'rivers of blood' speech about immigration. In 1981 Mr Benn was reprimanded by Shadow Cabinet colleagues for making a statement on energy which not only contravened collective policy but also transgressed the responsibility of the party's spokesman on the subject.

The parliamentary opposition has a number of resources at its disposal. It controls about a third of the parliamentary time, and may choose the subject for debate on opposition days. At present there are twenty Supply Days on which Labour (18) and Liberal Democrats (2) may choose the business of the day. It may also persuade the Speaker to grant an adjournment debate if an issue is considered to be an emergency. And, of course, it may at any time move a motion of censure on how a government or minister has conducted business. In 1940 a division by the Labour opposition, following an adjournment debate on the government's handling of the war, led to some eight abstentions on the government's benches and the eventual resignation of Neville Chamberlain. In March 1979 a Conservative motion of no confidence, carried by one vote, brought about the downfall of Mr Callaghan's Labour government. The opposition parties are represented on the select and standing committees in proportion to their strength in the House of Commons. Finally, it is traditional for an opposition MP to be chairman of the powerful Public Accounts Committee. In comparison to the time and opportunities allowed in legislatures elsewhere, arrangements in the British House of Commons must be judged as relatively generous to Her Majesty's opposition.

opposition activity in Parliament is limited by agreement on procedures and cooperation between the parties. For example, the programme of business for the following week is arranged by the parties' whips each Thursday, arrangements referred to as 'the usual channels'. Ministers who have to be absent from votes in the Commons because of government business are usually 'paired' with an opposition MP. Government business managers are aware that the opposition, by withdrawing goodwill and cooperation, can seriously disrupt their conduct of parliamentary business.

The mix of consensus and disagreement shows itself also in the division lobby. The consensus aspect is seen in the ultimate fate of bills which are still pending when Parliament is dissolved. In the new Parliaments of 1970, 1974, and 1979, two-thirds of these pending bills were reintroduced by the new government and enacted (Rose 1984*a*, ch. 5). Other non-contentious pending legislation is usually passed before the prorogation as a result of cooperation between the government and opposition. But a minister whose policy is unpopular with some of his own backbenchers may find opposition support embarrassing. And an opposition leader who offers some cooperation to the government may find that some of his backbenchers refuse to follow.

The essence of the relationship is not accurately captured in such phases as 'the Cabinet versus Parliament' or 'the executive and the legislature'. Because many members of the executive (about a third of the governing party) sit in Parliament we have to extract the government (around a hundred MPs hold a government post) from Parliament to appreciate its role. Anthony King (1976) has pointed to four distinct groups of actors in the 'normal' two-party-dominated Parliament. These are the government frontbenchers, government backbenchers, opposition frontbenchers, and opposition backbenchers. The variation in the role of Parliament may be analysed in three broad styles.

The first is the *opposition* style, in which the main actors are *the* government and *the* opposition. This is the view of Parliament conveyed by the mass media and the rhetoric of general elections. It is the classic adversary system which overlaps with a two-party system, and the arguments and votes are often predictable. The opposition can achieve various limited objectives: it can deny the government time in Parliament; it may expose the administrative or political weakness in a government's case; or it may undermine public and parliamentary confidence in a minister or a government, and so promote its own chances of winning the next general election. Above all, the opposition is poised to take advantage of a government's mistakes; its activities remind voters that there is an alternative government.

A second mode is what King calls the *non-partisan* one, when MPs act in relative freedom from the party whips. The work of select committees often shows MPs in a bipartisan mood; the desire to gain information, or expose policy failures by the executive, is found across the political spectrum of backbenchers. The other major opportunity for the relaxation of partisanship is on a free vote in the Commons. The government is relatively safe from defeat in both of the models outlined so far. In the former, party loyalty will usually secure its majority; in the latter, its position is not, by definition, at stake.

What does matter for the government, however, is King's third type, the *intra-party* model. For the ministers the most important MPs are their backbenchers. The leaders listen to the backbenchers, partly because they are in the same party, partly because they may represent opinion outside Parliament or in the party in the country (the trade unions or the Labour Party Conference, for example, or small businesses or farmers for the Conservatives), and partly because the executive is not monolithic. Divisions among backbenchers may reflect divisions in the Cabinet. The government may take the opposition of other parties for granted but, as John Major found when struggling to get the Maastricht Treaty ratified in 1993, its security depends on keeping its own party majority intact.

There are many examples of the government having to bow to pressures from its own supporters and either losing or modifying a bill. They include: abandonment of *In Place of Strife* (1969); abandonment of the reform of the House of Lords (1969); acceptance of the 40 per cent amendment in the referendum clause of the Scotland and Wales devolution bills (1978); defeat on immigration rules (1982); withdrawal of proposals to increase sharply the parental contribution to students in higher education (1984); defeat of the Shops Bill (1986); withdrawal of proposals to sell off parts of British Leyland to General Motors (1986); changes to proposals affecting housing benefit (1988); withdrawal of proposals to introduce identity cards for football spectators (1990), and concessions on the bill to ratify the Maastricht Treaty (1993). On Maastricht there was a core of two dozen Conservative Euro-sceptics, who could combine with Labour to threaten the government's majority. The motion to ratify the Treaty was carried in July 1993 only after the government declared that it was a vote of confidence—and that defeat would lead to a general election. Warnings of Conservative dissent cause the postponement of plans to privatize the Post Office in 1994.

Dissent

Until late in the nineteenth century many members were men of independent means and views and were willing to take decisions for themselves on the issues of the day. The professionalization of political life—the longer hours of parliamentary sittings, the pressure of work, and the increasing demands on MPs from constituents and pressure groups—has made Parliament more of a full-time career. For a number of Conservative and Labour MPs, even before 1945, politics was a matter of *noblesse oblige*, an involvement in public life which was motivated by a sense of duty or tradition. It was not a full-time role but something that could be combined with another career.

Increasingly, however, the new generation of MPs appears to look to politics as a career, is ambitious for political advancement, and, perhaps being largely dependent on a parliamentary salary, has been less likely to ignore the advice of the party whips. The decline of the 'independent' member and the rise of party has rendered the results of parliamentary votes predictable and made the government's position more secure. Such an analysis is highly plausible and seems to fit the party unity in the Commons for much of the post-war period, even though backbenchers were sometimes able to force concessions from the party leaders away from the floor of the House.

What is clear, however, is that MPs in the two main parties have become more rebellious, even to the point of voting for another party. This feature flies in the face of claims that as MPs become more 'professional' and ambitious for and dependent on ministerial advancement, so they would be more loyal to the party whips. Philip Norton's (1980) study of the division lists in Parliaments since 1945 shows that in the Parliaments from 1970 the proportion of divisions in which cross-voting (or a member defying his whip) occurred has increased. A consequence of this rebelliousness has been that governments have been defeated more frequently on the floor of the House of Commons and in standing committees.

It was no surprise that the Labour governments of October 1974 to 1979 were defeated on forty-two occasions, for they were in a minority for much of the time (Norton 1980). Nor should there be surprise over John Major's losses, given his slender majority and his party's divisions on Europe. The rebellions, however, continued under Mrs Thatcher, particularly with her greater majorities since 1983. The second reading of the Shops Bill was lost in 1986, when seventy-two Conservative MPs entered the opposition lobby, the first time this century that a government with an overall majority has lost such a vote. The Conservative government of John Major withdrew the whip from nine persistent rebels in 1994, an action unprecedented in the post-war Tory party—although Rupert Allason also had it withdrawn in 1993.

There appear to be four possible causes of the changes in behaviour. One is that a number of issues—incomes policies, Europe, devolution, and industrial policy—have divided both parties, while defence was an additional source of trouble for Labour. A second is that dissenting members rarely have the whip withdrawn; compared with the sanctions visited on left-wing Labour rebels or Conservative dissenters in the 1950s, the party leaders in the 1970s have been more tolerant. Alternatively, party leaders may be unwilling to make concessions to head off backbench dissenters. Mr Heath, in particular, was a goal-oriented leader, and Norton (1980) claims that his wilful style of leadership actually provoked dissent. Thirdly, the Labour governments between 1974 and 1979 and John Major's 1992 government had to be more relaxed about the consequences of a defeat in the division lobbies; the government lost votes on a number of important issues and then invited the rebels to support the government on a vote of confidence. The ultimate attempt by government leaders to defuse the impact of a backbench rebellion on a crucial issue is the refusal to issue a three-line whip, because the party is too divided. Finally, it may be that the more 'professional' MPs have become, the more they are frustrated by the conditions of parliamentary life and with lack of promotion. The long period of Conservative government has resulted in a large number of disappointed and potentially troublesome MPs—what an exasperated John Major was overheard to call 'the dispossessed and the never-possessed'.

House of Lords

For much of its history the House of Lords possessed co-equal powers with the Commons. In the nineteenth century the gradual extension of the suffrage for elections of the Commons emphasized that body's democratic credentials. The government of the day was also clearly responsible to the Commons because a defeat on an issue of confidence entailed its resignation; such a consequence did not follow from its defeat in the Lords. But before 1909 a constitutional clash over the powers of the two Houses was avoided. Late nineteenth-century Cabinets still drew half of their members from the Upper House and Lord Salisbury sat as Prime Minister in the Lords until 1902. In 1900, apart from the supremacy of the Commons on the matter of finance bills, the two Houses were more or less co-equal in their powers.

The sharper conflict between the Liberal and Conservative parties—particularly over the former's willingness to grant Irish Home Rule—and the preponderance of Conservatives in the Lords led to the clash. In 1909 the House of Lords took the unprecedented step of rejecting Lloyd George's Budget and paved the way for the Parliament Act (1911). This contained two important provisions which clearly established the subordinate position of the Lords. The first was that bills passed by the Commons and certified by the Speaker of the Commons as money bills were to receive the Royal Assent a month after being sent to the Lords, whether or not approved by the latter body. The second was that any other public bill (except one extending the life of a Parliament) passed by the Commons in three successive sessions and rejected by the Lords would receive the Royal Assent, provided that two years had passed between the bill's second reading in the first session and the third reading in the third session of the Commons. In 1949 the Lords' power to delay was further reduced to one year by the Labour government. At present one year must elapse between the bill's first and second reading in the Commons and the date on which it is passed by the Commons in the second session.

The largely hereditary composition of the

BOX 12.5. LORDS vs. COMMONS: THE CASE OF THE WAR CRIMES BILL

- In March 1990 the War Crimes Bill was introduced into the Commons. Its purpose was to permit alleged war criminals, who are now British citizens or are resident in Britain, to be prosecuted in the British courts.
- It was rejected by the Lords in June 1990.
- It eventually received royal assent in May 1991 after the government invoked the Parliament Acts to secure its passage in the face of the Lords' opposition.

CHARTING THE WAR CRIMES BILL

1989/90 Session	*1st Reading*	*2nd Reading*	*Standing Committee 1*	*Standing Committee 2*	*Report and 3rd Reading*
House of Commons	8.3.90	19.3.90	29.3.90	3.4.90	25.4.90
House of Lords	26.4.90	Negatived 4.6.90			

1990/91 Session	*1st Reading*	*2nd Reading*	*Standing Committee 1*	*Standing Committee 2*	*Report and 3rd Reading*
House of Commons	7.3.91	18.3.91			25.3.91
House of Lords	26.3.91	Negatived 30.4.91			

Invocation of Parliament Acts	1.5.91
Royal Assent	9.5.91

Source: John Dickens, 'War Crimes: A Case Study in the Use of the Parliament Acts', *Talking Politics*, 5/2 (1993), 105.

Other occasions on which the Parliament Acts have been invoked:

- 1914 Welsh Church Act
- 1914 Government of Ireland Act
- 1949 Parliament Act

BOX 12.6. HOUSE OF LORDS, 1994

Composition

- Hereditary peers: 773
- Life peers: 374
- Archbishops/bishops: 26
- Law Lords: 21

Total: 1,194

Of this total 77 are women, comprising:
- 17 hereditary peeresses
- 60 life peeresses

Payment for peers

Members of the House of Lords are unpaid but they are allowed to claim for travel on parliamentary business within the UK and for other expenses incurred when attending the House:

- £70.00 maximum per day for overnight subsistence
- £31.50 for day subsistence
- £30.50 for secretarial assistance and postage

Lords has been modified by two pieces of legislation. The *Life Peerage Act* (1958) made it possible for people to be given a peerage during their own lifetime without the title passing to their heirs; they were life peers. In the *Peerage Act* (1963) peers were given the option of disclaiming their titles within one year. Those who did this included Lord Stansgate (now Tony Benn) and Lord Home (who became Sir Alec Douglas-Home and Prime Minister in 1964). The House of Lords is at present composed of Church of England archbishops and bishops or the Lords Spiritual (26), the Law Lords (21), life peers (over 300), whose titles lapse with them, and hereditary peers (over 800). The actual work of the House is done by about 300 members, with the life peers being particularly active. Some 60 per cent of peers are hereditary. Nearly half of the peers take the Conservative whip and around 30 per cent are cross-benchers. The figures mean that a Conservative majority is not always assured, but also that a majority for the opposition is rare. Creating peers involves considerable Prime Ministerial patronage.

The present-day *powers of the Lords* are *relatively minor*. Its powers of delay may sometimes amount to obstruction if the government's term is nearly complete, its legislative timetable is already crowded, or it is unable to get the defeated measure through the Commons again. Since 1945, the Lords have generally felt free to reject or offer substantial amendments to a government bill which has not been mentioned in the government party's election manifesto. It frustrated the minority Labour government (1974–9) and was an irritant to the Thatcher government, defeating it 155 times between 1979 and 1990. In 1994 the government was forced to make concessions on the Police and Magistrates' Courts bill, particularly over the Home Secretary's assumption of greater powers to appoint members to police authorities. John Major has found that the Lords contains two of his

BOX 12.7. PARTIES IN THE HOUSE OF LORDS

House of Lords 1994
The state of the parties for the whole of the House of Lords

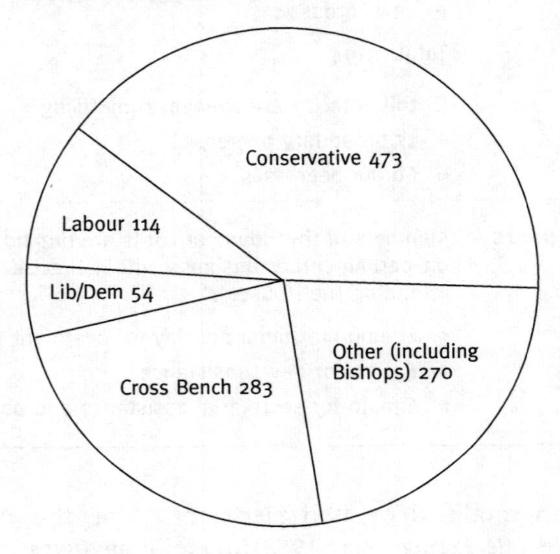

Life Peers 1995
State of the parties

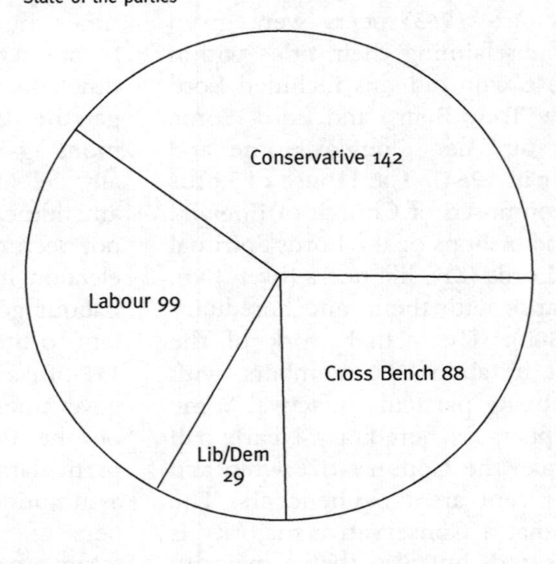

BOX 12.8. REFORM OF THE HOUSE OF LORDS

Plans to reform the House of Lords are frequently mooted, especially by Labour and the Lib Dems, but they are rarely high on the agenda. It was a Liberal government in 1911 that changed the Lords' veto into a two-year delaying power, and a Labour government in 1949 that reduced the delaying power to one year.

The last major attempt to reform the Lords was introduced in 1968. It was proposed that

- **hereditary peers be phased out**
- **members to be appointed**
- **voting rights be dependent upon attending 30% of sittings**
- **delay be reduced to six months**

It was defeated by the combined efforts, for different reasons, of the Labour left and the Conservative right.

Pressure for reform

Source	*Proposed reforms*
Liberal Democrats	Replace with a Senate elected by PR on a regional basis; two-year delaying power (except money bills)
Tony Benn	Commonwealth of Britain Bill 1991: members should be elected proportional to populations of component nations of Great Britain; one-year delaying power
Labour	In 1995 Labour proposed to end sitting and voting rights of hereditary peers; build in a majority for the government of the day; leave powers as at present

Some of their lordships' responses to Labour's proposals:
A group of peers led by the Earl of Carnarvon — a cross-bencher whose family peerage was created in 1793–laid out their objections in a paper. It warned that if hereditary peers were excluded the House would lose its wider sympathies:

> **'. . . the product of the arbitrary operation of the hereditary principle — and its capacity to espouse unusual causes, to express attitudes learned in a wide range of ordinary walks of life . . . to act as the small boy in the crowd who points out that the emperor has no clothes.'**

The Earl of Selborne (Conservative) warned that a House stripped of hereditary peers would be full of placemen, turning the Lords into the largest quango in the land.

most powerful Euro-sceptics, Lady Thatcher and Lord Tebbit.

Defenders of the second chamber claim that it has a number of uses. Governments may introduce relatively non-controversial legislation in the Lords. In revising and amending legislation it eases the legislative burden on the Commons and the great majority of amendments which it passes are accepted by the government—indeed ministers increasingly introduce amendments in the Lords. In comparison with the Commons the Lords also has more time and, since the introduction of life peers, more members with expertise in various walks of life, who can make informed contributions to debate. Indeed because leaders of business, finance, culture, the Civil Service, and trade unions are members, the House has been called a meeting place for yesteryear's élites (Shell 1988). The House also has a judicial function in that the Lord Chancellor and the Law Lords act as the highest court of the land. It may have a constitutional role, as a check (however fragile) against the elective dictatorship of a temporary majority of MPs in the Commons.

Critics of the Lords divide into two schools. One, found largely on the Labour left, would like to abolish it, and this was proposed in Labour's 1983 election manifesto. Some other liberal democratic countries, such as Sweden, Finland, Denmark, and New Zealand, manage without a second chamber. It is worth noting, however, that these countries, except New Zealand, also have proportional representation, a Supreme Court, and a Bill of Rights; single chambers are more representative of the electorate and operate in a system of formal constitutional checks and balances.

The other school accepts that the functions of the second House are necessary but argues that they should be performed by a reformed chamber. Proposals for changing the composition include: the direct election of members on a regional and/or proportional electoral basis (Labour's policy since 1989), the appointment of members in proportion to party strength in the Commons, with voting rights restricted to them, or some other mixture of nominated and elected members. The reformers have usually wanted to introduce the elective principle, reduce the role of hereditary peers, and not replicate the party balance in the Commons. For example, the Webbs in 1920 and Winston Churchill in 1930 favoured the creation of a second chamber which would more directly represent the major economic interests. The proposals for changing the powers have ranged from reducing the delaying power to six months to increasing it to two years. To date, disagreement among would-be reformers has prevented change of a fundamental nature.

Reform (or abolition) of the Lords presents several political and administrative difficulties, which is one reason why there has been no 'fundamental' reform since 1911. Since 1970 party leaders have also appreciated that constitutional measures, which must be considered on the floor of the whole House, are usually time-consuming, involve the sacrifice of parliamentary time that could be used for other measures, and are rarely popular with the public.

Conclusion

The House of Commons remains a focus of attention on major occasions—a Budget Statement, a motion of censure on a minister or the government, or an emergency, such as the Falklands debates in April 1982, or a finely balanced but important vote, as on the Maastricht bill in 1993. 'Winning the argument' in the House is important in maintaining the morale of the party in Parliament and transmitting this mood to the party in the country. Political leaders are recruited and socialized in the Commons and politicians' reputations and career prospects are affected by how they perform in parliamentary debates. The resignation from Cabinet of Leon Brittan (1986), David Mellor (1992), and Norman Lamont (1993) all followed evidence of a lack of sympathy for each of them among Tory backbenchers. Memoirs of former ministers frequently report the sense of relief or despondency—depending on the author's performance—following a parliamentary debate. A minister's rising stock in the Commons may be reflected in a relaxed working atmosphere among his senior officials, poor performance in a gloomy one. Paradoxically, if it is because so many ministers are members of the Commons that the latter has lost some of its influence and independence *vis-à-vis* the executive, by the same token this has made it a more important forum than might be the case with a strict separation of personnel.

Yet Parliament plays little part in formulating policy, or changing legislation, and has not come to terms with the loss of power to Brussels, quangos, or the bodies which regulate the utilities. Given the widespread assumption of the 'decline of Parliament' it is not surprising that there have been many proposals for reform. Advocates of reform often have different objectives in mind when they talk of parliamentary reform. Proposals have fallen into three broad categories:

1. Proposals to make the Commons more efficient. These include better pay and facilities for MPs, greater use of morning sittings to avoid late-night sittings, and reform of the House of Lords or the creation of regional assemblies to ease the burden on an overloaded House of Commons.
2. Proposals, usually by the government, to speed the passage of legislation. These involve some easing of the procedural checks which backbenchers or the opposition may impose on the executive, an increase in the number of enabling bills, or bills which would leave ministers to fix the details subsequently, or the introduction of more bills with timetables attached to them, so limiting the opportunities for discussion and amendment. Such ideas reflect the outlook of many ministers and civil servants who either want 'a quiet life' from Parliament or their legislation passed with the minimum of delay and embarrassment.
3. Proposals designed to improve the House's ability to scrutinize the executive more thoroughly. For some fifty years the most popular suggestion has been the extension of the role of standing and select committees.

Some reformers have been disappointed with the progress so far. They have therefore looked beyond the reform of Parliament itself as a means of seeking checks on the executive and promoting more accountable government. They have turned to proportional representation, a Bill of Rights, or devolution as levers of desired change.

If the committee system or other institutional reforms are to produce important changes they are likely to do so only as part of a broader package. The decline in party discipline and of members' willingness to obey the whips, already apparent, is one means of restoring influence to the backbenchers. Some backbench pressure is already exercised informally and privately, and on some issues the threat by backbenchers to withhold support, or even oppose, has frustrated government plans. Once the predictability of support is weakened there is more of a bargaining relationship between backbenchers and party leaders. The development of multi-partyism and the likelihood of deadlocked parliaments is one possible source of change. The period of the Lib–Lab pact (1977–8), minority government (1974 and 1978–9), and the troubles of the Conservative government (1993–5) showed that ministers have to bargain policies for support in the Commons, or accept defeat in the division lobbies.

Far from restoring the vitality and independence of the House, however, *insecure government* gives power to small groups of potential dissenters, maintains the opportunity for 'backstairs' deals, and gives scope for pressure groups, while making 'strong' government unlikely. It is not clear how such a government would be more representative of the public mood. The fragility of the Major government in the Commons is as much a reflection of the party's division on Europe, resentment of frustrated Conservative MPs, and the bargaining power of a small number of rebels because of the government's narrow majority.

There has been much criticism of workings of the Commons, of the damaging effects of the adversarial party battle and of its lack of power *vis-à-vis* the Cabinet or the European Union. Yet any study of the legitimacy and adaptability of British political institutions must award a high place to the House of Commons. The institution has a long, continuous history, well-established traditions, and a strong sense of institutional identity. It commands a deep-seated loyalty among most MPs (few leave it willingly), and defenders of the 'right' of Parliament strike a sympathetic chord in most parts of the House. A British politician who is a 'poor House of Commons man' is unlikely to succeed politically.

Summary

- Parliament's functions and role have evolved over several centuries. Its present-day functions include representation, debating, converting proposed legislation into law; it is a forum of recruitment for, and scrutiny of, the executive. The Law Lords in the House of Lords constitute the highest court.
- Democracy gave rise to the convention of House of Commons supremacy. This was given formal statutory recognition by the 1911 and 1949 Parliament Acts.
- Disciplined parties, the dominance of the two-party system, and the emergence of interventionist government have contributed to a 'decline' in Parliament in the sense of it being an institution independent of party control. Pressure groups and the EU have weakened Parliament's role as a consenting forum.

- The government is usually able to use its majority and legislative procedures to convert most of its desired legislation programme into law. An increasing tendency towards rebelliousness amongst back-benchers has produced some notable exceptions since the 1970s.

- There are three main types of committee in Parliament: standing committees which scrutinize and amend legislation; select comittees which scrutinize spending and the executive departments; and committees of the whole House which deal with short bills, and financial and constitutional bills. There are also various party committees.

- The party system, giving all but a few governments an assured majority in the chamber and in the committees, has weakened Parliament's control over the executive.

- There is formal recognition of the role of the opposition in Britain. The Leader of Her Majesty's Oppositition and the Oppositition Chief Whip receive an additional salary. Specific opportunities in the parliamentary timetable are allocated to the opposition. There are expectations that the opposition will be constructive and will cooperate with government through the 'usual channels'. Governments are more worried by the opposition on their own side than by criticism from the other side of the House.

- The House of Lords is still semi-feudal in that hereditary peers constitute two-thirds of its membership. Approximately 25 per cent of legislation begins its life in the Lords — mainly that of a non-controversial nature. The House has no power over money bills and only a one-year delay over other legislation. There have been many proposals to reform its compositition, its powers, or both, but insufficient support for any particular porposal has meant that fundamental reform has so far failed to materialize.

- Parliament remains an important political institution. Parliamentary performance is a crucial factor in recruitment to government office.

CHRONOLOGY

1265 De Montfort's Parliament

1689 Bill of Rights: creates parliamentary monarchy

1832 Reform Act: golden age of Parliament; the House of Commons makes and unmakes governments

1861 Public Accounts Committee established

1867 Reform Act: party rule begins

1909 Lords defy convention and reject the budget

1911 Parliament Act: powers of Lords reduced to two years delay; no power over money bills. Maximum life of Parliament reduced from seven to five years. Payment for MPs

1937 Ministers of the Crown Act: payment for Leader of the Opposition

1949 Parliament Act: delaying powers of Lords reduced to one year

1958 Life Peerage Act: life peers can be created and women peeresses can inherit a title and sit in Lords

1963 Peerage Act: permits hereditary peers to resign a title for their own lifetime

1969 Lords reform proposals defeated

1972 European Communities Act: Parliament accepts supremacy of EC law in areas of EC competence

1978 Report of the Select Committee on Procedure leads to setting up of select committees to scrutinize the work of government departments

1979 Callaghan government defeated on a vote of no confidence

1990 Lords reject the War Crimes Bill

1991 The War Crimes Bill becomes an Act of Parliament under the provisions of the 1949 Parliament Act

ESSAY/DISCUSSION TOPICS

1. How effective are the House of Commons select committees as a means of scrutinizing the executive?

2. Was Bagehot correct when he consigned Parliament to the dignified part of the constitution?

3. The sovereignty of Parliament is a meaningless concept which should be abandoned.

4. Why does the House of Lords survive in what claims to be a modern democratic state?

5. Discuss the view that political parties have transformed backbenchers into lobby fodder.

RESEARCH EXERCISE

1. You are a Conservative Euro-sceptic. Explain:

   (a) the kinds of pressure that might be applied to you before an important Commons vote on British participation in a single European currency, and

   (b) the kinds of arguments you could use for voting against your party.

2. Choose a Parliamentary select committee. Describe its composition, its functions, how frequently it meets, and give a brief outline of any reports it has produced or any controversy connected with its work.

FURTHER READING

For an introduction see P. Norton, *Does Parliament Matter?* (Hemel Hempstead: Harvester Wheatsheaf, 1993) and A. Adonis, *Parliament Today* (Manchester: Manchester University Press, 1990). On the Cabinet see Peter Hennessy, *Cabinet* (Oxford: Blackwell, 1986). On the House of Lords see D. Shell, 'The House of Lords, Time for a Change?', *Parliamentary Affairs*, 47 (1994), 721–37.

The following may also be consulted:

Dickens, John, 'War Crimes: A Case Study in the Use of the Parliament Acts', *Talking Politics*, 5/2 (1993).

Ludlam, Steve, 'Parliamentary or Executive Sovereignty? The Ratification of the Maastricht Treaty', *Politics Review*, Apr. 1994.

Norton, Philip, 'Independence without Entrenchment: The British House of Commons in the Post-Thatcher Era', *Talking Politics*, 6/2 (1994).

—— 'Resourcing Select Committees', *Talking Politics*, 8/1 (1995).

13 CIVIL SERVANTS AND ADMINISTRATION

Reader's guide

The British Civil Service is the permanent wing of the executive branch of government. On the one hand, it is acclaimed for being a 'well-oiled', efficiently run machine, a model bureaucracy envied and admired for its integrity and traditions of public service. On the other hand, it has been criticized for its conservatism, caution, and preference for compromise, and for being a caste of self-perpetuating gifted amateurs, out of touch with changing technology and modern management techniques. Politicians of both the right and left have argued that it is these latter characteristics that have blocked radical initiatives and contributed to policy failure associated with Britain's relative economic decline throughout the twentieth century.

This chapter examines the organization and underlying principles of the service and the effects of successive waves of criticism and reform. Will the cost of reforms be measured in the loss of the traditional virtues of the service? Are we witnessing the emergence of a new theory of the state?

THE British Civil Service has undergone far-reaching and continuing change in the 1980s and 1990s. Many of the changes—contracting out, performance pay, devolved budgets, management and information systems—have been imported from the private sector and introduced across much of the public service. Similar trends are at work in the civil services of other countries, particularly the USA and New Zealand. Governments everywhere are trying to cut costs, improve the management skills of civil servants, and provide public services more economically. Although claims that the service is in 'a state of crisis unprecedented in its history' (Plowden 1994: 1) are exaggerated, the effect of the changes has been to raise questions about many traditional assumptions about the British Civil Service.

This chapter examines the organization and underlying principles of the Service and of the effects of successive waves of criticism and reform.

Civil servants are, in the words of the 1929–31 Royal Commission on the Civil Service (Cmnd 3909), 'those servants of the Crown, other than holders of political and judicial offices, who are employed in a civil capacity, and whose remuneration is paid wholly and entirely out of money voted by Parliament'. This definition excludes such other servants of the Crown as judges and MPs, members of the armed forces (who are not employed in a civil capacity), and employees of local government and public cooperations (who are not servants of the Crown or paid out of parliamentary funds).

Organization and development

In April 1995 there were some 524,000 civil servants employed in forty-nine government departments. These are, in Peter Hennessy's graphic phrase, 'the government's direct labour force'. More than two-thirds of them worked in just five departments: Defence, Social Security, Inland Revenue, Employment, and the Home Office. The great major-

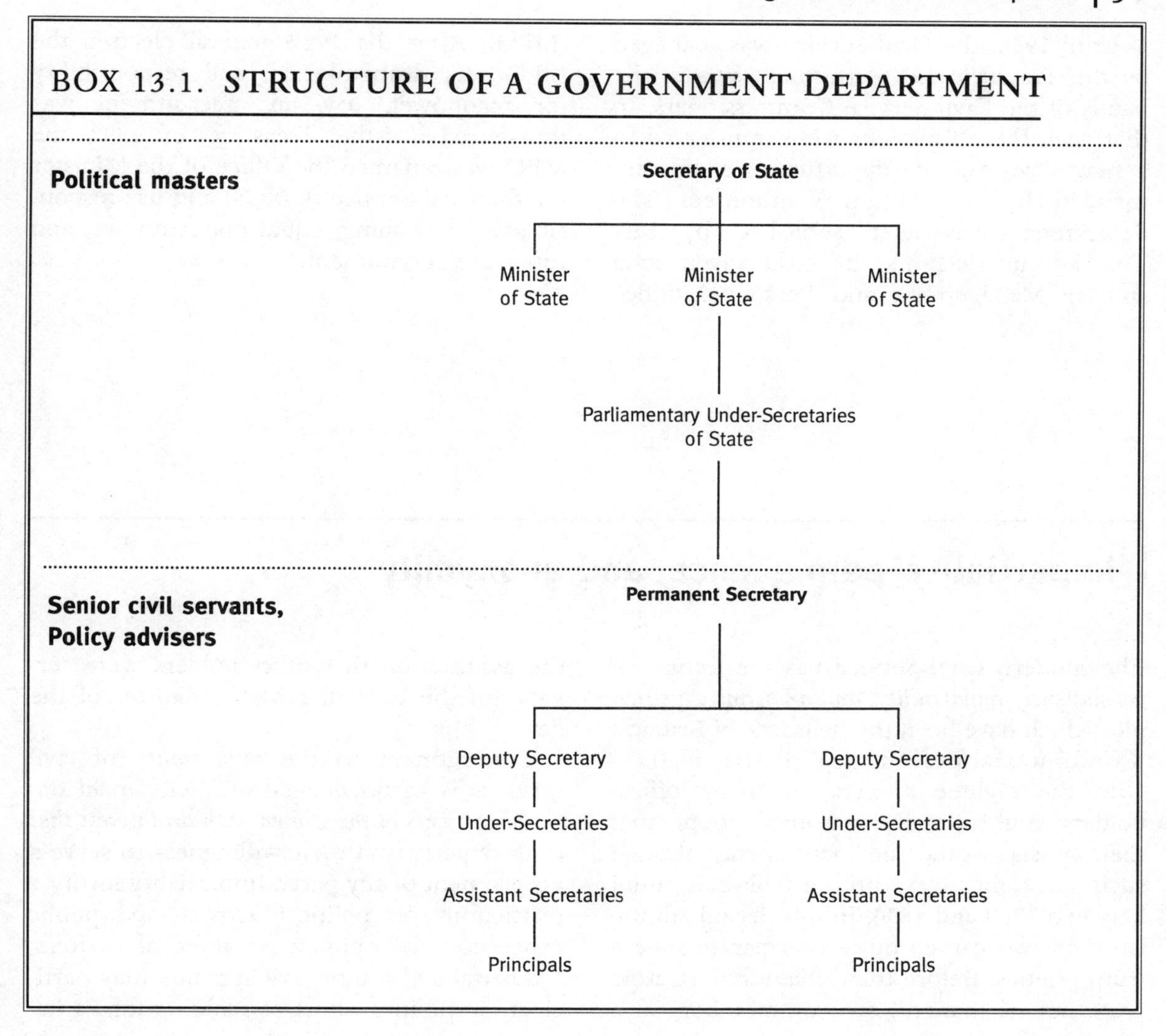

ity, over 95 per cent, deliver services and collect revenue—paying benefits and pensions, collecting taxes, and running prisons and courts. Only about 5 per cent of civil servants work in the London headquarters of their departments.

The official head of each government department is normally *a permanent secretary* or *a permanent under-secretary.* The former is usually also the accounting officer of the department and is responsible to the minister for all departmental activities. Some larger departments also have a second permanent secretary and, along with Deputy Secretaries, Under-Secretaries, Assistant Secretaries, and Principals, these officials are the main policy advisers to ministers (see Fig. 13.1). They assist the minister with parliamentary answers, speeches, and briefs for debates and appearances before House of Commons committees. They also help to prepare legislation and draft memoranda for the Cabinet and its committees.

Until 1968, the Civil Service was managed by the Treasury and recruitment was in the hands of the Civil Service Commissioners. In that year, these functions were transferred to a new Civil Service department, as recommended by the Fulton Committee. The department was later abolished by Mrs Thatcher in 1981 and its tasks given to a smaller Management and Personnel Office (MPO). After the 1987 general election the MPO was dismembered, and responsibility for manpower, pay, and recruitment was transferred to the Treasury. In 1992 the MPO was renamed the Office of the Minister for the Civil Service (OMCS) and has responsibility for training, equal opportunities, and top-grade recruitment.

Impartiality, permanence, and anonymity

The modern Civil Service has the *features of permanence, impartiality, and anonymity,* principles which have been the hallmark of Britain's 'Constitutional Bureaucracy' (Parris 1969).

In the eighteenth century, many office-holders could already reasonably hope that their posts would be permanent, though such a right was not established until between 1780 and 1830. In this period administration was carved out as a separate sphere from politics. Before then, the administrators were not permanent, civil or members of a service. The development of these features turned on the recognition of a distinction between administration and politics: the crucial difference in career terms was between those who remained and those who went when there was a change of government. The growth of government business made it increasingly difficult for people to combine political and administrative posts and they had to choose between them. Parliament also grew jealous of the Crown's power of patronage, and sought to reduce it by making posts permanent and salaried. The idea of civil or non-political appointments rested on the assumption that office-holders were servants of the Crown, not the minister of the day.

Commitment to the *impartiality* of civil servants is acknowledged by their *formal status as servants of the Crown with an interest that is above party,* and their willingness to serve a government of any party. Impartiality involves restrictions on political activity and public expressions of opinion on political matters. Industrial and junior civil servants may participate in politics, short of standing for Parliament, and senior officials are free to join a political party but may not stand for political office. In some other countries, France and West Germany, for example, senior civil servants may take leave of absence from their posts and enter politics. A British senior civil servant, in contrast, is required to resign his post once he is adopted as a prospective political candidate, and very few have followed this course. *A career in the British Civil Service is an alternative to and not a preparation for entry to party politics.*

Widespread acceptance of the impartiality of the senior Civil Service has depended to

BOX 13.2. THE CIVIL SERVICE AND THE CONSTITUTION

Civil servants

- servants of the Crown
- employed in a civil capacity
- paid out of taxes

MPs, judges, members of the armed forces, local government officers, and employees of public corporations are not civil servants.

There were 524,000 non-industrial civil servants in 1995. Of these, the 'open structure' consisted of approximately 3,000 senior policy advisers:

- Permanent Secretaries
- Deputy Secretaries
- Under-Secretaries
- Assistant Secretaries
- Principals

Civil servants are the confidential advisers to ministers.

- permanent
- politically impartial
- anonymous

Ministers are the political heads of governmental departments.
They receive the credit or blame for the advice they accept. They are:

- transitory
- politically partial
- publicly accountable

This relationship is reflected in the constitutional doctrine of Ministerial Responsibility.

Example

In 1986 Leon Brittan resigned as Secretary of State for Trade and Industry. He accepted responsibility for the actions of a civil servant in his department, Colette Bowes, who had leaked a confidential letter with the intention of discrediting Michael Heseltine, Secretary of State for Defence.

some extent on there being a large area of agreement between the two main parties and broad continuity in policy as one party succeeded another in government. It was therefore relatively easy for civil servants to appear, as Professor Ridley notes, as chameleons, 'changing colour as governments change'. In the 1980s, however, the growing ideological distance between the two main parties made it more difficult for the Civil Service to be seen as 'above party'.

The principle of impartiality is linked with that of *anonymity*. In theory, ministers decide and are answerable for policy. The idea of ministerial accountability to Parliament rests on the assumption that the civil servant acts as the agent of the minister. *The doctrine of ministerial responsibility means that, however fairly or unfairly, the minister receives both the praise and the blame for his departmental activities.* In its modern form the doctrine dates from 1873. In that year Mr Scudamore, an official in the Post Office, accepted personal responsibility for a misappropriation of funds and the Chancellor of the Exchequer indicated to the House of Commons that he accepted Mr Scudamore's responsibility in that matter. Then, Bernal Osborne, a backbencher, retorted: 'This House has nothing to do with Mr Scudamore. He is *not* responsible to us. We ought to look at the heads of departments.' As long as the minister is held to be responsible, and civil servants carry out his wishes, then it follows that the civil servants have no need to speak for themselves.

Ministers have jealously guarded this convention. In establishing the Fulton Committee in 1965, to report on the structure and management of the Civil Service, Mr Wilson excluded from its terms of reference any consideration of relations between ministers and civil servants. He claimed, in the House of Commons:

Civil servants, however eminent, remain the confidential advisors to ministers, who alone are answerable to Parliament for policy; and we do not envisage any change in this fundamental feature of our parliamentary system of democracy.

The convention limits the ability of parliamentary select committees to question ministers about the reasons why they took particular decisions and what advice they received from civil servants. It may therefore act as a shield for ministers and civil servants. The Westland affair in 1986, which led to the resignation of two Cabinet ministers, raised questions about ministerial responsibility, relations between ministers and civil servants, and the responsibility of both to Parliament. The government refused permission for the House of Commons Defence Select Committee in 1986 to question civil servants: 'The Government proposes to make it clear to civil servants giving evidence to Select Committees that they should not answer questions which are or appear to be directed to the conduct of themselves or of other named officials.' Instead the Committee had to make do with two sessions with Sir Robert Armstrong, head of the Cabinet Office, who interviewed the key officials.

Two of the above principles have come under strain in recent years. The unified Civil Service has been steadily dismantled and there is pressure for senior posts not only to be opened to outside competition but to be limited to three- or four-year contracts. Permanence is no longer thought to be such a good idea. Developments have also undermined the anonymity principle in practice. As early as 1971 a tribunal of inquiry into the collapse of the Vehicle and General Insurance Company blamed and named senior civil servants, in the then Board of Trade, for their negligence. Since then a more inquisitive mass media, publication of the alleged views and advice of civil servants in retired ministers' memoirs, reports of the Parliamentary Commissioner of cases of maladministration by civil servants, and more searching interrogation of civil servants by select committees and other bodies, together with greater openness by retired senior civil servants themselves, have all played a part in eroding anonymity. The public inquiry in 1994 under Sir Richard Scott into the behaviour of Cabinet ministers and senior civil servants over the Churchill-Matrix affair (on the illegal sale of arms to Iraq) was remarkable for the tough questioning of civil servants.

BOX 13.3. ARE THE TRADITIONAL ASSUMPTIONS ABOUT THE CIVIL SERVICE UNDER THREAT?

Permanence

Civil servants do not change with the changing political complexion of governments.

This principle has come under attack in recent years. In 1994 the White Paper 'Continuity and Change' set out proposals to reduce the job security of senior civil servants; pay was to be linked to performance and the movement of staff between the public and private sector was to be encouraged.

Impartiality

Civil servants are formally servants of the Crown, not of the government of the day; they are, therefore, assumed to be like the monarch, above party politics.

- they may not stand for political office
- they may not publicly express opinions on political matters

But

- they can vote
- they can join a political party

It has been argued by some that Mrs Thatcher endangered the political neutrality of the Civil Service. Between 1979 and 1985 there was almost a complete turnover of the two senior grades in whose appointments the Prime Minister has a direct say (permanent secretaries and deputy secretaries). Mrs Thatcher claims that, although she took an interest in senior appointments in the Civil Service, 'In all these decisions . . . ability, drive and enthusiasm were what mattered; political allegiance was not something I took into account.' Her claims were corroborated by Sir Robert Armstrong, Cabinet Secretary: 'There is no question of political consideration entering into the choice . . . she is not concerned with and . . . she does not seek to ascertain the political views or sympathies . . . of those recommended.'

The danger remains, however, that traditional Civil Service virtues of impartiality and caution may have been sacrificied as a result of the preferment of 'Thatcherite' 'can do' candidates, who enthusiastically embrace the government's policies.

Anonymity

Civil servants are the *confidential* advisers to ministers. Ministers take public credit or blame.

This principle has been eroded in recent years:

- public inquiries, Parliamentary Select Committees, and reports of the ombudsman are more willing to apportion blame to named individuals
- the media is more inquisitive and less sensitive to the traditional 'rules of the game'
- there is greater openness in the memoirs and diaries of former ministers and those of their political aides

Recruitment

In discussing the political and policy-making roles of the Civil Service we are primarily involved with the higher civil servants, that is, those at the level of principal or above. Most of these work in central London (except for a few who are employed in Edinburgh and Cardiff in the Scottish and Welsh Offices respectively) and are closely involved in formulating policy. They consist of direct entrants from university, promotees from the lower ranks of the service, and some appointments from outside the service. In addition, there are also specialists, like engineers, doctors, scientists, and lawyers, who play a largely technical role, giving specialist advice to the administrators.

Because the work of a higher civil servant demands high intellectual abilities, entry is usually through the form of a competitive examination and interview. The *Northcote–Trevelyan Report* (1854) took the first steps towards eliminating political patronage and improving standards by recruiting through open competitive examination. The main principles of the Report shaped appointments for the next century. Recruitment is still handled by an independent Civil Service Commission, thus keeping appointments out of the politicians' hands. Since 1969 external candidates (aged between 20 and 27 and possessing a university or polytechnic degree) have had to undergo a three-stage selection process for entry to a 'fast stream' of recruits. The first consists of written tests lasting one and a half days. Those qualifying proceed to the Civil Service Selection Board and take tests which cover written and oral exercises and interviews. The last stage is held by the Final Selection Board and consists of open-ended interviews on general topics.

Criticism and reform (1)

For most of the century following the publication of the Northcote–Trevelyan Report the British Civil Service was widely praised for its efficiency, fairness, and integrity. But from the late 1960s onwards it has been subject to sustained criticism (as have many other political institutions). In part, the growing national mood of self-criticism and dissatisfaction with so many policies made the Civil Service an inevitable target for attack. The different criticisms need to be distinguished. They included:

1. Complaints about Britain's comparatively poor economic performance, which focused on the Treasury, the main economic department.

2. Attacks from the political left about the middle-class backgrounds of senior administrators (alleged 'proof' that this made them anti-Labour) and from business about the Service's lack of understanding of its work.

3. Criticism that civil servants were 'amateur', lacking both the managerial and 'relevant' skills which a modern bureaucracy required.

4. Charges that civil servants had excessive influence over ministers, demonstrated in the continuity of a department's policy, regardless of the changes of government. This was a 'conspiracy view' of the British constitution, usually found on the left, but later on the political right also.

5. Suggestions that civil servants obstruct a minister, again, more frequently from the left. It is alleged that briefings may arrive so late that ministers have difficulty in rejecting the advice, or the minister may be overburdened with paper and minor matters, or that if officials disagree with a minister's policy their less than wholehearted cooperation hinders its implementation. The Crossman *Diaries* portray a minister's image of himself as being at war with officials who sought to win him over to the 'departmental view' and Mr Benn felt a similar tension when he was in government between 1974 and 1979 (Sedgemore 1980).

Some of the demands for reform were mutually inconsistent, such as that civil servants be more expert and take more initiatives, yet remain subordinate to the non-expert minister, or that they take part in a more public debate on policy, yet respect the principle of ministerial responsibility. Contradictory complaints about officials having too much influence and being too subordinate led two critics to reflect: 'Superman and supermouse make odd bedfellows' (Heclo and Wildavsky 1974: 28).

The Fulton Committee reported in 1968 and endorsed many of the conventional criticisms. Its report (Cmnd 3638) expressed regret that the Civil Service was dominated by the notion of the 'amateur' and the 'gifted layman'. It called for the creation of planning units in each department and the promotion of more specialists and the recruitment of graduates with 'relevant specialism'. It also proposed:

- the unification of grades and classes
- the establishment of a Civil Service College, to provide post-entry training in management, data processing, economics, and other skills
- the creation of a Civil Service department;
- the appointment of more outsiders
- the 'hiving off' of work to semi-autonomous bodies

Fulton did not lead to fundamental changes. Nothing was done about the planning units, 'hiving off' of work, or the recruitment of graduates with 'relevant' skills, although the Civil Service College was set up to provide post-entry training. A Civil Service department was formed out of the old Civil Service Commission and part of the Treasury and given responsibility for recruitment, training, and pay of the service as well as for the organization and efficiency of government. Mrs Thatcher, however, became increasingly dissatisfied with its work and eventually abolished it in 1981. More political advisers were recruited but, apart from some appointments at principal level, there was little recruitment of outsiders or late entrants. There was little change in the social and educational background of recruits and no recruitment of people on account of their 'relevant' skills. Some of the Fulton recommendations—including 'hiving off' some functions (the forerunner of contracting out and privatization) and management by objectives—found more favour in the 1980s.

Promotion

Most 'fast stream' entrants can look forward to promotion up to assistant secretary level. Promotions above that level are dealt with by the Treasury and Head of the Civil Service and depend on ability and seniority. Appointment to the top posts of permanent secretary lies in the hands of the Prime Minister of the day, who is advised by other senior permanent secretaries (who form a Senior Appointments Selection Committee), and the minister of the department. Mrs Thatcher appeared to take a greater interest than other premiers in the promotions. Civil Servants who have spent time in the Treasury have a better chance of gaining a top appointment, in part because the Treasury has traditionally recruited a disproportionate share of entrants who are 'high flyers'. A survey of twenty-three permanent secretaries at the end of 1979 found that eighteen had served at some time in the Treasury, the Cabinet Office, or the Prime Minister's Office (Theakston and Fry 1989).

Because an official's promotion depends in large measure on the approval of seniors and the accumulation of experience, there have been strong forces for continuity in role performance. The skills that are emphasized—drafting papers, formulating policies, presenting a case in committee, and getting agreement—have called for the qualities of a 'sound', 'safe' person rather than one who is innovating, risk-taking, or committed to a particular line of policy. The process of role-learning was probably facilitated by a certain broad similarity in the education and social backgrounds of senior civil servants. In the past decade there has emerged a new model, one based on the *New Public Management* (NPM) outlook (see below, p. 33). This has more in common with a dynamic business, is customer- or market-driven, and is more concerned with managing services and resources than providing policy advice for ministers.

Senior civil servants, clearly, are not generalists in the sense that they have had previous experience in other fields. In many ways the Service resembles a closed corporation, with most direct entrants still recruited straight from university, at the age of 21 or 22, and remaining until retirement. Few enter over the age of 30 and although some may retire early and take up positions in private industry and public corporations, there is little traffic in the reverse direction. In Britain the *generalist* approach to administration has attached more weight to taking a broad view of the policy process and its setting in a world of Cabinet, Parliament, and pressure groups. The traditional defence for the British practice is that there is a political side to administration which technocrats and specialists may find distasteful or unimportant. Many senior administrators in France or Germany are likely to have trained in administrative law, in the United States to have had a training in management or public administration.

Whitehall has not welcomed 'outsiders', apart from the exceptional circumstances associated with the two great wars in the twentieth century (Hennessy 1990). In part this resistance rests on a conception of a career service and fears that most outsiders would be political appointees who might act as a barrier between permanent officials and the minister and politicize the Service. There has also been an expectation among senior civil servants that merit allied to seniority, rather than party loyalty, should determine promotion. These views are changing and what might be termed the new model service is prepared to dispense with the idea of a lifelong career in Whitehall and, particularly for the new agencies, recruit outsiders with relevant expertise.

The Labour government in 1964 introduced a small number of temporary appointments from outside the Service, mainly journalists and economists, particularly at the short-lived Department of Economic Affairs. The experiment has been expanded and virtually all Cabinet ministers now have a special adviser, in some cases two. The advisers are usually party supporters and help the minister with research, constituency matters, liaison with backbenchers and party headquarters, and speech-writing. In some cases, the adviser may have expertise in the policy field and give opinions on the papers prepared by civil servants. There is still, however, no effective counterpart to the French minister's 'cabinet' of a dozen or so advisers.

Civil servants and ministers

Most departments work in a hierarchical mode, with papers being revised or approved as they move up the chain of command to the permanent secretary. Some ministers deliberately seek views from lower down the hierarchy and may encourage officials to debate policy issues in front of them. The brief which finally reaches the minister will list the pros and cons of various courses of action on a policy and the top paper will contain a recommendation. A minister who lacks energy or interest in the issue, or is 'managed' by officials, may choose to ignore the papers below the top one. Such a minister is more likely to be what Bruce Headey (1974) calls a 'policy legitimator' than a 'policy selector'. Roy Jenkins (1975) has written that the views of the Home Office were conveyed to him by the permanent secretary and in his first spell as Home Secretary (1965–7) he could not know to what extent various views had been expressed at junior levels. His fear was

> that the system effectively removed the point of decision from the Home Secretary. It also produced a certain reluctance on the part of officials to disagree with each other at meetings. Co-ordinated views on paper tend to produce co-ordinated silence round the table.

> Richard Crossman, at the beginning of the first volume of his *Diaries*, records the following description of the seductions and isolation of ministerial office:
>
> **I was appointed Minister of Housing on Saturday October 17th., 1964. Now it is only the 22nd., but, oh dear, it seems a long, long time. It also seems as though I have really transferred myself completely to this new life as a Cabinet Minister. In a way it's just the same as I had expected and predicted . . . already I realize the tremendous effort it requires not to be taken over by the Civil Service. My Minister's room is like a padded cell . . . they occasionally allow an ordinary human being to come and visit me . . . the Civil Service is profoundly deferential—'Yes Minister! No Minister! If you wish it, Minister!'—combined with this there is a constant preoccupation to ensure that the Minister does what is correct. The Private Secretary's job is to make sure that when the Minister comes into Whitehall he doesn't let the side or himself down and behaves in accordance with the requirements of the institution . . . I've only to transfer everything that's in my in-tray to my out-tray without a single mark on it to ensure it will be dealt with—all my Private Office is concerned with is to see that the routine runs on, that the Minister's life is conducted in the right way.**
>
> Crossman, *Diaries*, i. 21–2

The *private office* of a minister consists of a private secretary and up to three assistant private secretaries. It acts as a link between the minister and the department and organizes most of his working day—listening in on telephone calls, fixing appointments, and briefing him. The office sifts the papers coming into the department and decides most of which papers and delegations the minister should see. A decision to refer a matter to the minister depends on its sensitivity, demands on resources, need for a political judgement, and uncertainty about the minister's reactions. *An essential part of the policy-making role of senior civil servants is to know a minister's mind.* They should, in other words, seek to draft papers, deal with cases, and react to proposals from other departments in the way they think the minister would. They need to possess not only knowledge about but also a measure of empathy with the minister.

Senior civil servants spend perhaps a fifth of their time on work relating to Parliament, for example, preparing for parliamentary questions, drawing up legislation, drafting speeches, and handling the correspondence with MPs. It is not surprising that the civil servants, operating within such guidelines and exposed to possible criticisms from Parliament, courts, and the Parliamentary Commissioner, sometimes appear cautious. Demands for a more 'innovative' or 'political' Civil Service may be difficult to reconcile with such pressures.

Differing roles

Both the minister and his senior civil servant may agree on wanting a good press for the department, but there are important differences in their roles. *A minister's political reputation depends largely on the impact he has on the world outside—on the reactions of pressure groups, party activists, Parliament, fellow ministers, and the mass media.* He operates in a public world; doing good privately is not likely to bring him many political dividends. Because he may have only a couple of years to make his mark, his assessments about policy priorities, political timing, and risks may differ from those of his officials.

The civil servant's reputation and career prospects depend primarily on the esteem of fellow civil servants, not of ministers. Officials are concerned with the smooth working of the Whitehall 'machine' and maintaining the morale of the Whitehall 'community'. The admired administrator is one who is 'sound', 'reliable', and 'trustworthy', knows how to operate the machine, and gets on with colleagues in other departments. The work of the senior civil servant is largely introverted and 'in-house', with much time spent in consulting, negotiating, and attending to a variety of inter- and intra-departmental relationships. In other words, the civil servant is considered on Whitehall rather than departmental criteria. The trade-off is that the minister, not the civil servant, claims the credit for policy success and should get the blame if it goes wrong. The civil servant who is an innovator may create controversy, raise problems for other departments, or get his minister into trouble.

American observers have suggested that the much-prized impartiality of the British service does not encourage civil servants to be committed to a policy. Because they may have to serve different ministers and work

BOX 13.4. THE CIVIL SERVANT'S VIEW OF THE QUALITIES OF A GOOD MINISTER

A good minister is
- decisive
- defends his department's interests
- reads and digests his papers quickly
- able to win interdepartmental battles
- wins battles for resources
- able to gain the ministry a good press

A weak minister involves the department in damage-limitation exercises and adversely affects departmental morale.

Crossman reflects thus on a minister's role:

> 'In our new kind of civil service, the minister must normally be content with the role of public relations officer to his department . . .'
>
> R. H. S. Crossman, 'Introduction' to Bagehot's *English Constitution* (London: Fontana, 1963), 51.

with different policies, senior civil servants are more likely to be detached. A vehement critic of the British system is Sir John Hoskyns, the first head of Mrs Thatcher's Policy Unit (1979–82). He complains that a consequence of the cult of neutrality is that officials withhold the last ounce of commitment from the minister (1982, 1984). His remedy is to make senior posts in a department political appointments, drawn from outside the Service (mostly from business). But commitment by the official is not always an unmixed blessing. Officials working on the ill-fated poll tax did not consult experts on local government finance, study foreign practices, or adequately explore possible defects in the tax (Butler *et al.* 1995: 215). The government would have been better served if officials had urged caution—in the way their critics complain.

There is some agreement among civil servants about the qualities of a good minister. They like one who is decisive, makes decisions on time and follows them through, and defends the departmental interest. They also appreciate a minister who is able to read and digest papers quickly. A minister who is able to win a fair share of interdepartmental battles and present a good case in public and in Parliament improves morale in the department. Ministers' diaries (for example, those of Crossman and Mrs Castle) eloquently describe the diverse pressures on the minister—in battles with the Treasury over resources, with other departments over responsibilities, and with parliamentary managers for time in the legislative timetable. Conversely, officials in a department whose minister is weak, and regarded as such, have to spend time trying to limit the damage. Senior civil servants are political animals; they quickly learn to appreciate which policy line is emerging or fading and which ministers are marked out for promotion and which are not. The political standing of a minister reflects on the morale of the department in Whitehall.

Communication and coordination

Communication of solid information and gossip travels quickly in Whitehall. Knowledge of 'what's going on' across departments is eagerly sought, not least because of its possible impact on a department's work. The coordination and exchange of information (or gossip), often on an informal and 'need to know' basis, is necessary for the Whitehall machinery to operate effectively. Cooperation across departments stems from the acknowledged importance of personal relations in the small Whitehall community. In handling departmental disagreements, stratagems to preserve good relations and encourage cooperation include: delay (while seeking an agreement); ambiguity (to fudge or reconcile differences); and contradictions (to satisfy rival demands).

Most of the formal coordination in Whitehall is achieved through the interdepartmental or official committees, which parallel the Cabinet committees. Official committees consist of the senior officials of departments whose ministers sit on the Cabinet committees. They prepare the committee's agenda and try to reach some agreement on some of the matters going before ministers. The senior civil servants are expected to brief their departmental ministers according to the agreed line. There are also working groups of less senior civil servants which work with the official committees. The civil servants on these committees are, in principle, members of a team and not departmental spokesmen. Finally, there is a weekly meeting of permanent secretaries to discuss the business that is scheduled to come before the Cabinet.

The system contains an obvious source of tension between ministers and civil servants. An interdepartmental committee may go some way to foreclosing the options for ministers; once it has agreed a policy line, a Cabinet minister may hesitate to reopen the question. Crossman complained that his officials were loyal, firstly to the department, secondly to the Treasury, and then, if any loyalty remained, to himself. The committee system is designed to promote coordination and serves as a potential check on excessive departmentalism.

Duties and responsibilities

Civil servants are expected to carry out the wishes of their ministers. There is, however, a fine line between helping the minister with party matters and with managing the department. Help with the former conflicts with the principle of Civil Service impartiality. Civil servants are precluded from disclosing information to the opposition, although they may, with the permission of the government of the day, have discussions with opposition spokesmen, usually before a general election. They may also help with political tactics and stratagems which are designed to show the minister in a good light, but not with party business outside Parliament, for example, a party conference speech or election activities. A permanent secretary may also refuse to obey a minister's instruction if the intention is to obstruct the opposition in its attempts to question ministers in the House of Commons, or to make unauthorized expenditure. In 1985 such questions became topical with the trial of Clive Ponting, a civil servant in the Defence department being prosecuted for leaking sensitive material to an opposition MP. Ponting's defence was that his minister had lied to the public about the sinking of the Argentine submarine the *Belgrano* during the Falklands war and that by leaking a document he was exposing the falsehood and thereby furthering the public interest. The judge disagreed, stating that the public interest was something to be decided by the government of the day, not by an individual civil servant. As a result of the case, Sir Robert Armstrong, the Cabinet Secretary and Head of the Civil Service, issued 'Notes of Guidance' on the duties and responsibilities of civil servants in 1987. The key part stated that civil servants were servants of the Crown or, in effect, the government of the day, and their duty was first and foremost to the minister in charge of the department. In the 1987 Parliament the permanent secretary in the Department for Overseas Aid objected to the use of the aid budget to fund a dam in Malaysia, on the grounds that it was a misuse of the budget.

Civil Service power

A final question to consider is the broader claim that the civil servants effectively exercise great power behind the cloak of ministerial responsibility. One version is that the bureaucracy is conservative, that it has established methods of operating which, over the years, it comes to regard as satisfactory. Departments, for example, develop preferences about different policy options, which individuals and groups to consult, and a general policy stance, or what one permanent secretary called a 'stock of departmental wisdom' (Bridges 1950: 16). The latter comes from the 'slow accretion and accumulation of experience over the years', emerging from the exchange and blending of views in committees, informal discussions, and working with colleagues. It is because the civil servants are, according to one account, 'to some extent anaesthetized by the practicalities of (their) work and that (practical) view' that they sometimes feel 'that corrective should be applied' (Sir Brian Cubborn, quoted in Young and Sloman 1982: 31). Another senior civil servant has spoken of the importance of 'ongoing reality' and realizing the limits on what politicians and governments can do (Armstrong 1970: 21). Not surprisingly, in such an environment administrators may become sceptical about the reforming minister and hope to make him aware of 'the practical aspects of idealistic policies'.

A variant on the complaint is to argue that the conditions in which civil servants and ministers operate ensure that, over time, the balance of influence gravitates to the former. The sheer pressures of work on the minister, from reading parliamentary, governmental, and other papers, attending parliamentary,

BOX 13.5. THEORY AND PRACTICE: WHO EXERCISES REAL POWER: THE MANDARINS OR THE MINISTERS?

Theory

In theory:

- ministers set the policy agenda
- civil servants provide options for enacting policy
- civil servants execute policy decisions made by the minister
- civil servants administer the departments

Civil servants advise in private, ministers decide and take the public credit or blame.

Practice

In practice, systemic factors serve to weaken the position of the political masters *vis-à-vis* their official servants:

1. *Pressure of work*. Ministers have not only their departmental work but also parliamentary, party, and constituency demands on their time. *Ministers must delegate.*
2. Ministers are *temporary* outsiders in the department: they *lack expertise* and have to rely on Civil Service advice. Civil servants have the potential power to make the minister seem 'not up to the job' publicly.
3. Sheer weight of *numbers* is on the side of the civil servants: the ratio of ministers to senior civil servants is approximately 1:65.
4. *Information*. Civil servants have the task of sifting information, analysing options, and making recommendations; they decide what the minister will see and will not see. Ministers have limited access to alternative advice.

constituency, and governmental meetings, relative ignorance of at least some areas of the department's work, short spell in office, to the imbalance in numbers between the ministers and civil servants, mean that the minister may often have to struggle to keep his head above water. The amount of work means that civil servants have to exercise discretion and make decisions in the name of the minister. One report estimated that only 1 per cent of the work of the then Board of Trade in 1971 went before ministers. The minister may inform his officials of his political values, political priorities, and which issues he wants to study himself. But the process of sifting information, analysing options, and making recommendations also gives a good deal of initiative to civil servants. Complaints about the 'overload' of business on ministers reflect the tendency of most modern governments to expand their responsibilities, and are not unique to Britain. One calculation is that it takes a typical minister two years to learn about and master a department.

Ministers who appear to cope successfully with the burdens and who want to be in a position to initiate proposals have to limit their interests and objectives. Bruce Headey calls them 'key issues' ministers, who delegate as much as possible of the routine casework and non-urgent business to junior ministers and officials concentrate on a few issues. Inevitably, there will be some give and take, perhaps even tension, between the minister and the official. Some departments appear to

Marcia Williams, writing about her experiences in No. 10 as Harold Wilson's Personal and Political Secretary, 1964–70, reflects on Whitehall's power:

'In the offices of Whitehall and in the Clubs of Pall Mall lies immense power. The electorate believes that on Polling Day it is getting a chance to change history. The reality is that in many cases the power remains with the civil servants who are ensconced in Whitehall, rather than with the politicians who come and go at elections.'

Marcia Williams, *Inside Number Ten* (London: Weidenfeld & Nicolson, 1972), 344–5

Barbara Castle expresses surprise at the reality of the power relationship:

'In my innocence when I went into Cabinet government in October 1964, I still believed that governments worked this way: that Cabinets were groups of politicians who met together and said, these are the policies we are elected on, now what will be our political priorities? And they would reach certain decisions and then they would refer these decisions to an official committee to work out the administrative implications of what they had decided. I was soon disabused of that . . . I suddenly discovered that I was never allowed to take anything to Cabinet unless it had been processed by the Official Committee. In the Official Committee the Departments all had their interdepartmental battles . . . the departments did the horse-trading and having struck their bargains at official level they then briefed Ministers on it, and so at Cabinet meetings I suddenly found I wasn't in a policital caucus at all.'

Barbara Castle, 'Mandarin Power', *Sunday Times*, 10 June 1973

These impressions are not just a 1960s phenomenon. Kirsty Milne, writing in *The Times*, 29 July 1995, had the following to say about a Labour victory in the near future:

'. . . as some Labour MPs are beginning to point out, winning an election is not the same as running the country. Is the party ready for power? . . . The problem is that, for today's Shadow Cabinet, the last Labour Government is little more than folk-memory . . . Sir Peter Kemp, former Permanent Secretary to the Cabinet Office, told the *New Statesman* that ministers-in-waiting should be trained to cope with officials who may have already digested—and dismissed—Labour's policy proposals.'

Nor is it just a Labour problem. Alan Clark reflected on his early days at the Department of Employment thus:

'The office have found a new way of keeping me utterly exhausted. My time already curtailed by the demands of the Standing Committee, is being copiously allocated to an endless series of "Fit for Work" presentations.'

Clark, *Diaries* 65; see also 'civil service ambush' 13

have points of view which a minister has to come to terms with. But departments do change over time in response to events and political pressures and to the infusion of new officials and ministers with different ideas. In the Thatcher period, discontinuity was the order of the day in many departments. Compared with the period 1974–9 abrupt changes were made in employment, housing (particularly council house sales), education, and health. The Environment Department's traditional resistance to the ending of household rates was overcome when Mrs Thatcher pushed the poll tax.

If one looks at many of the alleged Labour policies which came to grief, the failures are open to explanations other than bureaucratic obstruction. The Land Commission, set up in 1967, bought available land with the object of increasing the supply for housing and slowing down the rate of increase in house prices. This plan foundered more through the sheer impracticability of the proposals than obstruction by officials. Mr Benn at Industry (1974–5) complained that the Civil Service frustrated his Industry Act, in particular weakening the section on compulsory planning agreements. In fact, the Prime Minister and most of the Cabinet simply did not support Benn's plans, and civil servants were aware of this.

Criticism and reform (2)

The role and ethos of the Civil Service has certainly been challenged as a result of government initiatives since 1979. Not surprisingly, relations between Downing Street and senior civil servants were sometimes tense. In understanding the relations and tensions between the Civil Service and the politicians four features are worth considering:

1. Mrs Thatcher's dislike of much of the public sector in general and the bureaucracy in particular. She was wedded to the promotion of the enterprise economy and regarded much of the public sector as unproductive ('pen pushers', spenders of taxpayers' money). By the end of her premiership the size of the Civil Service had been reduced by 20 per cent from April 1979, due partly to efficiency gains and partly to the transfer of tasks elsewhere.

2. As a critic of the post-war consensus—which she and her supporters regarded as enshrining social democratic policies—Mrs Thatcher viewed the Civil Service as favouring continuity, the so-called 'ongoing reality'. She was not alone. Critics from the left (see the Benn and Crossman diaries and memoirs by some of Mr Wilson's No. 10 staff) and the right have complained about the continuity in policy in many departments, for example, the pro-EC stance of the Foreign Office, the pro-incomes policy stance of the Treasury, the pro-comprehensive schools stance of Education, and complained that some departments had too close relationships with lobbies, for example, Department of Health with the health unions. It was tantamount to saying that officials had their own political agendas.

3. Mrs Thatcher, in contrast to many other Prime Ministers, had a policy agenda of change. She had little time for the balanced ('on the one hand, on the other') approach of senior civil servants. Her test of a minister's effectiveness was whether or not he or she could overturn the department's traditional line and curb its spending. The fact that there was substantial change in so many policy areas after 1979, as well as the substantial changes in the Service itself, weakens the credibility of claims that officials effectively decide policy.

4. Mrs Thatcher was unsympathetic to the culture of the Service. In 1983 Sir John Hoskyns (1982, 1984), a former head of Mrs Thatcher's Policy Unit, alleged that a number of senior civil servants were 'defeatist', associated with thirty years of failed policies and decline and lacked the convictions and energy necessary to implement Mrs Thatcher's policies. He favoured the appointment of more politically sympathetic administrators by the government of the day.

Changing the culture

Conservative criticisms of the Civil Service have traditionally referred more to 'waste', 'overmanning', 'red tape', and 'caution' than to political bias. Mrs Thatcher was no exception.

A number of steps were taken to transform the Civil Service to make it more cost-conscious and 'business-like' (Drewry and Butcher 1991, Dowding 1993).

Efficiency scrutinies

In 1979 Sir (now Lord) Derek Rayner of Marks & Spencer was recruited by Mrs Thatcher as her efficiency adviser, based in the Cabinet Office. His remit was to scrutinize departmental expenditure with a view to promoting efficiency, reducing costs, and eliminating waste in Whitehall. The scrutinies are now supervised by the Efficiency Unit, headed by the Prime Minister's Adviser on Efficiency.

The Financial Management Initiative

In 1982 a system of Financial Management Initiative (FMI) was introduced. Under this system, groups within departments assume the task of defining policy objectives and measuring performance in relation to objectives; managers have greater responsibility for managing their own budgets. Departments work within cash limits for their manpower and running costs. The aim was to promote among officials greater cost control and a more efficient use of resources.

Next Steps

The initiative was taken further when Sir Robin Ibbs, Raynor's successor as the Prime Minister's efficiency adviser, published the *Next Steps* report in 1988. This envisaged that the Civil Service would in future consist of a small policy-making core of 20–30,000, and that many tasks could be transferred to free-standing agencies under a chief executive who would be responsible for delivering services, for example, tax collection or driving vehicle licences. In other words, the strategic management could be separated from executive functions, with which some 95 per cent of civil servants were concerned. Crucially, it argued that 'The Civil Service is too big and too diverse to manage as a single entity'.

The government has supported the programme. Work concerned essentially with delivering services to the public is steadily being transferred to agencies, including the Customs & Excise, Prisons Service, and Benefits Agency (which has over 60,000 staff). The chief executives of agencies are recruited by open (including outside) competition, and appointed for three to five years. The agencies, rather than the Civil Service Commission, have responsibility for budgets, pay, manpower, and recruitment.

The agency programme clearly involves a fragmentation of the Service. But it also raises important constitutional issues (Bogdanor 1994). Because the agencies are semi-autonomous they are no longer subject, in theory, to departmental direction. Once the agency's budget and objectives have been set by the sponsoring department, MPs are encouraged to raise matters directly with the chief executive, not the minister. Ministers become involved only if the MP is not satisfied with the original response. It remains to be seen how managerial independence in practice is reconciled with ministerial accountability to Parliament. The dismissal in 1995 of the Director General of the Prison Service by the Home Secretary brought these concerns to a head. The Home Secretary drew a distinction between policy, which was his concern, and operational responsibilities, which belonged to the agency. But critics pointed out that in practice the distinction broke down and there was ample evidence that the

Home Secretary had intervened on operational matters. The large number of complaints to MPs about the work of the Child Support Agency was such that the minister could not deflect blame on the chief executive. By the end of 1995 some three-quarters of civil servants were employed in agencies with a chief executive, objectives, and fixed budgets. The *Next Steps*, like the FMI, is designed to change the culture and practice of the Service.

It is important to note that these changes are not confined to the British Civil Service or public services. Achieving better management and value for money and the introduction of private-sector disciplines have been recurring themes in reform of bureaucracies in many other states in the past decade. It is almost a new theory of the state.

Citizen's Charter

The Citizen's Charter was launched by John Major in 1991. It is a programme, expressed in a series of charters for all public services, designed to improve the services and make them more responsive to their users. Its essential principles require the services to provide:

- published standards of services, so that citizens can see how they are performing
- information about performances, sometimes including audit by independent agencies
- mechanisms for consultations with users of services
- complaints procedures, coupled with independent review where possible

Many services now have their own charters, for example Parents' Charter, Tenant's Charter, Passenger's Charter, etc. The government's hope is that consumers will put pressure on service providers to deliver them to a high standard (Pollitt 1993).

Market testing

Another initiative designed to improve public services has included the extension of competition. The 1991 White Paper, *Competing for Quality* (Cmnd 1730), set out a number of 'prior option' tests which needed to be applied before decisions were made about non-core government activities. The tests include:

- does the activity need to be done at all or can it be abolished?
- if it is to continue, must it be carried out in the public sector or can it be privatized?
- if it is to remain in the public sector can it be contracted out or should a market test be carried out?

Once a department has decided that one or more of its functions could be subject to competitive bids from other suppliers, it is invited to compare the costs of rival bids. Some services, such as catering and cleaning, are provided by outside bodies, others are provided in-house. In early 1995 government ministers claimed that about £2.6 billion worth of civil service work, covering 54,000 jobs, was subject to such bids. Market testing has got rid of half of those jobs and produced savings of £400 million.

Mrs Thatcher also encouraged her ministers to challenge the policy advice given by senior civil servants when this reinforced the *status quo*. She was willing to look for new ideas to sympathetic think-tanks like the Centre for Policy Studies and Adam Smith Institute and relied on her Policy Unit to promote new thinking. Many government policies were thrashed out in bilateral meetings in Downing Street when Mrs Thatcher was flanked by a Policy Unit adviser and negotiated face to face with a Cabinet minister and his permanent secretary. In many respects the post-1979 experience shows that the Civil Service does respond to ministerial direction—provided that a clear lead is given over time and that ministers are determined.

The early retirement of some senior officials, allegedly on grounds of a lack of enthusiasm for government policies, her close involvement in top appointments, and the radical change in a number of policy areas do not constitute proof that Mrs Thatcher politicized the service. She was more interested in rewarding senior civil servants who were 'doers' concerned with good management and value for money—rather than detached policy advisers. But concern has been expressed that the public science ethos of the service has been undermined. The all-party Public Accounts Committee in 1992 reprimanded senior Treasury officers over a concealed payment to help the Chancellor, Norman Lamont, with his expenses when removing a tenant from a property he owned in 1991. Justice Scott's report on Matrix-Churchill in 1995 showed a degree of collusion among ministers and senior officials. The Public Accounts Committee's 1994 report on the conduct of public business warned that the pace of change in the public services threatened traditional values of accountability and integrity.

The long period of one-party rule means that for a decade and a half officials have mixed with ministers who have favoured market over public-sector solutions to problems. Peter Hennessy (1990) speculates that civil servants may have become more willing to give advice which ministers wish to hear than that which is balanced. Senior officials in the Environment Department, according to Nigel Lawson (1992), cast aside proper caution in their enthusiasm for the poll tax—once it was clear that the Prime Minister was behind it.

Conclusion

The effectiveness of British government is a product of the abilities and efforts of both ministers and senior civil servants. The similarities between the two are striking: both are drawn largely from middle-class and university backgrounds, are generalists, and much of their training takes the form of 'on the job' learning.

Some of the alleged shortcomings of the Civil Service have been in large part a comment on the kind of bureaucracy that the politicians and the public have chosen to have. Complaints about slowness, caution, and lack of innovation or initiative are to some degree inseparable from the qualities of fairness, even-handedness, and thoroughness for which the Service is often praised. Demands have often been voiced for a different kind of Service, for

- one that is more sensitive to political considerations and more committed to the policies of the party in office (as with some top posts in the USA)
- one that is more self-confident and interventionist, a Service which assumes responsibility for long-term and innovative policies

BOX 13.6. BRITISH POLITICS TODAY: CONTINUOUS REVOLUTION IN THE CIVIL SERVICE?

The 1994 White Paper on the Civil Service: 'Continuity and Change'.

On 13 July 1994 William Waldegrave, Public Services Minister, unveiled the White Paper 'Continuity and Change', which contained the proposals for continuing the changes in the Civil Service initiated by 'Next Steps':

- 50,000 Civil Service jobs are to be lost at all levels over the next four years
- departments are to have greater autonomy over recruitment and salaries of senior posts
- a 3,500-strong senior Civil Service is to be created
- departments are to consider advertising senior posts outside the Service in order to inject 'new blood' into the organization.

Mr Waldegrave insisted that the core principles of the non-political Service remained unaltered.

Comments:

Michael Meacher, Labour's Shadow Public Service Secretary, claimed that 'This . . . is a menu for accelerated privatization of the Civil Service, which threatens to destroy the constitutional principles on which it has rested for 150 years.'

Sir Robin Butler, Head of the Home Civil Service, said that he would want to see a greater movement of staff between the private and public sectors.

Barry Reamsbottom, General Secretary of the biggest Civil Service union, the Civil and Public Service Association, said: 'This is the latest stage in dismantling the British Civil Service which the Prime Minister has described as the envy of the world.'

Simon Jenkins, political editor of *The Times*, asserted, 'The White Paper is the first substantive reform of the Civil Service for 30 years. Thousands of jobs will be hived off into self-managing agencies, with local contracts and no security of tenure. The ancient custom of centralized pay negotiation will end . . . Senior civil servants will have personal contracts and these may depend on performance bonuses.' Whilst he concedes that there is some room for improvement in Whitehall, he warns that these proposals might undermine the Civil Service virtues of incorruptibility, impartiality, and confidentiality. He claims: 'There is an art of government and it cannot be acquired overnight. The Civil Service corps d'elite might be less important were British government less centred on its political directorate in Downing Street. The independent Civil Service has become a surrogate parliament, subjecting policy to scrutiny and deliberation in advance of decision, less inane and corrupt than the real Parliament. To perform this function, it must have at its core a profession, one able to translate the language of politics into the language of government. Such a profession must be an independent discipline, not just a scatter of hired consultants.' (Simon Jenkins, 'Governing is an art', *The Times*, 13 July 1994)

In January 1995 the government published another White Paper on the Civil Service with new proposals:

- a new Civil Service Code will set out the essential values of the Civil Service and specify the duties and responsibilities of civil servants and ministers
- an independent line of appeal to the Civil Service Commissioners will deal with cases of alleged breaches of Civil Service Code
- a new Civil Service Commissioner is to be appointed who will monitor internal appointments

The former demand calls for a politically subservient Civil Service, the latter for a class of 'super' civil servants who, as it were, stand above party politics. The United States has achieved this through the recruitment of senior 'in and outers', persons accomplished in other fields who are appointed by the President or Cabinet Office to oversee a policy. France is often regarded as the model for the second.

In the 1990s a different model has emerged, one which encourages most civil servants to regard themselves as managers and deliverers of services to customers, with a small core left to provide policy and strategic advice for the minister. The cumulative effects of the introduction of market testing, cost centres, privatization, agencies, and the Citizen's Charter are transforming the role of many officials. The emphasis is to make governments smaller and more efficient by making greater use of markets, competition, contracting out, and setting performance targets. Governments, both local and central, are urged to become more like Marks & Spencer or Sainsbury, whose success depends on providing a satisfactory service to customers. This is a new public management (NPM) philosophy, and is not confined to Britain. In the United States, the best-selling book *Reinventing Government* (1992) by Osborne and Gaebler has propounded similar themes. In many states there is a collision between voters' demands for better public services and their resistance to paying higher taxes. Governments need other mechanisms to improve services, with the same level of resources. Better management is the new watchword.

Some reformers would like to go further. A British think-tank, the European Policy Forum, supports the introduction of the New Zealand system of putting senior civil servants on three- or five-year contracts, paying them a market rate for their work, and assessing them by the quality of their performance. Allocating blame for policy errors to officials would clearly dilute the convention of ministerial responsibility. It would, however, bring senior officials into line with the trends in the agencies and local government.

There is today a greater willingness to question whether the traditional model of the career Civil Service is adequate. The Service is no longer unified, but is divided between a small policy-making core and the great majority who are delivering services in agencies with their own budgets, management styles, and pay scales. The possible drawbacks to having an permanent impartial Civil Service have to be set against the costs of potential politicization in the form of 'jobbery', turnover of officials, discontinuity of policy, and lower morale with changes of government. Calls for officials to show more initiative and imagination must come to terms with the present principles of political control and accountability, and calls for a more politically responsive Civil Service with the principles of the neutrality and permanence of civil servants.

Summary

- The foundation of the modern Home Civil Service can be traced to the 1854 Northcote–Trevelyan Report, which established the principles of a permanent, politically impartial, unified civil service recruited and promoted on merit.
- Ministerial responsibility is the constitutional theory underpinning the relationship between ministers and mandarins. Senior civil servants are the confidential advisers to ministers.
- It is the background, recruitment, training, and power of senior policy advisers — assistant principals and above — that are of primary interest to students of politics.
- Policy failure associated with long-term relative economic decline inevitably invites criticism of, and attempts to reform, sources of policy advice.
- The Civil Service has been criticized by the left for its social exclusiveness and suspected Conservatism; by those of the right for its vested interest in big government associated with the policies of welfare and the post-war consensus.
- It has been criticized by all sides for its caution, its exercise of negative power in favour of continuity, and its tendency to block radical initiatives.
- Reform of the Civil Service is frequently on the political agenda, but Whitehall's skill at de-radicalizing policy proposals extends to the area of Civil Service reform itself. Since 1979, however, the Civil Service has been subjected to an almost continuous revolution in an attempt to create a leaner, more efficient, less 'bureaucratic' bureaucracy.
- Recent changes in the Civil Service can be seen to reflect the emergence of a new theory of the state.
- Change has brought benefits but it may not be without cost. A distinction continues to exist between the ethos of public service and that of private enterprise.

CHRONOLOGY OF IMPORTANT DATES IN THE DEVELOPMENT OF THE HOME CIVIL SERVICE

1066 Normans conquered Britain continued the process of organizing and centralizing the administration of the state. The origins of the Civil Service lie in the exchequer/treasury function, the collection of taxes

1830 By this time the principle of permanence had become established

1854 Northcote–Trevelyan Report:- permanent, professional bureaucrats, recruited and promoted according to merit

1873 Doctrine of ministerial responsibility and anonymity of civil servants established

1919 Warren Fisher reforms establish the unified Civil Service envisaged by Northcote–Trevelyan

1968 Fulton Report: recommended establishment of Civil Service Department; Civil Service College; departmental planning units; appointment of special advisers; increased management training; wider net of recruitment

1979 Sir John Hoskyns proposes 'de-privileging' the Civil Service
Sir (now Lord) Derek Rayner heads the Efficiency Unit and 'Rayner's raiders' scrutinize the Civil Service for waste

1982 Financial Management Initiative: departments are cash-limited for manpower and running costs; clearer management responsibilities are defined

1988 Next Steps, the Ibbs Report: envisages gradual reduction of the Civil Service to a policy-making core. Executive agencies are to take over the delivery of services

1991 Citizen's Charter: sets standards of performance and complaint procedures for public services

1994 Continuity and Change: greater departmental autonomy over pay, conditions, and recruitment of senior ranks; open market recruitment is encouraged

ESSAY/DISCUSSION TOPICS

1. Do the traditional assumptions about the British Civil Service have any contemporary relevance?
2. 'Civil servants advise, ministers decide'. Is this an accurate description of the relationship between officials and their political masters?
3. With reference to examples, discuss the kind of constitutional issues raised when tasks previously performed by civil servants in government departments are transferred to agencies under a chief executive.
4. In what respect can recent changes in the organization of central government be said to imply a new theory of the state?
5. 'Civil Service advice is impartial and informed by experience. It must not be sacrificed to a dogma.' Discuss.
6. Whitehall has succeeded in neutralizing or resisting all attempts to reform the Civil Service. Discuss.

RESEARCH EXERCISES

1. Who were Colette Bowes and Clive Ponting? What can we learn about the relationship between ministers and civil servants from the events in which these two people were involved?

2. In 1994 prison breakouts, the treatment of long-term offenders, and the availability of illegal drugs in jails were some of the issues that aroused widespread controversy about the administration of British prisons. In 1995 the Home Secretary dismissed the Director General of the Prison Service, Derek Lewis. What kinds of constitutional issues were raised by these events?

FURTHER READING

See K. Dowding, *The Civil Service* (London: Routledge, 1993) and Peter Hennessy, *Whitehall* (London: Fontana, 1990). For more up-to-date developments see J. Willman, 'The Civil Service', in D. Kavanagh and A. Seldon (eds.), *The Major Effect* (London: Macmillan, 1994), 64–82.

The following may also be consulted:

Burch, Martin, 'Civil Service Reforms', *Talking Politics*, 5/3 (1993).

Butcher, Tony, 'The Civil Service in the 1990s: The Revolution Rolls On', *Talking Politics*, 8/1 (1995).

Farnham, David, 'The Citizen's Charter', *Talking Politics*, 4/2 (1991/2).

Greenaway, J. R., 'The Civil Service: Twenty Years of Reform', in B. Jones and L. Robins (eds.), *Two Decades in British Politics* (Manchester: Manchester University Press, 1992).

Pyper, Robert, 'A New Model Civil Service?', *Politics Review*, 2/2 (1992).

Wass, Douglas, *Government and the Governed: BBC Reith Lectures 1983* (London: Routledge & Kegan Paul, 1983).

Yes Minister (videos of the TV series).

14 LOCAL GOVERNMENT AND DECENTRALIZATION

Reader's guide

Local government disburses approximately 25 per cent of public expenditure and is responsible for delivering a wide variety of personal services and municipal amenities. But the turnout for local elections is less than 40 per cent, many people have only a vague idea of which authority, central or local, county or district, is responsible for which service and which decisions, and few people can name their local councillors. The doings of the parish pump hold little local, let alone national, interest unless they can be captioned under the headline 'Loony Left' or attract the lable of 'sleaze' because councillors stand accused of, for example, manipulating council house sales for political advantage.

Throughout this century, for both main political parties, support for thriving local democracy has been the rhetoric of opposition. Once in office, the idea holds less appeal. Governments of either party have been intolerant of any attempt by local councils to challenge public spending policies and/or of failure to meet nationally set standards of service provision. A long period in office has enabled the current Conservative government to mount a more sustained effort to tackle local attempts to frustrate the nationally set agenda. Powers and responsibilies have ebbed away from local government in both a downwards and upwards direction. The result has been a notable reduction in local autonomy and democratic accountability.

Does this matter? Is local democracy important? How is local government organized and to what kind of controls is it subject? What kinds of people become councillors and local government officers, and what kind of relationship exists between the two? What kind of relationship exists between central and local government? This chapter addresses these questions and then goes on to examine the almost continuous revolution which local government has undergone since 1979. It explores the critique of the New Right and the New Left and examines the role of the parties in local politics. It explains the crucial part played by finance in central government's need to control local government and the means of exercising control. The chapter concludes by reviewing some of the ways in which powers and responsibilities have been slipping away from local government and by drawing a contrast between the centralizing tendency noted in Britain with the decentralization occurring in many of Britain's European partners.

IN the past two decades the role of local government in British political life has been transformed—and severely weakened. The Thatcher governments' agenda of 'rolling back' the state has certainly been felt by local government. This chapter reviews the case for local democracy, examines the present local government structure, discusses the state of central–local relations, and finally assesses the effects of the drastic changes in powers and resources of local government.

Local democracy

Arguments for local government have often been advanced on the grounds of either greater democracy and/or greater efficiency, two different but not mutually exclusive arguments. Since 1979, both have been challenged. Supporters of democracy claim that local government provides opportunities for people to participate in politics, develop a sense of citizenship and political responsibility, and promote a sense of community. Supporters of efficiency claim that central government and Parliament are already overloaded and that policies developed in a local context are more likely to be appropriate for and supported by the local population.

Critics have objected to the arguments for decentralization on the grounds that the likely variations in policies and quality of services may lead to undesirable inequalities in services between areas, or that local units will be less efficient than central government. They are also able to point to the evident lack of a sense of community in many larger conurbations and the typical turn-out of less than 40 per cent in local elections. Some of the distinctions between local and central government responsibilities are unclear to many people and local elections are often decided on national not local grounds. In the mid-1990s the many changes made to the structure and organization of local government have done little to make the distinctions clearer or to increase democratic accountability.

The role of local government usually figures prominently in discussions of the nature of democracy. Some political theorists see democratic vitality emerging from the existence of various decision points; pluralism allows many groups to have some influence and to check one another. According to this view local government may act as a buffer between citizens and central government, and as a check on the latter. An alternative view is that political democracy is enhanced when power is centralized, increasing the capacity of government to be responsive to the preferences of the majority.

In the 1960s and early 1970s, reorganization of central and local government reflected a belief that 'big is beautiful'. A large authority was able to provide a more specialized staff and service because of the greater scale of its operations. Demands for a more integrated approach to decision-taking also seemed to require the creation of larger local authorities. Interestingly, studies for the Royal Commission on Local Government in the 1960s did not find any clear-cut relationship between population size and quality of service provided by the local government. At the same time, neither did they support claims that a person's 'social attachment' and 'interest in local affairs' increased with the smaller types of authority (Newton 1982).

The decentralized dimension of British government is expressed in various ways. Some agencies are primarily *territorial*, as in local government, while others are primarily *functional*, such as education authorities which administer services on a local basis. Many central government departments also administer their own policies and services—pensions, national insurance, employment, etc.—through their local offices. Central government may lay down national policies on housing and education, for example, but they do not build houses or schools, or directly employ teachers and builders. The local authorities undertake these tasks for state schools and council housing. Teachers are employed by a local education authority or school managers rather than by central government; they are not civil servants, as they are, for example, in France. Parents, pupils, and teachers in the education service, and tenants and builders in the housing service, will be more aware of local councillors and officers as their authorities. Yet local authorities derive their powers from Parliament, the main lines of policy for local services are laid down by parliamentary legislation, and a large part of the central government grant is given for the authorities to carry out particular services.

Local government structure

The origins of local government in England can be traced back to the Middle Ages, when a number of boroughs gained royal charters from the monarchy and were allowed to run their own affairs. In the nineteenth century a new system of local government was established and it lasted until 1974. By then changes in patterns of employment and residence, development of new forms of transport, and demands for new types of services all combined to undermine the logic of the system drawn up in the late nineteenth century.

After much debate, the Local Government Act (1972) came into effect in 1974. This established a two-tier system of thirty-nine counties and six metropolitan counties for the major conurbations outside London. Within the metropolitan counties was a lower tier of thirty-six metropolitan districts. Wales was divided into eight counties and thirty-seven districts. In Scotland the proposals of the Wheatley Commission (1969) were implemented in 1975. The Scottish structure has seven top-tier regional authorities and thirty-seven second-tier districts, with special arrangements for the three 'island' authorities.

Under the basic division of functions, the counties were awarded services for which uniformity of policy is considered desirable, and the districts were to provide more local services. The counties deal with the major services of planning, education, traffic, police, fire, and highways, for example, and the districts with housing, cemeteries, local planning, parks, and refuse collection. The powers of the six metropolitan counties included strategic planning and passenger transport but not education or personal social services, and were fewer than those of the shire counties. The lower-tier metropolitan district councils were given control of education and of some services administered by the counties elsewhere.

In the 1980s, local government became increasingly politicized and there were frequent battles between central and local government and between the Labour and Conservative parties. The Conservatives abolished the Greater London Council (GLC) and metropolitan authorities—all Labour-controlled—and devolved their duties to London boroughs and metropolitan districts respectively. The history of local government reform is one in which negotiations, inquiries, commissions, and discussions have been coloured by the partisanship and the self-interest of different political parties and groups. It has long been 'part of a bargaining process, extended over almost fifty years by a perceived need to proceed by consent and to avoid major upheavals' (Alexander 1982: 3). Since 1979, the Thatcher government unilaterally changed the structure, responsibilities, and financial basis of local government.

At the time of writing, local government in England is being reviewed by a Local Government Commission under the chairmanship of Sir David Cooksey. Legislation passed in 1994 abolished the two-tier structure in Scotland and Wales with effect from April 1996 and replaced it with a single tier of unitary authorities: twenty-two county and county borough councils in Wales and twenty-nine county councils in Scotland. Elections for these new authorities took place in May 1995. England seems likely to be little altered by the proposals, due to be finalized at the end of 1995. The Commission has recommended only ten new all-purpose unitary authorities from a list of twenty-one it was asked to consider by the government. The underlying logic of the proposals is that only districts with strong local identity and economy on the fringes of counties are viable as unitary authorities.

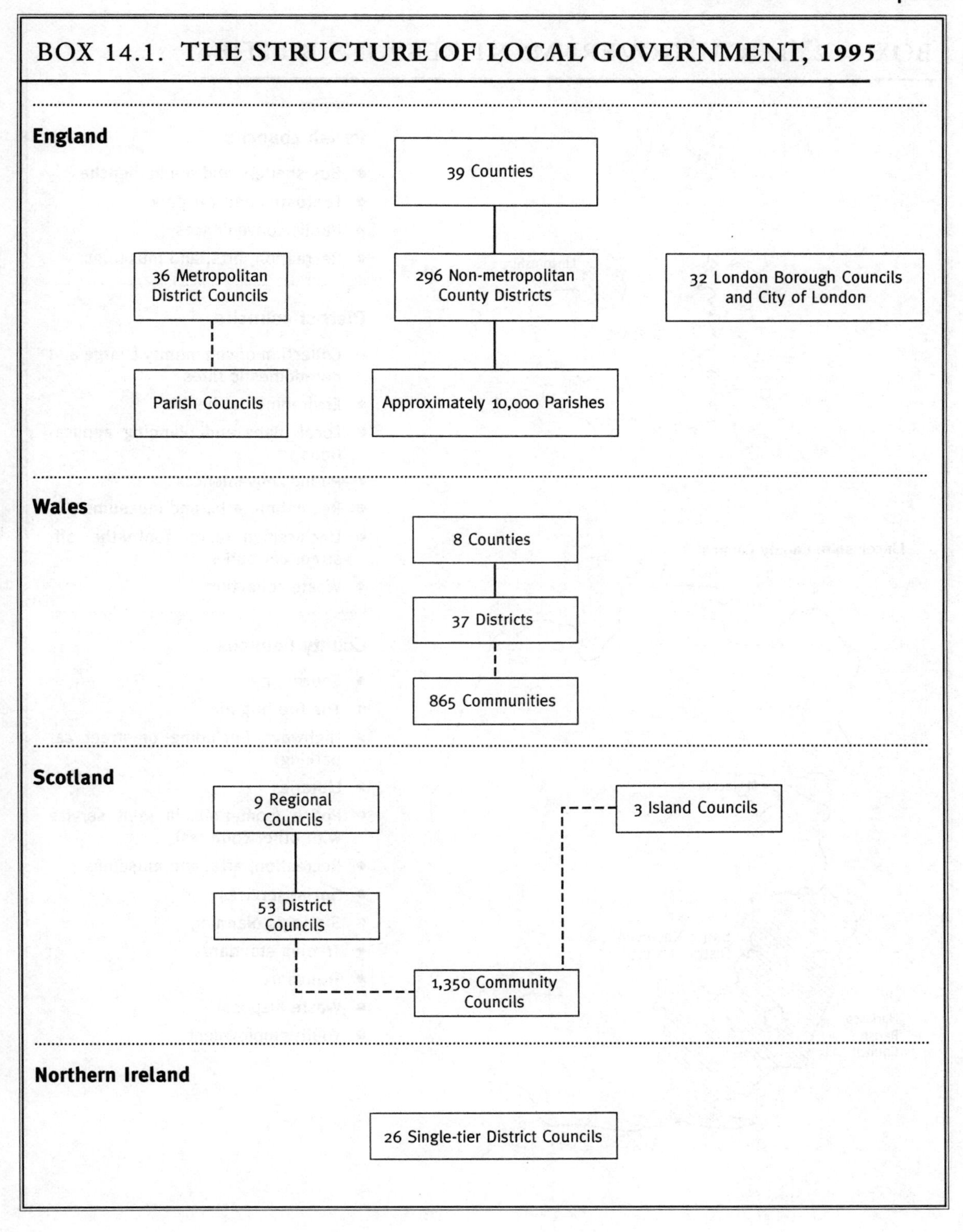
BOX 14.1. THE STRUCTURE OF LOCAL GOVERNMENT, 1995
England
39 Counties
36 Metropolitan District Councils
296 Non-metropolitan County Districts
32 London Borough Councils and City of London
Parish Councils
Approximately 10,000 Parishes
Wales
8 Counties
37 Districts
865 Communities
Scotland
9 Regional Councils
3 Island Councils
53 District Councils
1,350 Community Councils
Northern Ireland
26 Single-tier District Councils

BOX 14.2. LOCAL GOVERNMENT RESPONSIBILITIES

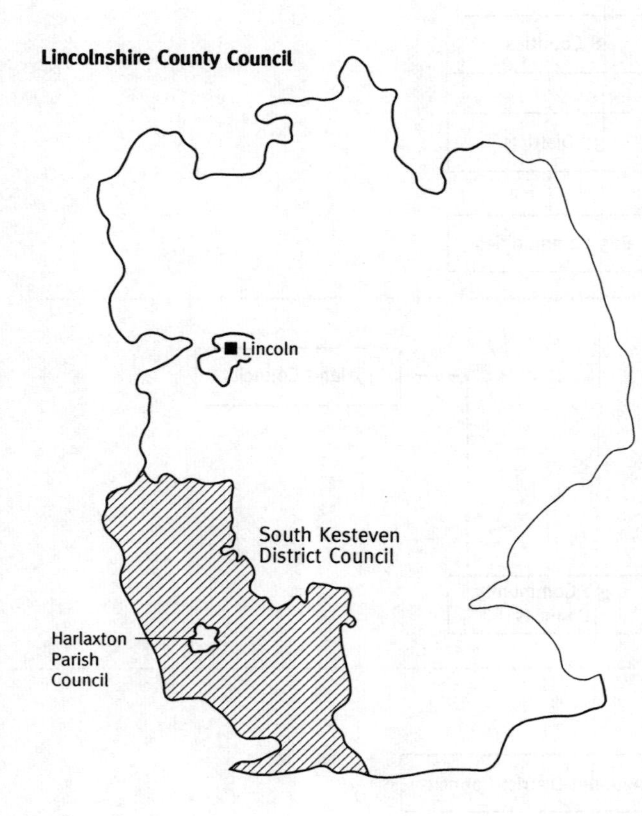

Parish councils

- Bus shelters and public benches
- Footpaths and car parks
- Public conveniences
- Recreation, arts, and museums

District councils

- Collection of community charge and non-domestic rates
- Environmental health
- Local plans and planning applications
- Public conveniences
- Recreation, arts, and museums
- Unclassified roads, footpaths, off-street car parks
- Waste collection

County Councils

- Education
- The fire brigade
- Highways (including on-street car parking)
- Libraries
- Police (sometimes a joint service with other counties)
- Recreation, arts, and museums
- Social services
- Strategic planning
- Trading standards
- Transport
- Waste disposal
- Youth employment

BOX 14.3. BRITISH POLITICS TODAY: CHANGES IN LOCAL GOVERNMENT

At the time of writing, local government in England is being reviewed by a Local Government Commission under the chairmanship of Sir David Cooksey. Legislation passed in 1996 abolishes the two-tier structure in Scotland and Wales with effect from April 1996 and replaces it with a single tier of unitary authorities. Elections for these new authorities took place in May 1995.

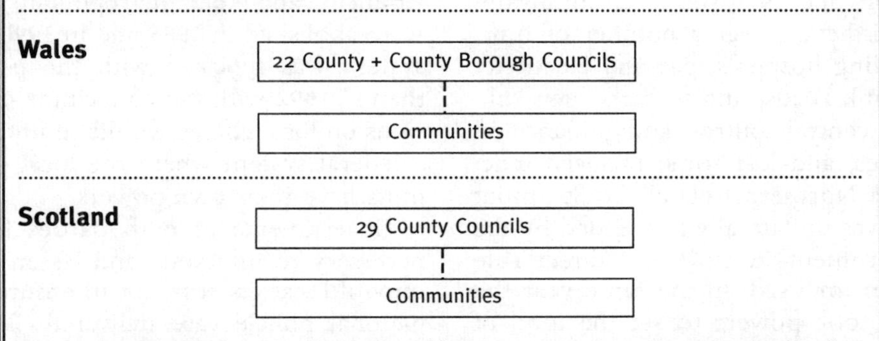

England seems likely to be little altered by the proposals — due to be finalized at the end of 1995. The commission has recommended only ten new all-purpose unitary authorities from a list of twenty-one it was asked to consider by the government. The underlying logic of the proposals is that only districts with strong local identity and economy on the fringes of counties are viable as unitary authorities.

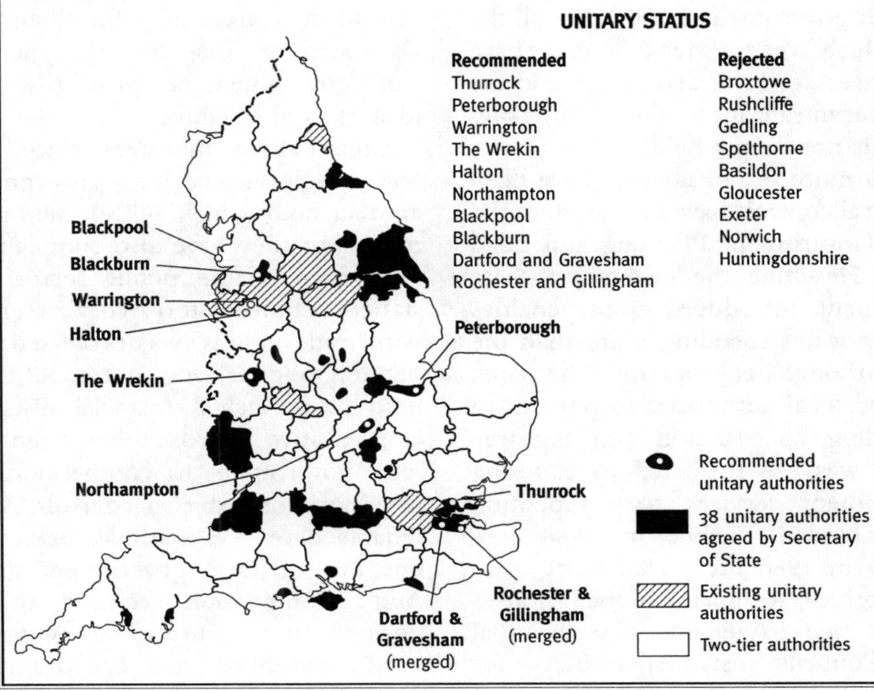

Central controls

The British system of government is unitary, with sovereignty concentrated in Parliament. Such powers as a local authority has are conferred and may be withdrawn by Parliament. Between 1945 and 1979 the local authorities had lost jurisdiction over a number of functions (including hospitals, gas and electricity supplies, trunk roads, and water), been subject to more central controls and guidance on other services, and lost some financial independence. In Northern Ireland the Stormont Parliament was unilaterally suspended by the British government in 1972 and direct rule from London imposed. In the same year the government took powers to set the level of rents for local authority housing (subsequently repealed). In 1976 a Labour government instructed all authorities to reorganize their systems of secondary education on non-selective lines—an instruction repealed by the new Thatcher government in 1979. For all the lip-service which Westminster politicians have paid to the need for local democracy and the virtues of decentralization, the record suggests that it is not deeply held.

Since 1979 more responsibilities have been lost and central controls have increased. In the 1980 Local Government Planning and Land Act, Michael Heseltine, the Secretary of State for Environment, introduced grant penalties for local authorities spending more than the government thought appropriate. The same act compelled local authorities to put out to tender building construction and highway repairs; this was the first step to exposing local government services to competition and was extended further in 1988 (see below). Also in 1980 the government compelled authorities to make council houses available for sale to tenants at substantial discounts. Councils lost responsibility for further education and polytechnics, and schools were allowed to 'opt out' of local authority control and be funded by central government. The GLC (Greater London Council) and the metropolitan authorities were abolished in 1985 and in 1990 the rating system was replaced with the poll tax and then in 1992 with the council tax. Such incursions on local choice would be unthinkable in a federal system where the local or regional units have their own powers.

Governments of both parties have felt it necessary to intervene and lessen the autonomy of local government to ensure that *their* national policies are delivered. Before 1979 Conservatives usually intervened on grounds of macroeconomic policy, particularly to control public expenditure, Labour on grounds of redistribution and equality. Both parties when in government have used central funding to reward their areas of political support. It has also been the case that the party excluded from government has paid lip-service to the idea of local freedom.

Conservative ministers since 1979 have increasingly viewed local government as *just* another body which delivers service to customers. But they have also complained that it, like much of the public sector, was overstaffed, enjoyed too 'cosy' a relationship with trade unions, was protected from competition, and, as long as it could raise household rates, lacked financial discipline. The Conservative approach has been to subject local government to competition in service provision and external controls. Schools and tenants have, for example, been allowed to opt out of local government control, the Audit Commission acquired an important position in measuring the performance of local authorities and encouraging greater economy, effectiveness, and efficiency, and private-sector management techniques are used increasingly. Above all, a consistent theme of the central government interventions has been to curb spending by local authorities.

The 'top-down' model

Both parties, when in government, have found the 'top-down' model of British administration convenient. The key elements of the model are the supremacy of Parliament as lawmaker, ministerial responsibility, Whitehall control, and the absence of formal checks and balances. Conservatives, with a belief in 'strong' government, have understandably found the model congenial, as has Labour, which, traditionally, has regarded centralized decision-making as the most effective way to promote equality. The central government lays down the framework and minimum standards for services, provides most of the funding, and tries to ensure uniform standards throughout the country, while leaving delivery of the service to the local authority. Entitlement to and levels of welfare benefits, or the school-leaving age, for example, are the same wherever one lives in the United Kingdom, though there are differences in the *quality* of service across regions.

By contrast, a system with less central control would allow for more regional and local variation above and below a minimum or 'norm' standard of service. Provision of hospital beds or medical consultants, the extent of home ownership, or proportions of children staying on at school and going to university vary markedly across the country. On most indicators of socio-economic development and social provision, the South-East of England is better off than other regions, a symptom of the 'Geography of Inequality' (Sharpe 1982).

Councillors and officers

To be eligible to serve as a councillor a person must live or work in the seat or have some connection with it. Holding a paid office with the local authority disqualifies one from election. Also ineligible are undischarged bankrupts, lunatics, and anyone who has been sentenced to more than three months' imprisonment over the previous five years, or convicted of illegal electoral practices. Many councillors work long hours and may spend twenty hours or more a week on council duties. The reduction in the number of authorities and number of councillors (to about 26,000, or half of the earlier numbers) has led to an increase in the councillors' typical workload (Widdicombe 1986). Proposals for

BOX 14.4. LOCAL ELECTIONS—A COMPLICATED SYSTEM?

1. Local elections take place annually on the first Thursday in May.

2. Not everyone votes every year, however:
    - County councils are elected *en bloc* every four years.
    - Districts can opt for either *en bloc* four-year elections at the mid-term point of county elections
      or
      'by thirds', one-third of councillors standing for re-election each non-county election year

3. Eligibility to vote:
    - 18 years or over
    - citizen of Britain, Ireland, or the Commonwealth
    - resident in the locality
    - registered

4. Eligibility to stand as a candidate:
    - 21 years or over
    - citizen of Britain, Ireland, or the Commonwealth
    - on the local electoral register
    - resident or employed in local authority for at least twelve months prior to standing
    - support of ten registered electors in the ward

Turnout at local elections is approximately 40 per cent. The turnout in May 1995 was 38 per cent.

Local elections are now portrayed by the media, and to some extent accepted by politicians, as a verdict on national government.

the payment of salaries have been resisted, but the attendance and other allowances have been increased. These may have done something to attract younger and more economically active councillors; particularly assiduous councillors may collect the equivalent of an average wage in allowances. Yet the tradition of unpaid public services is still important in local government.

Councillors, like MPs, are sociologically unrepresentative of the population, being disproportionately male, middle-aged, and middle class. Unpaid council work is most easily combined with certain types of employment—particularly such self-employed professions as the law, estate agency, and small businesses—and trade union and political organizers, housewives, and the retired.

People stand for council for a variety of motives. For some it is a step on the road to national politics—in recent parliaments about 45 per cent of Labour MPs and 25 per cent of

Conservative MPs had been councillors before election. Some mention personal reasons for participation: council membership is seen as an alternative to humdrum jobs, or members enjoy the feelings of enhanced social status and membership of 'a good club'. Many recruits are already community activists and typically belong to an average of seven or more local voluntary 'organizations'. Nomination by a political party is now a virtual precondition to pursuing a local political career.

As local government acquired more functions during the nineteenth century, so Parliament insisted that officers be appointed. The 1972 Local Government Act specified the categories of officials whose appointment by local authorities is compulsory. The chief officer of a department, in contrast to the councillor, is expected to possess professional training and qualifications. An officer is also expected to be politically impartial and debarred from standing for political office until a year after his retirement as an officer. The officer's task is to provide the best professional advice to the councillors and then carry out the policy decisions of the council, short of breaking the law.

In theory, the relationship between councillors and officers resembles that between ministers and civil servants. But there are differences: a councillor, in contrast to an MP, is part of the executive; there is closer contact between officials and councillors than between civil servants and MPs; each authority decides its own staff structure, compared with the uniform structure of Whitehall departments; most chief officers are appointed for their relevant professional expertise in a specialist area (engineering, accountancy, planning, and so on), while Whitehall administrators remain generalists, and whereas the officer is appointed by the council, senior civil servants are promoted by colleagues and the Prime Minister. Relationships between the chief officer and the councillors may vary. In some cases an officer's personality and skills may enable him to 'lead' the councillors; in others, a committee chairman or a small group of councillors will provide the direction. The legal position is clear enough. Officers are employees and servants of the *council* and, in the last resort, take instructions from it. Perhaps as a consequence of the greater partisanship between Conservative and Labour councillors in the 1980s the work of officials came under greater challenge. Some Labour-controlled London boroughs in the 1970s and 1980s openly advertised for officers who were sympathetic to their radical policies. Rather than respecting the professionalism and expertise of the officer such councils looked for political commitment.

In the 1980s new concerns prompted a fresh look at the management of local government. Ministers expressed alarm at politicization or, more specifically, radical left-wing policies being provided 'on the rates'. More efficient and business-like management was seen as a way of delivering services more economically. Councils could and should cut back the number of committees and concentrate on strategic management. They should also cease to be regarded as local *government* (with the potential for conflict with central government) and more as an *enabling* authority, purchasing services from a variety of providers and monitoring quality and value for money.

Council leaders

The leader of the council is invariably also the leader of the majority party and, more recently, of the policy committee. The leader's emergence is largely a post-war development. Previously, political leadership was usually dispersed between the chairmen of the different committees; acting together, they might provide collective leadership. The ending of the aldermanic system and the more frequent turnover of councillors have probably reduced the possibilities for the emergence of such experienced influentials. There have been some prominent, even autocratic, local leaders, such as Herbert Morrison in London (in the 1930s) and Jack Braddock in Liverpool (in the 1950s and 1960s). In the 1970s and 1980s Ken Livingstone in London, Derek Hatton in Liverpool, 'Red' Ted Knight in Lambeth, and Lady Porter in Westminster attracted publicity, even notoriety. But as a rule local leaders do not gain the national media attention of the mayors of big American cities. Compared with the American or French mayor or German *Bürgermeister*, British local government lacks 'visible' local political leaders. Indeed, the fact that the American and West German leaders are directly elected gives them a function and standing separate from that of their parties. Michael Heseltine has long advocated the election of mayors to be high-profile leaders of local government, but fellow ministers have not supported the idea. In Britain, the leader is more likely to work through the party, be part of a collective leadership, and work closely with the chief executive.

Central–local relations

Relations between central and local government came under growing strain in the 1980s and as a result left local government a much enfeebled body. For long there has been a debate about whether the central–local relationship is or should be one of *partnership*, or if local government should be an *agent* of the centre. It has been a question of finding the right balance between local autonomy which, if pushed to the limit, may result in marked disparities between authorities in the provision of services, and central guidance and control which, if pushed to the limit, might make local government merely the executive arm of central government. Few people in local government are satisfied, however, that the relationship between central and local government at present strikes a proper balance.

Different models have been employed to describe the central–local relationships (Rhodes 1988). In support of the *agency* model, it can be noted that central government has the power to create or abolish the units and powers of local government. In this model, the national framework of a policy is established and largely financed centrally and local authorities carry it out, with little scope for discretion or variation. The *partnership* model claims that the authorities do have

BOX 14.5. THE RESOURCES OF CENTRAL AND LOCAL GOVERNMENT

Central government

- Control over legislation and delegated powers
- Provides and controls the largest proportion of local authorities' current expenditure through the Revenue Support Grant
- Controls individual authorities' total expenditure and taxation levels by 'capping'
- Controls the largest proportion of local capital expenditure
- Sets standards for and inspects some services
- Has national electoral mandate

Local government

- Employs all personnel in local services, far outnumbering civil servants
- Has, through both councillors and officers, detailed local knowledge and expertise
- Controls the implementation of policy
- Has limited powers to raise own taxes and set own service charges
- Can decide own political priorities and most service standards, and how money should be distributed among services
- Has local electoral mandate

Source: Wilson and Game (1994), 111.

some scope for local choice. Local government has its own political legitimacy, finance (from the council tax) resources, and even legal powers, and the balance of power between the centre and locality fluctuates, according to the context. There is too much variation in local services to sustain the agency model, even though local authorities are clearly subordinate in the partnership. Yet we need to remember that, for all the well-publicized conflicts, the normal pattern is one of cooperation, consultation, and even bargaining.

The relationship between the two levels of government is based on a mix of controls, dependence, and cooperation. Local authorities have few legislative powers themselves; all local powers are conferred by Parliament, and may be amended or withdrawn by it. If a local authority exceeds its powers then its actions will lack legal validity, and it may be held *ultra vires* by the courts. Powers are conferred in private or general Acts relating to a particular function, like education. Some responsibilities are mandatory, such as the compulsory sale of council houses to tenants, and some are permissive, such as whether or not local authorities reorganize secondary schools on comprehensive lines. Responsibilities may move from one category to the other—for instance, the reorganization of secondary education was made compulsory by a Labour government in 1976, and then permissive by a Conservative government in 1979.

There are also controls which stem from central government's involvement in many services. The provision of primary and secondary education is the responsibility of local authorities. However, the scope of the curriculum and training and the supply of teachers lie largely in the hands of the Department of Education, and the school-leaving age is decided by Parliament. Under the Education Reform Act (1988) the government assumed greater power over the curriculum and introduced nationwide assessment of pupils. Another source of central influence is the power to approve the appointment and dismissal of certain chief officers. The approval of the Home Secretary, for example, is required for the appointment of a local Chief Constable, and the approval of the appropriate minister is necessary for appointments of chief education officers or directors of social services. Inspectors in education, police, fire, and social services are all employed by central government to report on the standards and efficiency of the services. A local authority's building plans for schools and the making of by-laws, require the approval of the appropriate department. Central government circulars also contain advice and guidance to local authorities. In finance the government has the power to reduce a local authority's central grant if the latter is deemed to have 'overspent', and it approves local authority applications to raise a loan for new capital expenditure.

Finally, there are the legal controls which the government, organizations, or even individuals may invoke if an authority has broken the law. Local authorities may provide only those services which they are empowered to by statute. The Audit Commission appoints auditors to inspect the local authority accounts each year to establish that all expenditure has been duly authorized. Individual councillors may be held responsible and surcharged for any unauthorized expenditure and even disqualified from membership of a local authority. In 1986 Labour councillors in Liverpool and Lambeth, for example, were surcharged and disqualified for refusing to set a legal rate.

One has to beware of portraying local government in general, or even a particular local authority, as monolithic and unified. The continuing importance of committees, departments, parties, and factions may make an authority appear pluralistic to people who work in it. Local authorities also vary in their orientations to Whitehall; differences in size, resources, powers, problems, and needs affect their relationships with government departments. There is no single spokesman for the local authorities; the task is shared by the various local authority associations (such as the Association of County Councils and the Association of Metropolitan Authorities), which may be further divided on party or political grounds. Although there is a Minister for Local Government—in the Department of the Environment—Whitehall is still effectively departmentalized. Government departments negotiate with functional bodies or with the professional associations of chief officers—the Department of Education, for example, negotiates with the Association of Education Committees. In their dealings with local authorities about particular services, Whitehall departments before 1979 varied between a *laissez-faire* approach at one extreme to highly interventionist strategies at the other. Under Mrs Thatcher most became interventionist.

A decade of change

In spite of the relentless pressures from the centre in the 1980s to limit their autonomy and resources, local authorities still account for nearly a quarter of total public spending and still deliver many local services, particularly in schools and housing. It can also be said that the trend towards more control from the centre has been a long-standing one. All central governments, regardless of party, have used their substantial powers over finance (in the form of the government grants), structure, and policies to influence local government.

Yet since 1979 there have been major changes in the structure, financial basis, and internal workings of local government (see Box 14.6) as well as in the extent of local autonomy, and in relations between local authorities and the private sector and citizens. The changes have been largely imposed rather than agreed. The shift has been from consultation and bargaining to a much greater degree of central control. From merely complaining about the wasteful spending of some Labour authorities, the Conservative governments steadily increased controls, particularly financial, over local authorities, curtailed their activities, and set up new bodies to provide some of the services. They also developed a radical critique of their traditional role, dismissing the concept of a local mandate and arguing that many local services traditionally provided by local authorities can be better provided by other agencies.

The most important change in the 1980s was the disagreement between the national parties on the role of local government. In part this was a consequence of the breakdown of the consensus at national level and in part a consequence of the politicization of local government (see below). In both Labour and Conservative parties, groups have

BOX 14.6. MAIN LEGISLATIVE CHANGES AFFECTING LOCAL GOVERNMENT, 1979–1993

1980	Local Government Planning and Land Act: Powers to cut government grants to 'overspending' local authorites
1982	Local Government Finance Act: abolition of supplementary rates
1984	Rates Act: rate-capping
1985	Local Government Act: GLC and six metropolitan councils abolished
1986	Local Government Act: statutory prohibition of local authority publicity of a party political nature
1987	Rate Support Grant Act: local authority grant fixed to expenditure
1988	Local Government Act: local councils to put public supply services out to tender
	Housing Act: allowed council tenants to choose their own landlords and created Housing Action Trusts to take over bad council estates
	Local Government Finance Act: rates abolished, replaced by community charge
	Education Act: allowed schools to opt out from local authority control and be funded directly from Whitehall. National curriculum for state schools
	Local Government Act: extended compulsory competitive tendering (CCT)
1991	Poll tax abolished, replaced by council tax, to begin April 1993. 'Capping' power retained by centre

rejected the old model of *partnership* between central and local government. In this model the central government's essential right to determine policies was not challenged by the local authority, while the latter was left with a good deal of initiative in applying policy in many fields. Jim Bulpitt (1983) has called this a *dual polity*—as long as central government controlled policies of 'high' politics, essentially finance, defence, and foreign policy, it was prepared to leave what it regarded as 'low' politics to the localities and other groups. For the Conservative government of Mrs Thatcher curbing public spending, particularly by high-spending left-wing Labour authorities, became a priority.

The New Left

A younger generation of left-wing Labour councillors came to prominence in the 1970s. They were determined to expand services and increase spending, and defied the Labour government (1974–9). Not surprisingly, they were even more hostile to Mrs Thatcher. Labour councillors in Manchester, Liverpool, and a number of London boroughs were dismissive of what they regarded as right-wing parliamentary Labour leaders, and were determined to use local government as a base from which to challenge Thatcherism (Gyford 1985). They pursued policies of local economic development, equal opportunities, and defied government guidelines on spending and raising local domestic and business rates. The conflict was one reason why the GLC and other metropolitan authorities were abolished by an impatient Conservative government in 1985.

The New Right

The New Left attitude was matched by a *New Right* critique of local government among some Conservatives. Traditionally, Conservatives have praised local government as a means of checking interventionist (probably Labour) central government, encouraging diversity, and providing an opportunity for experimentation with different policies. This was one reason why, for example, the party opposed mandatory comprehensive secondary education, and insisted that it was for each local authority to choose its own system. Yet the New Right critique, developed in the 1970s and 1980s, dismissed the legitimacy of local government. Electoral turnouts were low, local elections were usually fought on national not local issues, and electorates could be dominated by a public sector workforce (for example, teachers, direct labour workers and transport employees), council house tenants, and many who did not pay rates. In other words, the claims by local government to possess a local mandate were bogus. Moreover, more than half of the money spent locally actually came from central government. This New Right critique argued that because local government was a creation of Parliament, such systematic defiance of Whitehall was illegitimate.

The critics therefore urged the importation into local government of greater competition and a business approach to management. Local government should be forced to compete with other groups in the delivery of school education or refuse collection. It would be better, the New Right concluded, if local government administered contracted-out local services, with funding and standards of services determined by central government. Such reforms would make local services more efficient and more responsive to consumers (N. Ridley 1989).

Local party politics

Party politics in some form or other has long been a part of local government. In the mid-nineteenth century many councils were divided between rival political factions and even parties. In the 1870s Joseph Chamberlain used the Liberal party and a party programme in Birmingham to great effect, showing what could be done to give political leadership to local government. The Liberals and Conservatives became dominant in local politics towards the end of the nineteenth century, especially in the North and the Midlands. The major expansion of the Labour party in the 1920s in many parts of the country formalized and accentuated partisanship.

Since the 1974 reorganization party politics has virtually taken over local government. The number of 'Independents', or non-party councillors, has been drastically reduced and the proportion of uncontested seats has fallen from 40 per cent to 16 per cent (Game and Leach 1995). Some 90 per cent of councils are now controlled by one or a coalition of the three main national parties.

The arguments for and against party politics in local government, particularly where it duplicates the national rivalry, are well known: partisanship may frighten off able people who would otherwise offer their services to the community; many local services require good administration rather than the application of party ideology, and the sense of community identity is weakened as national parties with national programmes move in. Against this, it may be argued that parties provide a focus and coherence for local political debate; they stimulate political interest and turnout in elections; and they are important in enabling voters to pin political responsibility at elections.

Yet, viewed in a wider perspective, particularly in comparison with the United States, France, or West Germany, it is the disjunction between local and national politics in Britain that is so striking. In the countries cited, parliamentarians are expected to act as local spokesmen and the advancement of a political career depends on having a local base. In Britain, by contrast, many local councillors have indeed been 'localists', and regarded Westminster politics as remote and not very interesting. They may also regard themselves primarily as members of a local political community rather than a national political party, and behave accordingly. Apart from Northern Ireland, national political élites are usually securely based in London. National politics is 'higher' than local politics. Nevil Johnson (1977: 128) has commented on the dominance over a long period of national political elites and their relatively sharp separation from local politics, understood as the maintenance of local positions of political influence. As a result, interest and influence do not flow up continuously from the base to the top of British political life.

Finance

Local government's financial activities are so significant in the national context that central government cannot turn a blind eye to them. If we aggregate expenditure and employment by local authorities, then we are talking about big government. They spend about a quarter of total public expenditure and employ nearly three million people (about a tenth of the working population), own a quarter of the national housing stock, and provide primary and secondary education for over 90 per cent of children.

There are three main sources of local finance. The first is the grant from central government. The second is revenue from rents, fees, and charges from or on housing, sites for markets, services in parks, leisure centres, and so forth. For long the third source was the rates, a tax levied on the occupier of premises, including houses, offices, and factories. Rates were a form of local property tax and were replaced by the poll tax in 1990, which in turn was superseded by the council tax in April 1993.

The domestic rating system dated back to a time when only property owners had the vote. Rates were paid only by householders, and became increasingly unpopular at a time of inflation. In 1974 the Conservatives promised that they would abolish rates if they were elected. They did not repeat the pledge in 1979, though Mrs Thatcher's opposition to the rating system was well known. Various committees examined alternatives to the rates and proposals for central funding of some services. The Treasury opposed any form of local income tax, and the general Whitehall view was that a local source of finance was essential if local independence and accountability were to be maintained.

The largest source of finance for local authorities is the grant from central government. As a proportion of local authority spending, the central grant steadily grew from 29 per cent in 1938–9 to 65 per cent in 1976–7 and has since fallen back to just over 40 per cent. The Treasury has consistently tried to reduce or hold steady public spending as a share of GDP and to restrain both local spending and the proportion financed by the central government. In 1981 the government took powers to make an assessment of each authority's grant-related expenditure (GREA). The grant is fixed each year after negotiations have taken place between associations of the local authorities and the departments most concerned with local government. As already noted, this grant may be reduced if the Secretary of State decides that an authority's expenditure

BOX 14.7. LOCAL GOVERNMENT FINANCE: KEY TERMS

Term	Definition
Rates	A tax on rentable value of business and domestic property. It is paid by the owner or occupier. It is a regressive tax because it takes no account of income. Property values are periodically revised. The last revision in England was in 1973. The 1985 revision in Scotland brought rate rises of 50%, and fear of likely repercussions of a similar rise in England prompted the government to seek an alternative. Many people were exempt from rates, and this weakened the link between voting for increased local spending and responsibility for footing the bill. Rates were abolished in Scotland in 1989 and in England and Wales in 1990.
Community charge	A 'poll tax' introduced by the Local Government Finance Act in 1988. It is a flat-rate tax to be paid by every adult with a few exemptions. It has the advantage of simplicity and it revived link between voter, charge, and service. But it is regressive, a very visible tax, and difficult and costly to collect. In the face of its extreme unpopularity, the government abandoned this tax as of 1993.
Council tax	A tax introduced in 1991 to replace the community charge, effective in 1993. It combines a personal and property element and is levied on households; it assumes two adults, and single occupants can claim 25 per cent rebate. It reflects property values: each property is allocated to a valuation band set by the Inland Revenue. The tax level is set by councils.
Uniform business rate	Replaced rates in 1990. It is based on business property values and therefore similar to the old rating system, but set and collected nationally and redistributed to councils through government grants.
Rate capping	In 1982 the Local Government Finance Act removed the right of councils to levy a supplementary or extra rate in the middle of the financial year. The 1984 Rates Act gave the government the power to cap the level of rates charged by a local authority. It still applies in the form of Council Tax capping.
Revenue support grant	This is the main contribution of central government to the current income of local authorities. The government determines the **Total Standard Spending** for local government. It then decides what proportion will be met from national taxation. The remainder has to be met from locally determined sources. *Example*: 1992–3 • Government-determined TSS would be £70 billion • 85% would be provided nationally (£59.4 billion) • local authorities would have to raise 15% (approx. £10 billion)

exceeds its GREA 'target', that is, what it should spend.

Some local councils reacted to the reduction in the real value of central government grants by increasing their rates, or levying a supplementary rate when the initial one proved insufficient for their expenditure. The government then took power to outlaw supplementary rates. The same act allowed the Secretary of State to adjust (i.e. reduce) the block grant to individual authorities to encourage reductions in their expenditure. In 1990 the Rate Support Grant became the Revenue Support Grant. It produced a broadly similar amount of money but used fewer criteria for assessing the grant.

The poll tax

Yet for all the battles and extra controls imposed by the centre, local government current spending continued to increase in real terms. In some frustration, the government finally decided to abolish the rates, replacing them with the Community Charge, or poll tax, as it became known. The tax was a flat-rate charge for local services levied on all adults over the area, although the government reserved the power to cap it in each authority. In addition, central government set a uniform charge for businesses.

An attack on the rating system was an important part of the *New Right critique*. As long as only a minority of voters paid rates voters had little incentive to protest at wasteful spending or the imposition of high rates by the council. The Thatcher government calculated that if all adult residents had to pay a fixed charge, then they would be more likely to vote and be critical of a 'high' spending and 'high' taxing local authority. In other words, the government hoped that by restoring the link between taxation and voting, people would vote in local elections to punish 'wasteful' councils. Government spokesmen argued that in the past, local authorities had frightened business away from cities through high business rates. By imposing a *uniform* business rate across the counties, local authorities would be able to increase local spending only by increasing the poll tax. Control would remain in the form of an enhanced role for the Audit Commission to scrutinize local spending and ensure value for money.

The poll tax has been described as the single most disastrous democratic policy decision made by a post-war British government, and a detailed study of its passage is titled *Failure in British Government* (Butler *et al.* 1995). The principle of charging virtually every adult a flat-rate charge was bitterly opposed. It led to riots, a slump in Tory electoral support, and widespread evasion, and was a factor in Mrs Thatcher's downfall in 1990. Many Tory MPs were convinced that the party would lose an election badly unless the tax was repealed—and that was out of the question, as long as Mrs Thatcher was leader. A new council tax, based partly on property values and partly on a personal element, replaced the poll tax with effect from 1 April 1993. The government has retained the power to cap the tax—so much for local choice—and the tax now raises only 20 per cent of local spending. The fact that the government ended up with a form of property tax, not dissimilar to the old rates, is a tribute to the complexity of the problem and the failure of a decade of reform.

Local spending

The growing controversy concerning local government finance shows how difficult it is to separate questions of financial control and political autonomy. As Labour and Conservative governments in the past two decades have tried to control the growth of central public expenditure by imposing cash limits on many programmes, so they have tried to 'export' similar controls to local government. Some three-quarters of local authority spending is effectively dictated in the policies laid down by central government, particularly in transport, education, housing, and social services. And global assessments about the 'correct' level of local authority finance may overlook the differences between frugal and high-spending authorities. Variations in levels of local expenditure on particular services are partly a consequence of the different levels of resources available to authorities, as well as of different local needs and demands for particular services. Inner-city areas, for example, with high levels of social and economic deprivation, face a greater demand on housing and family support services, But the different political values of Labour and Conservative matter also. Central government increasingly shapes and finances much of the current spending by local authorities. Under both Labour and Conservative administrations, the central government has used the grant as an instrument of national economic policy, parrticularly on public expenditure and public-sector pay, and as a means of shifting the allocation of resources between different areas, or altering the support for different services.

A changed balance

The major changes affecting the role of local government have been financial. But there have also been significant changes in other areas. Since the abolition of the GLC London is now the only major city in Europe without a directly elected authority. The centre has also taken power to control the internal workings of local authorities. Following the Widdicombe Report (1986) it has excluded a number of local officers from political authority and specified rules for achieving party balance on committees.

The Government has *bypassed local authorities* in many policy fields. In *education* it expanded the role of the Manpower Services Commission in further education, established City Technology Colleges free from local authority control, abolished the Burnham Committee on teachers' pay, on which local authorities were represented, and imposed a core curriculum and regular testing of pupils in schools. It removed the polytechnics (now universities) and sixth-form colleges from local control and allows secondary schools to opt out of local authority control and become grant-maintained schools funded by Whitehall. School governors have rights to hire and fire teachers and control budgets.

BOX 14.8. LOCAL GOVERNMENT SPENDING BY SERVICE, 1993

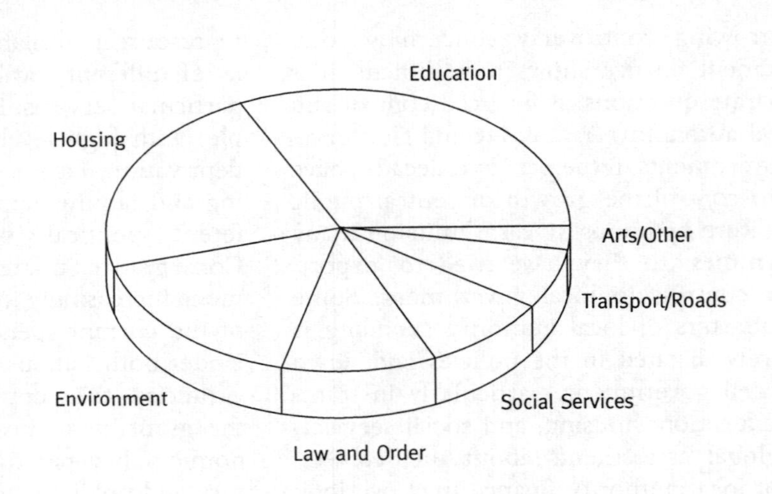

Total Local Government Spending	**£70 billion**

Local government accounts for approximately 25% of public expenditure but 75% of local authority spending is effectively dictated by the policies laid down by central government.

Education	33%
Housing	22%
Environmental services (including refuse collection and disposal, health, recreation, and trading)	17%
Law and order, protective services	11%
Personal social services	9%
Transport and roads	7%
Arts and libraries	1%
Other	1%

In *housing*, the government has allowed tenants to purchase their council houses at a substantial discount (Housing Acts, 1980 and 1984) and over one million have done so since 1979. Under the Housing Act (1988) tenants are also allowed to opt out of local authority control and form their own housing associations. In addition, it has set up Urban Development Corporations (UDCs), dominated by private businesses, to help in the regeneration of inner cities. The government has also made it compulsory for local authorities to put out to tender such services as street cleaning and refuse collection.

The Local Government Acts (1988 and 1992) extended the practice of Compulsory Competitive Tendering (CCT). The first act obliged local authorities to put such services as street cleaning, refuse collection, catering, and cleaning of buildings out to tender to the private or voluntary sector. The second extended competition to include professional, technical, and financial services. The importance of the policy is that local government is no longer seen primarily as a deliverer of services. Instead local government is seen as an *enabling authority*, which has a responsibility for ensuring the quality and delivery of services but with the provision to be contracted out to other bodies. Local government is now only one of a number of agencies providing local services, in competition with the voluntary sector and private agencies. It remains the case, however, that most of the contracts have been awarded in-house.

Such a massive reduction in financial independence and responsibilities has led some observers to question the role of local government. Does it, they ask, have any purpose beyond satisfying the administrative convenience of Whitehall? Local government has been squeezed between greater control from the centre, competition from other agencies which have assumed some of the responsibilities, and the role which has been given to parents, school governors, tenants, and consumers of services. Significantly, no local government representative participated in the formulation of the poll tax (Butler, *et al.* 1995).

Democratic deficit

The Conservative government's view of local authorities as 'mere' enabling bodies is defended on the grounds that many people's primary contacts with local government are as consumers of public services and that what matters to them is the quality of the services, not whether the deliverers are accountable. The idea has been to make the producers or deliverers of these services more accountable to consumers, which is accomplished not through the election of councillors or mediation of government ministers, responsible to Parliament, but through devolution to such institutions as grant-maintained schools and trust hospitals, contracts for the level and quality of services with suppliers, and the role of regulatory bodies.

Ministers claim that this approach complements their own electoral accountability to voters with local accountability to consumers. The new agencies provide fuller information about the services they deliver figures on, for example, on schools' truancy rates and examination results. Yet these bodies operate free from many of the constraints that bind elected local councillors. In addition to the discipline of regular elections, council meetings are open to the public, disclosure of councillors' interests is required, surcharges may be imposed on them for illegal or improper expenditures, and accounting officers are appointed to conduct audits of local spending. In so far as the new bodies are accountable, it is more often to the minister in Whitehall than to consumers. Britain is now perhaps the most politically centralized state in Western Europe. The ratio of elected officials to voters in Western democracies varies from 1:250 to 1:450. In Britain it is 1:2,200. And local government is the only body that can integrate policies and set strategy for the whole community.

Territorial government

British government works according to a mixture of territorial and functional principles (Hogwood and Keating 1982). Most ministries, such as Education, Health and Social Security, or Defence, are primarily *functional*. There are also *territorial* departments which cover Northern Ireland, Wales, and Scotland, each with its own Secretary of State, who sits in the Cabinet. The Ministry of Defence, the Treasury, or the Foreign Office make policies for the United Kingdom as a whole, while the territorial Secretaries of State are mainly interested in how the policies of the functional departments relate to their own territories. As members of Cabinet they are entitled to a say in any department's policy which affects their nation; equally, policies for their nations are subject to collective Cabinet decisions. The territorial ministries are not at the top of the ministerial hierarchy, but usually appear in the bottom half of the list of Cabinet ministers announced at the opening of each Parliament (Rose 1982*b*).

Britain to date has resisted attempts to introduce political devolution to Scotland and Wales, and any form of regional government. The system remains unitary. Federal systems, as in the United States, Australia, or West Germany, have another, regional, layer of government which has its own independent powers to make laws and raise money.

Richard Rose (1982*b*) has drawn a useful distinction between different types of territorial and regional policies. *Uniform policies* involve measures and laws which are applied similarly throughout the United Kingdom. These apply particularly to the defining policies of defence, foreign affairs, and taxation. *Concurrent policies* are laid down nationally but administered by different institutions, and make some allowance for local circumstances. Exceptional policies apply only to a particular area, for example, the security legislation in Northern Ireland.

Conclusion

The past two decades have seen much controversy surrounding the structure, role, and financial basis of local government. Local government clearly operates within an environment which is the product of each authority's social, historical, political, and economic conditions. It also operates within a national environment—councillors belong to national parties, officers to national associations, policies are made in Whitehall, and the state of the national economy is always important.

In Britain's 'top-down' model of political authority, local government is often regarded by the centre as a useful agency for the delivery of services. The unitary system combines the Treasury's determination to control public expenditure and taxation with Whitehall's officials' views of tidy administration to reinforce this model. In contrast, a number of other countries operate federal systems which guarantees independent powers to local units. In the past decade many EC member states have been moving in a different direction to that of the British government and devolved powers to regional and local governments. They have also applied the principle of *subsidiarity* to empower sub-national and not merely national governments *vis-à-vis* Brussels (see Ch. 4). When officials devised the poll tax, European experience was ignored and the verdict of the authoritative study on the subject (Butler, *et al.* 1995: 3) was:

> **The neglect, bordering on contempt for wider European practice in local government . . . is a telling commentary on Britain's rapport with its continental neighbours.**

Summary

- Local democracy is defended on the grounds of participation and efficiency; but these arguments are undermined by low turnouts for local elections, low interest in, or awareness of, what local government does, and unacceptable inequalities in standards of service provision.
- The structure of local government was reorganized in 1974 broadly into counties and districts. Six metropolitan counties in the major conurbations and the Greater London Council were subsequently abolished in 1985. In 1991 the government allowed counties and districts the option of continuing the present two-tier arrangements or creating single-tier authorities.
- Local government is subject to central control: the powers of local authorities are conferred, and may be withdrawn, by Parliament. A variety of Acts such as Education, Housing, etc. affect local government's powers and responsibilities and often specify particular central–local relationships. Both main political parties when in power favour a 'top-down' model of central control.
- Councillors have local connections; they are usually members of a party, are sociologically unrepresentative, are unpaid but receive expenses. They often regard local government as a stepping-stone to national politics.
- Local government officers are full-time officials usually having expertise in particular fields of local government work, e.g. planning, education. The relationship between councillors and officials is similar to that between ministers and civil servants.

- Party politics now dominate local councils and the leader of the majority party usually becomes the Council Leader.

- Central–local relations vary to some extent from department to department. In some ways it is more like a partnership, in others local government is more like an agent of the centre. The relationship is based on a mix of controls, dependence, and cooperation. Central government is closely involved in overseeing the provision of many services in an attempt to ensure minimum or equal standards.

- Since 1979 there have been major changes in the structure, financial basis, and the internal workings of local government. Increased central control has eroded local autonomy and the old partnership model has been superseded by a greater emphasis on the agency model.

- The New Right argue that local government lacks legitimacy and that defiance of Whitehall on grounds of a local mandate is bogus. Local government needs greater exposure to the market.

- The New Left saw local government as an opportunity for radical opposition to unwelcome Conservative policies.

- Finance comes from three main sources: grants, council tax, and income from rents, fees, and charges for local amenities. All are subject to central control. The poll tax was an unsuccessful attempt to link local spending to local control.

- Local government has not only been subjected to more stringent financial controls in the last decade, it has also been bypassed in many policy fields, e.g. housing and education. It has been forced to operate a Compulsory Competitive Tender system for a variety of previously 'in-house' services.

- Local government operates within a local and national environment. Central government intervention is inevitable when it bears electoral responsibility for the economy, and local government accounts for 25 per cent of public spending. Britain's increasingly centralized model, however, offers a contrast with the decentralization that is taking place in many of Britain's EU partners.

CHRONOLOGY

1964 Maud Committee on Management of Local Government

1966 Redcliffe-Maud Commission on Local Government

1969 Wheatley Commission on Local Government in Scotland

1970 'Fair Rents' Act: rent on council houses to be based on ability to pay

1972 Local Government Act: rationalizes the previous 1,400 local authorities into 420-plus parishes. The government sends commissioners to adminster housing in Clay Cross; councillors refuse to implement the 'Fair Rent' Act

1974 The 1972 Local Government reorganization comes into force

1975 Reorganization of local government in Scotland

1976 Local Education Authorities are instructed to reorganize education on comprehensive lines

1980 Local Government and Planning Act: grants penalties for overspending authorities. Housing Act: mandatory sale of council houses

1982 Local Government Finance Act: abolishes supplementary rates

1984 Rates Act: introduces rate-capping

1985 Local Government Act: GLC and metropolitan counties abolished. Transport Act limits powers of local authorities to subsidize and regulate public transport

1986 Widdicombe Report into the Conduct of Local Authority Business. Local Government Act prohibits local authority publicity of a party political nature

1987 Rate Support Grant: grants fixed to expenditure

1988 Local Government Act: introduces Compulsory Competitive Tending. Housing Act: tenants can form Housing Action Trusts. Local Government Finance Act: rates to be replaced by Community Charge and Uniform Business Rate. Education Act: introduces opting-out, national curriculum, local management for schools

1989 Community Charge (poll tax) introduced in Scotland

1990 Community Charge replaces rates in England and Wales

1991 Council Tax replaces poll tax as of 1993. Commission for Local Government to examine structure and make recommendations

1992 Education Act: every school subject to inspection at four-year intervals

1994 Local Government Acts for Scotland and Wales: new single-tier authorities to be introduced in 1996

ESSAY/DISCUSSION TOPICS

1. How legitimate is the local electoral mandate?
2. Why is local government subject to central government control?
3. British people attach little importance to local democracy. Discuss.
4. In what respects are the role and power of local government changed in recent years?
5. Why was the Community Charge introduced and then subsequently abolished?

RESEARCH EXERCISES

1. Undertake a research project on the local government of your area. Organize your work under the following headings:
   - overall structure
   - county responsibilities
   - district responsibilities
   - organization of work within the district council — committee structure, etc.
   - number of councillors; elections; party strength
2. What was the significance of Clay Cross in 1972 and Tameside in 1976–7?

FURTHER READING

For an introduction see T. Byrne, *Local Government in Britain*, 6th edn. (London: Penguin, 1994) or David Wilson and Chris Game, *Local Government in the United Kingdom* (Basingstoke: Macmillan, 1994). On the Poll Tax and much material about Whitehall's attitude to local government see D. Butler, A. Adonis, and T. Travers, *Failure in British Government* (Oxford University Press, 1995).

The following may also be consulted:

Conley, Frank, 'The Local Elections 1995', *Talking Politics*, 8/1 (1995).

Manton, Kevin, 'The Ideology of the New Right in Local Government', *Talking Politics*, 6/1 (1993).

15 THE JUDICIARY AND RIGHTS

Reader's guide

Britain has neither a codified constitution nor a Bill of Rights, and there is a fusion rather than a separation of powers. There is no judicial review, in the sense of judges ruling on the constitutionality of the government's actions or laws. The judicial branch of state in Britain, therefore, is less obviously political than, for example, its American counterpart. Nevertheless, headlines such as 'Judge criticizes Howard [Home Secretary] over IRA prisoners' (*The Times*, 29 September 1995) clearly demonstrate that the line dividing law and politics is a fine one, if it exists at all. Law can never be neutral: statute law is the outcome of a political process; common law and case law are based on the decisions of a socially exclusive group of professional lawyers; and it is common to refer to the judicial branch of government or branch of state.

What is the role of the judiciary and what safeguards, if any, exist to ensure its independence? Whislt there is no formal judicial review, is it nevertheless possible to identify judicial activism? What has been the impact of Europe on British justice? This chapter addresses these questions and then goes on to examine the somewhat precarious status of civil liberties in Britain. It reviews what is entailed by the doctrine of the Rule of Law and examines the arguments for Britain adopting a Bill of Rights. It explores the impact of the European Court of Human Rights on British politics. The chapter concludes by examining the growing body of delegated legislation and administrative tribunals which blur the traditional distinction between law and politics.

THE law is often regarded as a *set of procedures* which promotes order, justice, and predictability in relations between individuals, groups, and the state. In a free society law also underpins the freedoms of citizens by guaranteeing certain civil liberties and imposing limits on the authorities. The judiciary acts to uphold law and order and arbitrate in clashes. A second and narrower usage of law may be as a *means of resolving disputes*, which is ultimately backed by criminal sanctions imposed by the courts. Another is the *common law*, derived from precedent and judges' interpretations of the law. Finally, one may refer to *statute law*, that is, an Act which has been passed through Parliament and received the Royal Assent. In Britain, statute law takes precedence over common law, but even here judges have a role to play, in assessing criminal guilt or innocence, interpreting a statute, and determining its boundaries and application.

The noted law-abidingness of the British has rested on several planks. As long ago as 1865 Bagehot noted that the natural impulse of the English 'is to resist authority', to regard government as 'an extrinsic agency' and legislation as 'alien action'. Respect for law in Britain has depended, in large measure, on the laws being regarded as reasonable and on the restrictions on liberty being limited. There have been occasions when groups have felt it necessary to resort to direct action against a law, or to further a cause outside of the parliamentary process. In the nineteenth century, the Catholic Emancipation League, the Chartists, parliamentary reformers, Irish Nationalists, and then the suffragettes before 1914 all employed methods of direct action to promote their causes. The violent demonstrations against the poll tax in 1990 followed in a long tradition.

With a few exceptions British political scientists have not shown much interest in the judiciary. In part this derives from the lack of a written constitution, in part from the subordination of the judiciary to Parliament, and in part from the political independence of the judiciary. Politics and law have usually been seen as separate spheres. This attitude is changing, partly because of the impact of the European Community, partly because of the growth of judicial activism, and partly because of growing unease about the power of the executive.

This chapter considers first the judiciary and its relation to politics, and the forces making for judicial activism. It then examines civil liberties and the case for a Bill of Rights. Finally, it considers the citizen's opportunities for obtaining redress of grievances against the administration.

The judiciary's role

We observed in Chapter 3 that in the British Constitution there is no higher authority than statute law—with the notable exception of European Community legislation. The sovereignty of Parliament refers to Parliament's unlimited right to make law and the inability of the courts to overturn a statute, apart from matters pertaining to the EC. By contrast, in Germany and the United States, both of which have written constitutions, the courts

BOX 15.1. KEY CONCEPTS: THE JUDICIARY

Common law	Derived from ancient custom as interpreted in court cases, not codified.
Statute law	Acts of Parliament, which take precedence over common law, but even here judges play a role in assessing guilt or innocence and in interpreting statutes.
European Union law	Takes precedence over British law in areas of EU competence and requires no parliamentary action to become operative in Britain.
Public law	Law which covers the relationship between the state and the citizen, including both civil liberties and administrative matters, for example, the compulsory purchase of private property or entitlement to benefits. Britain does not have a formally identified body of public law.
Criminal law	Concerned with wrongful acts harmful to the community such as murder and arson.
Civil law	Concerned with rights, duties, and obligations owed by individuals towards each other, for example, divorce, debts.
European Court of Justice	An institution of the EU created to ensure uniformity in the interpretation and application of EU law. It sits in Luxemburg.
ECHR	European Court of Human Rights. Britain ratified the European Convention on Human Rights in 1951 but did not incorporate the convention into British law; therefore it does not confer legal rights enforceable in British courts but its decisions are influential, carry moral authority, and frequently embarrass the British government, for example, the ruling in September 1995 on the SAS killing members of the IRA in Gibraltar.
Ultra vires	The action of government must be based on law. Courts can declare actions to be *ultra vires*, or 'beyond the powers', if they are not covered by the terms of a statute.

play a crucial role in interpreting the Constitution and making judgements which affect public policy. The British judges are not, however, entirely passive. Judicial review of the actions of the executive allows judges to rule whether ministers or their agents have statutory authority for their actions. Judges are also competent to rule that an office-holder has been guilty of *ultra vires* (i.e. acting beyond authority) or of an error of law (i.e. an improper act). They have, however, been reluctant to go further and decide on the merits or constitutionality of a particular statute, since, by so doing, they might lay themselves open to the charge of attempting to usurp the legislative function. By tradition they have contented themselves with a literal reading of the statute to establish whether a minister or official acted within his statutory powers. Even here judges are bound to some degree by precedent and certain rules of procedure. This is one aspect of the informal separation of functions in the British constitution.

Since the case of *Pepper v. Hart* (1992) judges have also been able to consult the relevant parliamentary debates, reported in *Hansard*, when the bill was being discussed, in the hope of gaining a better knowledge of the intentions of the framers of the Act. There is, however, no guarantee that such study will uncover a clear legislative intention. Statutes are usually drawn up by civil servants, who remain anonymous, and voted by a Parliament, many of whose members will not have read the wording of the bill. An Act in many instances represents a compromise between different points of view and may be open to different interpretations by its supporters, and one cannot be certain which view or interpretation prevailed when it was enacted.

Judicial independence

The independence of the judiciary from political control or influence is a key principle of the British constitution. Judges as a rule refrain from public comment on contentious issues, they may not be MPs, and MPs who become judges are required to resign their seats. The salaries of judges are a standing charge on the Consolidated Fund and may not be altered through the annual Estimates or be the subject of parliamentary debate. Senior judges also have security of tenure; they may be removed from office only as a result of an address to the monarch by both Houses of Parliament. A final safeguard is that judges are, ostensibly at least, appointed and promoted on professional rather than political grounds. (Before 1914 this was less true.) This last feature is striking because the appointments of judges are made by the Lord Chancellor, himself a politician and member of the Cabinet. The Prime Minister also has a say in the appointment of the Lord Chief Justice and of the Lords Justice and Lords of Appeal. MPs, even though protected by parliamentary privilege, generally withhold comment on judges and on legal matters which are pending; these are regarded as being *sub judice* until the proceedings are completed. *There is a mutual exchange of self-restraint between politicians and judges* and a recognition of their separate spheres.

The separation of law and politics is not total, however. The Lord Chancellor, as noted, is a member of the government, sits in the Cabinet, and presides over the House of Lords. The Attorney-General and Solicitor-General, as well as the Law Officers for Scotland (Lord Advocate and Solicitor-General), are also members of the government. The Law Lords, sitting in the Lords, act as the supreme court of appeal—and usually confine their speeches to legal matters. The law is usually the best-represented profession in the Commons: in post-war Parliaments the number of MPs who were or are solicitors or barristers has rarely fallen below a hundred, or a sixth of the House. Judges have also found themselves involved in presiding over tribunals, commissions, and various inquiries into politically contentious issues, for example, Lord Wid-

BOX 15.2. LAW AND POLITICS: A FUSION OF POWERS

An assumption of liberal constitutional theory is that in order to guarantee individual liberty and avoid arbitrary government or tyranny, the three main branches of state should be separate. These are:

- legislature: those empowered to make laws
- executive: those responsible for implementing the laws
- judiciary: those tasked with applying the laws

The British constitution pays scant regard to this doctrine. There is some recognition of the need for judicial independence but the British arrangements represent a fusion rather than a separation of powers. The Lord Chancellor, as head of the judiciary, member of the House of Lords (legislature), and member of the Cabinet (executive), is at the centre of three overlapping branches of state.

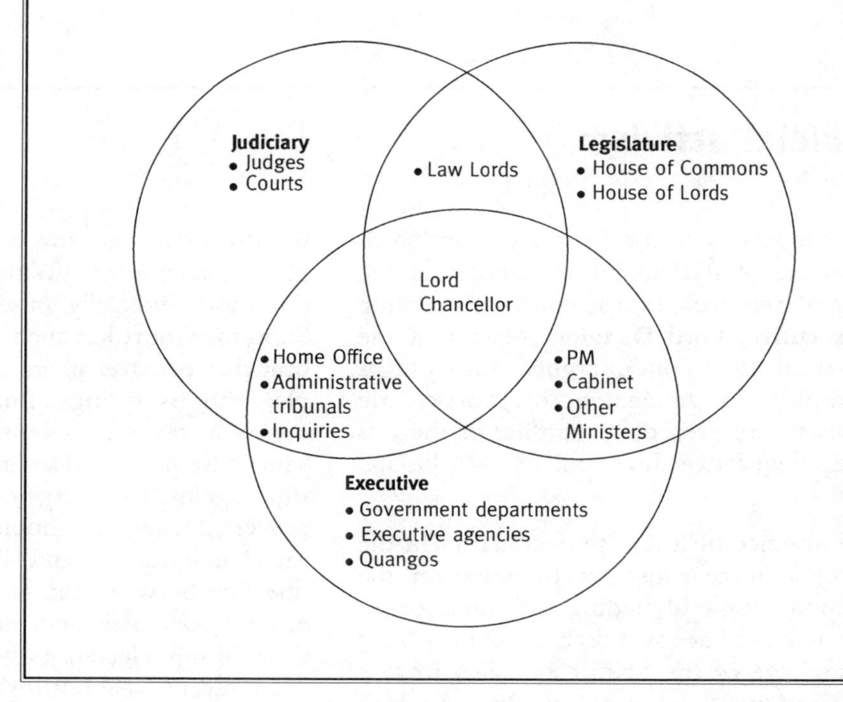

gery on 'Bloody Sunday' in Londonderry (1972), Lord Denning on the security aspects of the Profumo affair (1963), Lord Scarman on the Brixton riots (1981), Sir John May on the convictions of the Guildford Four (1989), Chief Justice Scott on arms for Iran (1995), and Chief Justice Nolan on standards in public life (1995).

In the 1970s and 1980s the role of judges was viewed in increasingly party political terms. For supporters the judiciary is an essential element in a liberal democracy, buttressing a system of checks and balances and the rule of law, and providing a safeguard against arbitrary government. But for critics it is part of a 'dominant' political order, usually making judgements in favour of the state or its own class. Judges are also frequently attacked for political 'bias' from their exclusive social and educational backgrounds and legal training. John Griffith (1991) is a vigorous proponent of the thesis that judges have frequently been partial to the interests of the authorities in cases involving property and law and order, hostile to trade unions, and too often have identified the public interest with the government's interest. Ministers, it is alleged, find it convenient to use judges as 'neutral' members of commissions, tribunals of inquiry, and other investigatory and advisory bodies, to reassure the public that an independent inquiry is being held into an alleged impropriety or disaster. A danger is that involvement in such extra-judicial activity will undermine the myth of judicial neutrality. As Lord Hailsham complained in 1973, 'You cannot keep independent judges in Britain if you constantly expose them to ordeal by public criticism which is not only inevitable but legitimate and proper whenever you ask them to preside over tribunals of inquiry' (cited in Drewry 1975: 60).

Judicial activism

Some supporters of the judiciary, pointing to the decline of Parliament in relation to the executive, see a role for the courts in checking the executive. Lord Denning, Master of the Rolls until 1982, often argued that judges should play a more creative role, particularly in cases where precedents conflict or there is no clear legislative direction. In 1949 he stated:

> In the absence of it [i.e. perfect clarity in the Statute] a judge must set to work on the constructive task of finding the intention of Parliament, and he must do this not only from the language of the Statute but also from a consideration of the social conditions which gave rise to it, and of the mischief which it was passed to remedy, and then he must supplement the written word so as to give 'force and life' to the intention of the legislature. (Cited in Marshall 1971: 88)

There is, however, no written constitution against which the courts may review acts of the executive, nor any tradition of the courts playing such a role. Within the United States the more judicially (and politically) activist Supreme Court has such a role, and the other branches of government are expected to comply with its rulings. But the Court's judgments in several contentious areas in recent years have produced a reaction on the political right against its rulings on civil rights, school prayers, busing of schoolchildren to promote racial integration, and abortion. There is a fine line between the law-making powers of elected politicians and the judicial interpretations of non-elected judges.

In recent years British courts have been more willing to strike down the actions of ministers who have acted on the basis of too indulgent an interpretation of their powers. In 1975, the Court of Appeal ruled that the Home Secretary had no right to revoke the television licences purchased in advance by people who wished to avoid paying an increased fee later on. In 1976, the same

BOX 15.3. THE COURT SYSTEM IN ENGLAND AND WALES

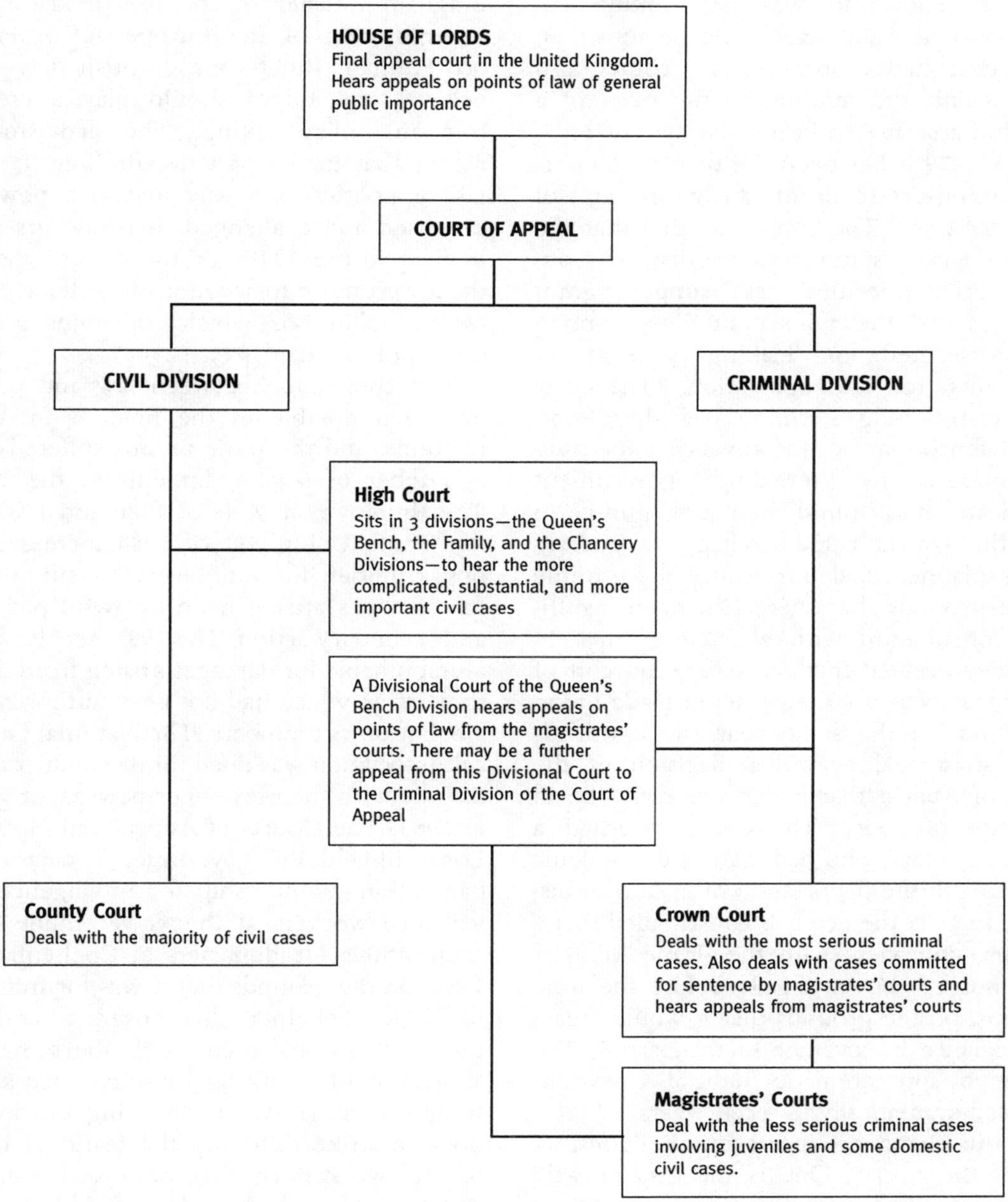

HOUSE OF LORDS
Final appeal court in the United Kingdom. Hears appeals on points of law of general public importance

COURT OF APPEAL

CIVIL DIVISION

CRIMINAL DIVISION

High Court
Sits in 3 divisions—the Queen's Bench, the Family, and the Chancery Divisions—to hear the more complicated, substantial, and more important civil cases

A Divisional Court of the Queen's Bench Division hears appeals on points of law from the magistrates' courts. There may be a further appeal from this Divisional Court to the Criminal Division of the Court of Appeal

County Court
Deals with the majority of civil cases

Crown Court
Deals with the most serious criminal cases. Also deals with cases committed for sentence by magistrates' courts and hears appeals from magistrates' courts

Magistrates' Courts
Deal with the less serious criminal cases involving juveniles and some domestic civil cases.

court decided that the Education Secretary was not empowered to prevent Tameside Education Authority from overturning a plan for comprehensive secondary education and restoring academic selection but was only entitled to intervene if the authority could be shown to have acted 'unreasonably'. The courts also ruled against the minister in the cases of a refusal of a licence to Laker Airways (1976).

Since 1979 it has been the turn of Conservative ministers to be on the wrong end of court decisions. The courts decided that the minister lacked statutory authority for reducing local authorities' rate support grant (1981). In 1985 the civil servant Clive Ponting was prosecuted for 'leaking' government documents to the opposition. The jury rejected the judge's ruling, refusing Ponting's defence that he was advancing the public interest by revealing government dishonesty, and found him not guilty. In 1992 the courts ruled against the government's planned coal pit closure programme on the grounds that there had been insufficient consultation with affected groups. In 1993 they decided that a teachers' boycott of school tests was not illegal under trade union legislation. In the same year the law lords heard an appeal against a decision of the Court of Appeal that a minister could be in contempt of court. The case concerned a Home Secretary who had disregarded a court order to halt the deportation of a Zairean dissident. In 1995 the Scottish courts ruled that a *Panorama* interview with the Prime Minister could not be shown shortly before the local elections, on the grounds that it would create an imbalance in coverage of the parties. The number of applications for judicial review has increased dramatically in recent years.

Various factors are increasing the 'political' role of the courts. One is the greater willingness of groups, aggrieved at legislation, to resort to the courts, and the expansion of law into new areas, particularly in employment, discrimination, and central–local government relations. A second is the changed outlook of the judges. A number of recently appointed senior figures, including Lord Taylor (the Lord Chief Justice) and Sir Thomas Bingham (Master of the Rolls), favour the incorporation of the European Convention on Human Rights into British law—and believe that judges should play a creative role in policy-making. The activism has meant that in the past decade 'Law is overtaking politics as a way in which power is exercised and challenged. It is lawyers more than opposition MPs or the media who call the government to account, the brief as much as the ballot box which constrains government policy' (Lee 1994: 123–4).

The connection between law and politics has been notable in the field of industrial relations, and the trade unions suffered from a number of legal judgments in the 1980s. The Employment Acts of 1980 and 1982 and the Trade Union Act of 1984 increased the opportunities for employers to sue unions for damages arising from unlawful picketing and secondary action. The 1984 Act also made a union liable for damages arising from industrial action which had not been authorized by a ballot of its members. The National Graphical Association was fined for its conduct of the dispute with the Messenger newspaper group in 1983. The Courts of Appeal and House of Lords upheld the government's decision to ban union membership for intelligence surveillance workers at the Government Communications Headquarters at Cheltenham in 1984, on the grounds that it was for ministers to decide whether this posed a potential threat to national security. In the same year a group of working miners successfully sought a court ruling declaring unconstitutional a strike called by the National Union of Mineworkers (NUM), because a ballot of NUM members had not been held before the strike was called. The union was then fined for contempt of court because it insisted that the strike was official.

European influences

British membership of the European Community is another force giving judges a more creative role. In cases where British legislation has an EC element, or in which there are disputes about the EC treaties or related matters, or in those which reach the House of Lords, there must be reference to the European Court of Justice (ECJ) for a definitive ruling. Where British domestic law conflicts with EC laws, British judges are obliged to give priority to the latter. The British government had to change social security policy after an ECJ ruling that it should not differentiate between full-time and part-time workers. More far-reaching is the impact of the *Factortame* case, by which British courts were empowered to suspend provisions of an Act of Parliament which might appear to breach EC law, pending a ruling by the ECJ. One should not make too much of actual or potential conflict, for Britain has one of the best records of member states for compliance with EC laws and directives. But membership clearly limits the sovereignty of Parliament.

If a Bill of Rights, based on the European Convention, were to be passed, and entrenched, this would provide the judges with another document to which they could refer in deciding the 'constitutionality' of the actions of the executive.

Civil liberties

The British have traditionally prided themselves on the security of their individual liberties. In practice, the liberties and rights of individuals and groups are qualified because they may conflict with other rights or with a broader social goal, such as public order. On the whole, individual rights and liberties in Britain have emerged from, and been sustained by, a largely negative view of freedom: individuals may do what they want unless and to the extent that they are forbidden to do it. Even today there are few statutory guarantees of rights, but instead a general freedom to do as one wishes, as long as it is not forbidden, that is, as long as the action does not transgress the law or interfere with the rights of others.

British freedoms of speech, organization, and demonstration are supported more by tradition and the political culture than by law. For example, laws against blasphemy, defamation, and obscenity qualify freedom of expression. The Race Relations Act (1976) also makes it a criminal offence to utter or publish statements which are likely to incite racial hatred. For long the dominant view was that British liberties, emerging from the common law and the decisions of the ordinary courts, were better protected than the rights enshrined in a constitution or statute as in other states. What could be so easily given by the state could with equal facility be removed.

There is no absolute freedom of meeting or assembly for British citizens. Meetings and marches are subject to the laws prohibiting obstruction of the highway, public nuisance, and trespass, and to local authority by-laws. For example, no march may be held within a mile of Parliament when it is in session. If the local police have reason to fear that a march may provoke disorder, they may insist on rerouting or even banning it. These powers were given to chief constables under the 1936 Public Order Act, which was prompted by disturbances surrounding the marches and meetings of the British Union of Fascists.

In Britain the police are bound by laws and rules, and are subordinate to the civil power. For example, the police may not, unless invited, enter a person's premises without a search warrant. A set of 'Judge's Rules' provides guidelines for police conduct; they require that a person be brought for trial shortly after his arrest, that he have access to a friend or solicitor, and that no force or other pressure be used to produce an involuntary confession. They uphold the principles of freedom from arbitrary arrest and imprisonment, a person's presumed innocence until proven guilty, the right not to be detained without a trial (that is, if a person is detained he or she has to be charged with a specific offence), and to be brought before a magistrate within 24 hours if charged, and to a fair trial.

These rights are generally observed but there have been exceptions. During the Second World War, the Defence Regulations 18B empowered the Home Secretary to detain any person whom he had reasonable cause to believe was hostile to the state. It was under this regulation that Oswald Mosley, the leader of the British Fascists, was detained without trial between 1940 and 1943. The violence in Northern Ireland has weakened the rule of law in that province. Under the Internment Act (1971) the authorities were allowed to imprison suspected terrorists or their protectors for an indefinite period and without a trial. In 1974, following bombings in Birmingham, Parliament rapidly passed the Prevention of Terrorism (Temporary Provisions) Act, which made the Irish Republican Army illegal. Under the Act, any UK citizen born in Northern Ireland may be arrested, detained, and then deported to Belfast without any charge being made against him and without a court hearing. The Northern Ireland (Emergency Provisions) Act, reviewed annually, also allows those accused of terrorism to be tried by a 'judge sitting alone, and not by jury'.

An important new legal constraint on personal conduct in recent years has been the growth of anti-discrimination legislation. Traditionally, no special protection has been given to particular sections of the community. But the growth of 'women's rights' movements, and the influx of immigrants in the 1950s and 1960s from the new Commonwealth states in East Africa, the Indian subcontinent, and the Caribbean challenged that tradition. The evidence of sexual and racial discrimination, and demands that they be tackled, provided the stimulus for legislation.

The Equal Pay Act (1970) required that women receive the same rates of pay as men when doing similar or equivalent work. The Sex Discrimination Act (1975) forbids discrimination on the grounds of sex and established an Equal Opportunities Commission. The 1965 and 1968 Race Relations Acts, later replaced by the 1976 Act, prohibit discrimination on grounds of race, religion, or national origin in various public places, in the provision of services and sale of goods, and in housing, employment, and membership of associations. A Race Relations Board was established to assist conciliation between races. Paradoxically, while the role of the law has been relaxed in many other areas of personal relations, such as divorce, abortion, and sexual behaviour, anti-discriminatory law has been used in a *tutelary sense*, that is, designed to influence values and conduct.

Advocates of a Bill of Rights complain that civil liberties have been restricted in recent years. Critics point to the banning of broadcast interviews with supporters of terrorism in Northern Ireland, the ending of trade union membership for employees at GCHQ, the imposition of a national curriculum on schools, the 'anti-gay' section of the 1988 Local Government Act, the increase in police powers, and the legislation which permits, without reference to a court, telephone tapping on such grounds as 'the economic well-being of the country'. There have also been several cases of miscarriage of justice: the 'Guildford Four', convicted of bombings in Guildford in 1975, were released in 1990, as were the 'Birmingham Six' in 1991, who had been convicted of IRA bombings in 1974. The sheer number of miscarriages of justice prompted unease about the ability of the judicial process to protect individual liberties. John Major's government appointed a Royal Commission on economic justice which made several recommendations for protecting the rights of defendants.

Rule of Law

One set of protections for citizens is enshrined in the *Rule of Law*. This concept was famously expounded by the nineteenth-century constitutional laywer A. V. Dicey. The essence of the concept is that rulers as well as the governed should be subject to law, and that governments should not act in an arbitrary manner. Dicey argued that the British Constitution rested on three major principles which amounted to a 'rule of law'. These were:

(1) No person is above the law, all are equal before the law, and disputes are decided in ordinary courts.
(2) No person is punishable except in the case of a distinct breach of the law.
(3) The principles of the Constitution, especially the liberties of the individual, are the result of judicial decisions. The rights of the individual precede and do not derive from the Constitution.

But Dicey's view no longer stands unchallenged. It is, for example, hardly compatible with the absolute sovereignty of Parliament. Parliament, as we have seen, has given arbitrary powers to the government in the exceptional circumstances of wartime. Moreover, many public authorities (such as ministers acting for the Crown) have special and discretionary powers. Dicey's statement may be more important today as a set of ideals rather than as a guide to actual practice.

Bill of Rights

Several countries have a Bill of Rights or some formal statement of the civil liberties of citizens and groups. These provide constitutional guarantees of freedom, for example, speech, freedom from arbitrary arrest, the right to privacy, or the freedom to emigrate. In

Britain there is no such document. The 1689 Bill of Rights deals largely with protecting the rights of Parliament against the monarchy and the Protestant succession to the throne. The British have placed less reliance on such formal safeguards and trusted more to a pluralistic political system, independent judiciary, acceptance of limits on what a government may do, and sense of 'constitutional morality' among the population. Britain has long been regarded as a country in which civil liberties were secure because of these cultural factors.

In recent decades, however, as the power of the executive and the scope of legislation appear to have increased, concern has grown about Parliament's ability to control the executive, not least to protect individual liberties. There is, of course, an element of party calculations to this, just as there is over the constitution. Under the 1974–9 Labour government the Conservative Lord Hailsham was eloquent about the dangers of executive dictatorship. Since 1979 it has been the opposition parties, including Labour, which have expressed alarm about an over-mighty government and the need for an overhaul of the British constitution, which would curb the power of the executive. A party's concern about the dangers of the absolute sovereignty of Parliament seems to be uppermost when government is controlled by another party.

There is, however, a non-partisan case to be made also. For the protection of individual rights to be effective, there is a need for a reviewing body which is independent of the government. Critics argue that the traditional safeguards of individual liberties—public opinion, shared political values, and self-restraint by the executive—no longer suffice. They claim that a Bill of Rights might stiffen the resolve of the courts to defend liberties and promote popular awareness of the issues involved. Lord Scarman (1974) has been the most notable spokesman for the view that the common law is now a fragile shield against a sovereign Parliament and the more interventionist scope of legislation. On a number of occasions a Bill of Rights, modelled on the *European Convention on Human Rights* (ECHR), has passed through all stages in the House of Lords, but no time for debate was provided in the Commons. If enacted, it would have the effect of allowing people to bring an action against the government for breaches of civil liberties. Various pressure groups and minor parties have promised support for the measure and it is part of the constitutional reform agenda, but the Conservative and Labour parties, apart from individual MPs, have shown no inclination to introduce such a measure.

European Court of Human Rights

One possible approach to safeguarding human rights would be for Britain to incorporate into British law the ECHR, to which it is a co-signatory. The Convention forbids a wide range of discriminatory practices and provides for an equally wide range of freedoms and rights (such as the right to life and to a fair trial). A complaint that a signatory state has violated the Convention may result in the government being taken before the Court.

The Human Rights Commission, to which the complaint is initially addressed, has to decide if there is a case to answer and satisfy itself that other possible remedies have already been tried. Only when domestic procedures have been exhausted and the Commission's efforts to conciliate have failed is the case referred to the Court. This means that it may take five years or more before a case is resolved. The Commission, not the complainant, takes the decision to refer a case to the Court.

The work of the Commission differs from that of the European Court of Justice (ECJ). Under Article 164 of the Treaty of Rome, the ECJ supervises the interpretation and application of Community law in the member states. So far nineteen of the twenty-three member states of the Council of Europe have ratified the Convention, and the majority of these have recognized the right of citizens to petition the Court and recognized its jurisdiction. Parliament has regularly renewed the Convention, so allowing individuals to bring a complaint against the United Kingdom to the Human Rights Commission in Strasburg. Indeed, there is widespread cross-party support for incorporating the Convention into British law. But since Britain is still the only co-signatory not to incorporate the Convention into its domestic law, citizens may not use it to appeal to British courts.

Since 1966, when Britain allowed individual citizens to bring cases to the Court, there have been successful appeals by citizens about the closed shop on British Rail (1981), the use of torture on prisoners by the security forces in Northern Ireland (1976) (although the concurrent appeal against internment failed), and corporal punishment in schools (1982), although the Court upheld the UK government's broadcast restrictions on Sinn Fein. By 1995, over a hundred cases against the UK government had been judged admissible, and the court has decided thirty-seven cases against the British government, one of the worst records of the signatories to the European Convention. Paradoxically, the European Court is able to act as a quasi-Supreme Court and judge governments on the criteria of the Convention. Supporters of the Court point out that its judgments declaring unlawful such practices as police telephone tapping, restrictions on prisoners' access to lawyers, inadequate legal protection for detained mental patients, and discrimination against foreign husbands of British women have expanded individual liberty.

There remains influential resistance to the development of public law in Britain. The supremacy of Parliament, in Lord Scarman's words, 'makes it difficult for the legal system to accommodate the concepts of fundamental and inviolable human rights' (1975: 15). Yet the European Convention, in recommending that the limits on human rights should not be 'unreasonable' or 'arbitrary', leaves much subjective judgement to the judges. The growth of judicial activism, already referred to, may encourage some judges to use the Convention as a guideline in cases before them. Simon Lee (1994: 138) has claimed that the 'twin streams of European law' (from Luxemburg and Strasburg) are working to curb the autonomy of the UK government and that in the 1980s and 1990s, law as much as politics had become the chief means of effectively challenging the government.

BOX 15.4. BRITAIN AND THE EUROPEAN COURT OF HUMAN RIGHTS

In 1951 Britain was a co-signatory of the European Convention on Human Rights, which was designed to improve the observance of human rights by European governments. Britain did not incorporate the convention into British law; British citizens, therefore, cannot appeal to British courts on the basis of the Convention. Since 1966, however, governments have granted British citizens the right to appeal to the ECHR in Strasburg. The British government is not bound to act on the Court's verdicts but usually does so since they carry considerable moral authority.

An Example

'Outrage over Death on the Rock verdict by Euro Court'. This headline appeared in *The Times*, 29 September 1995, the day after the European Court of Human Rights (ECHR) ruled against the British government in a case brought by the families of three members of the IRA—Daniel McCann, Sean Savage, and Mairead Farrell—killed by the SAS on Gibraltar in 1988.

The Verdict

In this particular case the original inquest in September 1988 found that the SAS had acted lawfully when it gunned down the three members of the IRA. In 1994 the European Commission on Human Rights upheld the inquest's verdict by 11 votes to 6. A year later, however, the ECHR rejected these rulings by 10 votes to 9. Although the Court accepted that the executions were not part of a shoot-to-kill policy, it condemned some aspects of Britain's anti-terrorist operations. The Court declared that there was no justification for the use of such force on the basis of the evidence and found the British government guilty of breaching an article of the Human Rights Convention which guaranteed the 'right to life'. Britain was ordered to pay costs of £38,700.

The Response

The British government's immediate response was to threaten to review the right of British citizens to appeal directly to the Court.

A Downing Street spokesman said that the judgment, overturning a series of verdicts in favour of Britain in the lower courts, defied common sense and was incomprehensible.

Some Conservative MPs noted that it was a 'rag tag and bobtail' group of states that voted against Britain and that they were countries with questionable records on human rights themselves.

How Judges Voted

For Britain	*Against Britain*
Britain	Italy
Norway	Luxemburg
Germany	Greece
Iceland	Spain
Turkey	Portugal
Sweden	Malta
Finland	Poland
Hungary	Slovakia
Slovenia	Lithuania
	Estonia

Administrative justice

An Act frequently delegates to a minister (or his officials) or a local authority certain law-making powers which are implied in the substance of the Act. The authorities are given powers to work out details, amend legislation to bring it up to date, or create machinery to administer it—all within the framework of the Act. These delegations obviously assist the speed and flexibility of the legislative process, but there have been frequent complaints that the growing use of such *delegated legislation* amounts to a form of executive dominance of Parliament.

In a powerful tract, *The New Despotism* (1929), Lord Hewart argued that the delegation was a form of lawlessness, because of the absence of 'known rules and principles, and a regular course of procedure'. The government subsequently set up the Donoughmore Committee to review the powers of ministers. The Committee's report (1932) did not wholly share Hewart's alarm, and only in 1944 did Parliament set up a Statutory Instruments Select Committee to review such Acts. There is now a joint committee of the House of Commons and House of Lords whose task is to review the operation of instruments and draw the attention of Parliament to any instrument that appears to make unusual use of the power conferred by the original statute.

In the twentieth century, the greater role the government has assumed in the economy, the rise of the Welfare State, and the complexity surrounding the duties and rights of citizens have added a further dimension. Disputes arise about administrative decisions in areas ranging from planning permission for building motorways and extending houses to slum clearance, hospital treatment, allocation of housing and level of rents, dismissal from work, and pensions, unemployment, and national insurance benefits. Such disputes raise problems of administrative justice, but if they were left to the ordinary courts they would be overwhelmed with work.

Tribunals, inquiries, and the Parliamentary Commissioner

Public inquiries and tribunals provide a quasi-judicial review of the actions of administrators and the possibility of redress for aggrieved persons. The tribunals and inquiries were set up to examine a class of disputes which were not referred to courts of law. Public inquiries are set up ad hoc, as and when the government considers they are needed or when there is an appeal against a government decision. In a public inquiry into, for example, the use of land, an inspector from the Department of the Environment will hear the views of different parties—the house owner and the local authority, say. Eventually, the inspector submits a report and makes a recommendation to the minister, who makes a decision.

Tribunals deal with more specialized matters. For example, an industrial tribunal may consider a worker's complaint of unfair dismissal and award compensation or even order

BOX 15.5. PUBLIC LAW: METHODS OF APPEAL

A network of semi-judicial bodies adjudicate in an informal system of 'administrative' or 'public' law. There are two main types:

Administrative tribunals

These are numerous and cover a wide range of topics, for example, industrial tribunals hear claims of unfair dismissal, redundancy, etc.; education tribunals hear claims against allocation of school places, expulsions, etc.

Composition: usually a legally qualified chairman plus two lay members representing the interests in question; for example, industrial tribunals include one representative of unions and one representative of employers.

Appeal: some tribunals allow appeals to higher court; in others, e.g. NHS tribunals, there is no appeal.

Statutory inquiries

These are usually ad hoc, often set up in connection with planning and compulsory land purchase.

Composition: the chairman is appointed by a minister and advises the minister of his findings. The relevant departmental minister usually takes the final decision. Procedures are regulated by the 1971 Tribunals and Inquiries Act.

the reinstatement of the complainant. In contrast to inquiries, these are standing bodies and make decisions.

Continued dissatisfaction with the tribunal system and concern at the lack of effective redress against administrative injustice led reformers to look abroad. In France a powerful *Conseil d'État* sits as a court of administrative justice to which aggrieved citizens may appeal and French ministers may be summoned and forced to justify their conduct. In Scandinavian countries and New Zealand, a Parliamentary Commissioner or Ombudsman is available to examine complaints of maladministration or cases of harsh or unreasonable decisions. *In 1967, a Parliamentary Commissioner for Administration (PCA) was established in Britain*; the Commissioner is appointed by the Crown but is the servant of the House of Commons.

The PCA deals with private citizens' complaints that the authorities have failed to carry out the law, observe proper standards of conduct, or follow established procedures. Examples of maladministration by officials include bias, incompetence, delay, and arbitrariness in making a decision. The Commissioner was precluded from investigating maladministration by local authorities, hospital boards, the police, armed forces, and nationalized industries. Many of the complaints originally received concerned these groups and thus were outside his terms of reference. Subsequently, Commissioners were established for Northern Ireland, local government (1974), and the National Health Service (1977), and an independent element was introduced into the police appeals procedure (1977). Under the Citizen's Charter most public services also have complaints procedures for customers.

If the Commissioner decides that a case falls within his remit, the first step is to invite comments on the complaint from the department. The Commissioner is empowered to call for the relevant files of the department concerned, although he lacks any executive authority of his own. If he finds a case of maladministration, the department is invited to rectify it. If the department refuses to act on his report then he lays it before Parliament and a select committee will consider the case and issue a report. Of some 300 cases examined each year, some maladministration is found in about 20 per cent of them. Most of the complaints upheld have concerned the departments of Inland Revenue and Health and Social Security, departments heavily used by members of the public.

The Commissioner's effectiveness has been limited by the terms of reference. Alleged cases of maladministration must have been committed in the United Kingdom and by a department of central government. The complainant has no right to appeal to a court or tribunal against the verdict. The Commissioner is also precluded from questioning policy, or the merits of discretionary decisions, as long as these were taken legally. Concern with maladministration alone, that is, with cases of unfairness which arise from the official not following rules and procedures, has disappointed those who feel that injustice may also result from an official following the law. The indirect method by which citizens' complaints are made to the Commissioner probably reduces the public perception of his role. Citizens' complaints must be forwarded in the first instance to the Commissioner through an MP. The Commissioner reports the results of his investigation to the MP and is required to submit an annual report on his work to Parliament. The number of cases referred to the Commissioner has not grown and most MPs prefer to write directly to the minister concerned if there is a problem. In Sweden administration is more open to public scrutiny and citizens have the right of direct access to the Commissioner. But such a move is likely to be resisted by those MPs who would resent being bypassed.

Citizens' rights against public authorities are upheld by various other non-political and non-legal mechanisms. The use of the Citizen's Charter is one such remedy (see p. 328). Greater use is also being made of regulatory bodies in the formerly nationalized utilities, which are now privatized. Because the scope for direct competition in those services has been limited, the government established independent regulators. The most significant are Ofgas (for gas), Ofwat (for water), Oftel (for telecommunications), and Offer (for electricity). These bodies exercise quasi-judicial powers in regulating prices, profits, quality of service, and consumer satisfaction for their respective utilities. For schools, Ofsted is charged with maintaining standards in education.

Conclusion

The connections between politics and law have grown in recent years and are likely to grow in the future, not least because of the growing impact of Europe on British politics. Applications for judicial review have increased, and public authorities, including ministers, departmental officials, and quangos, are aware that judges are metaphorically looking over their shoulders. The courts themselves are likely to be drawn into political controversy. Interestingly, if judicial intervention has traditionally been attacked by the left, in the past decade it has been the political right that has more often expressed concern.

Summary

- In Britain judges do not review the merits or constitutionality of the government's actions or laws but they can judge that a minister or official has acted beyond his statutory powers.

- In an attempt to ensure judicial independence a mutual exchange of self-restraint exists between poiticians and judges. Judges refrain from public comment on contentious issues; MPs avoid commenting on cases before the courts; judges may not be MPs; their salaries are not debated in Parliament; they have security of tenure and are appointed and promoted on professional grounds.

- The separation of law and politics is not total: the Lord Chancellor is appointed by the PM, is head of the judiciary, a member of the Cabinet (the executive), and a member of the House of Lords (the legislature). The PM is also involved in appointing other senior lawyers. Law is the best represented profession in Parliament. Judges often preside over commissions, tribunals, and inquiries that come under the auspices of the executive.

- Some argue that the judiciary is a safeguard against arbitrary government. Others argue that the judiciary buttresses the dominant political order and can never be impartial.

- Since the 1970s there have been signs of increasing politicization of the judiciary: a notable increase in judicial activism; a greater willingness for those aggrieved by legislation to seek redress through the courts; and the nature of some legislation, paticularly in relation to industrial law, has drawn judges into a more conspicuously political role. Determining the compatibility of British law and EU law has also contributed to this process.

- In Britain civil liberties are residual: they are not enshrined in a Bill of Rights, and citizens are free to do anything not specifically forbidden by law. The police are expected to treat those taken into custody in accordance with a code of conduct, Judges' Rules.

- The Rule of Law entails a set of ideals rather than a constitutional principle or a guide to practice.

- Some groups, particularly politicians in opposition, argue that the traditional safeguards of individual liberty in Britain are no longer sufficient and that the time has come to adopt a Bill of Rights.

- Britain is a co-signatory to the European Convention on Human Rights, but to date its provisions have not been incorporated into British law. The British government has on several occasions suffered defeat and embarrassment in the ECHR.

- Parliament delegates much detailed law-making to the executive in the form of enabling statutes. The Statutory Instruments Select Committee reviews the operation of these laws.
- Public inquiries and tribunals allow the executive to act in a quasi-judicial manner.
- The Parliamentary Commissioner for Administration provides citizens, via their MPs, with a means of redress in specifically defined cases of maladministration. But the PCA can only rely on moral suasion to enforce his rulings.
- The line between politics and law is becoming increasingly blurred and the role of the judiciary has become more controversial in recent years.

CHRONOLOGY

1951 Britain ratifies the European Convention on Human Rights

1967 A Parliamentary Commissioner for Administration is appointed

1971 Tribunals and Inquiries Act: regulates procedures of administrative judicial bodies

1972 European Communities Act: EU law takes precedence over British law in areas of EU competence. Lord Widgery chairs an inquiry into 'Bloody Sunday'

1974 Parliament passes the Prevention of Terrorism Act in 24 hours. The local government ombudsman is appointed

1975 The Court of Appeal rules that the Home Secretary had no right to prevent people from buying TV licences in advance to avoid increased fees

1976 The Court of Appeal rules that the Home Secretary for Education could not force the Tameside Local Education Authority to adopt comprehensive schooling

1977 The National Health Service ombudsman is appointed

1981 Lord Scarman's inquiry into the Brixton riots

1982 The ECHR upholds the case against corporal punishment in schools

1984 The Court of Appeal and House of Lords uphold the government's decision to ban unions at GCHQ

1989 Sir John May's inquiry into the conviction of the Guildford Four

1990 Courts and Legal Services Act: deregularizes some rules in respect of solicitors and barristers; creates Legal Sevices ombudsman

1991 *Factorame* Case: Lords rule that the 1988 Merchant Shipping Act is incompatible with EU law

1994 Criminal Justice and Public Order Act: curtails some civil liberties

1995 Scott Inquiry into arms for Iraq. Nolan Committee inquiring into standards in public life. Scottish courts rule that the *Panorama* interview of PM cannot be shown before local elections. The ECHR rules against the British government over killings of members of the IRA in Gibraltar

ESSAY/DISCUSSION TOPICS

1. Examine the case for and against Britain adopting a Bill of Rights.
2. How effectively are individual rights protected in Britain?
3. Discuss the view that the Rule of Law is a theoretical concept that has no practical application.
4. 'Judicial independence is guaranteed by a mutual exchange of self-restraint between politicians and judges'. Discuss.
5. Examine the factors that have contributed to an increasing politicization of the judiciary in recent years.

RESEARCH EXERCISES

1. You feel that you have been unfairly treated by the officials in your local social security office. You have complained to the head of department but this has brought no redress. What other courses of action could you try?
2. You have received a letter informing you that the Department of Environment intends compulsorily to purchase your property in order to construct a new bypass. What can you do to resist this action?

FURTHER READING

For an overview, see J. Griffith, *The Politics of the Judiciary*, 4th edn. (Oxford: Clarendon Press, 1991), and S. Lee, 'Law and the Constitution', in D. Kavanagh and A. Seldon (eds.), *The Major Effect* (London: Macmillan, 1994), 122–44.

The following may also be consulted:

Davis, Howard, 'The Political Role of the Courts', *Talking Politics*, 6/1 (1993).
Dowdle, John L., 'The Factortame Case', *Talking Politics*, 6/3 (1994).
Puddephett, Andrew, 'The Criminal Justice and Public Order Act and the Need for a Bill of Rights', *Talking Politics*, 8/1 (1995).

16 CONCLUSION

NO political system is immutable and in the British case there has been a good deal of change over the past two decades. It is appropriate that the conclusion to the book tries to strike a balance between the continuities and changes. Inevitably, deciding whether the direction has been one of progress or decline is a matter of controversy. Conservative apologists claim that the net effects of the changes have been beneficial, while their opponents deplore many of the changes.

Continuities

Much continuity is evident.

1. Viewed in comparative perspective, the degree of persistence in political institutions is remarkable. A citizen in no other European state, comparing his present-day political system with that, say, in 1922, would note the same degree of continuity which exists in the British system. Outwardly, the roles of the monarchy and House of Lords remain the same, and there are the same borders, electoral system, and the two dominant political parties.

2. The vast majority of public spending programmes and laws on the Statute Book pre-date not only 1979 but also 1945. As shares of GDP, public spending is virtually the same as in 1979 and the share taken in taxation is slightly higher. So much for a decade and a half of 'cuts' in spending and taxes under Conservative administrations.

3. In spite of at least two decades of intense debate about the need for constitutional change, little has been achieved. The starting-point for the debate is appropriately Lord Hailsham's Dimbleby Lecture on *Elective Dictatorship* in 1976. Yet twenty years later, for all the sound and fury, there has been no move to proportional representation, devolution, reform of the House of Lords, or Bill of Rights. The most significant political change has been the whittling away of the powers of local government, which has only added to the centralization.

4. Similarly, in spite of the rise and fall of the Social Democrats and the Alliance, the period 1982–7 now appears to have been a potential turning-point when the party system failed to turn. The Liberal Democrats' share of the vote in the 1992 general election was the same as that for the Liberals in 1974.

Changes

Yet there have been changes, including:

1. The shift in the style and role of government since 1979. Particularly significant has been abandonment of many features of the post-war social democratic consensus. These include the moves away from government

intervention in the economy, the large-scale privatization of state utilities, extension of market principles into the Welfare State, attack on trade union immunities and rights, and the acceptance of high levels of unemployment.

2. The contemporary vocabulary of political debate is increasingly about markets, competition, choice, consumer rights, value for money, and improving standards in public services. In central and local government the emerging model is of a core of decision-makers who contract out services and then set standards and monitor performance. Many of these ideas (associated with *Reinventing Government*), and the policies designed to encourage them, have been promoted by the political right. So-called modernizers in the Labour party have also accepted them and therefore a large part of the Thatcherite agenda. Many of the concerns and responses of policy-makers in Britain are found in other industrial societies. It is interesting that the right seems to have won this argument at a time when the Conservative party since 1992 has suffered a sharp fall in electoral support and is divided about its future direction.

3. The retreat of local government. This includes a loss of powers, imposition of far-reaching controls by central government, and more stringent financial constraints.

4. A consequence of the decline of elected local government has been the rise of the so-called new *magistracy*, the membership of which is not elected but appointed by ministers or organizations to quangos, which makes key decisions in education, health, and other areas. The role of these bodies raises important questions about democracy and accountability. Another consequence has been a further centralization of decisions in the hands of Whitehall.

5. There has been a sharp decline in the influence of trade unions and of corporatist styles of economic policy-making. Governments have been markedly less willing to negotiate with major interests about economic policy. They have also been eager to challenge the monopoly position and practices of the professions. Pressure groups seem less confrontational than they were in the 1970s.

6. There has been a fragmentation of the party system, from a largely two-party one to a three-party one. This has coincided with the long-term electoral decline of the Labour party, already apparent in the 1970s but partly masked by the unrepresentative electoral system. Since 1979 the division of the non-Conservative vote has weakened the Opposition in Parliament and allowed the Conservative party to amass large parliamentary majorities with only 42 per cent of the popular vote. The electoral decline of Labour has also produced, until 1992, a major imbalance in the party system, in terms of parliamentary representation.

7. In the mid-1990s there has been a reversal in the fortunes of the Labour and Conservative parties. The new Labour leadership has carried further the organizational and policy changes that Neil Kinnock began and even talks of a 'new' Labour party. At the same time the Conservative party has been bitterly divided over Europe. By 1996 the Labour party had enjoyed a lead in the opinion polls for over two years and by an unprecedented large margin.

8. Europe impinges increasingly on Britain's domestic politics. More and more, policies are shaped by guidelines from Brussels. The handling of the Maastricht Treaty and debates about whether Britain should join a single currency showed the weaknesses of John Major's government and added to them. For Euro-sceptics, largely found on the political right, a federal Europe threatens not only Britain's sovereignty but also its identity and place in the world. Proponents of a 'pooling' of Britain's sovereignty point to the 'real' world limits on the contemporary nation state's independence. These limits have

led to the creation of the EC and Britain's decision to join it. External forces, for example, the power of international finance to force Britain from the ERM in September 1992, were a brutal reminder of the limits of state independence. Whether or not Britain is a member of the EC, its main economic policy has to satisfy the 'good housekeeping' expectations of international markets.

9. The judiciary has played a more important role in domestic policy, in part as a consequence of Britain's membership of the European Union.

BIBLIOGRAPHY

ABRAMS, M., and ROSE, R. (1960), *Must Labour Lose?* (London: Penguin).

ADDISON, P., *The Road to 1945*, 2nd edn. (London: Pimlico).

ADONIS, A. (1990), *Parliament Today* (Manchester: Manchester University Press).

ALDERMAN, R. (1976), 'The Prime Minister and the Appointment of Ministers: An Exercise in Political Bargaining', *Parliamentary Affairs*, 29: 101–34.

——— and Cross, J. (1979), 'Ministerial Reshuffles and the Civil Service', *British Journal of Political Science*, 9: 41–66.

ALEXANDER, A. (1982), *Local Government and Britain since Reorganisation* (London: Allen & Unwin).

ARMSTRONG, W. (1970), 'The Role and Character of the Civil Service', talk given to the British Academy, London, 1970.

BAKER, D., GAMBLE, A., and LUDLAM, S. (1993), 'Whips or Scorpions? The Maastricht Vote in the Conservative Party', *Parliamentary Affairs*, 46: 151–66.

——— ——— ——— (1994) 'The Parliamentary Siege of Maastricht 1993: Conservative Divisions and British Ratification', *Parliamentary Affairs*, 47: 37–60.

BARNETT, C. (1986), *The Audit of War* (London: Macmillan).

BARNETT, J. (1982), *Inside the Treasury* (London: Andre Deutsch).

BEER, S. (1964), *Modern British Politics* (London: Faber).

——— (1982), *Britain against Itself* (London: Faber).

BENN, T. (1980), 'The Case for Constitutional Premiership', *Parliamentary Affairs*, 33: 7–22.

BERRINGTON, H. (1968), 'Partisanship and Dissidence in the 19th-Century House of Commons', *Parliamentary Affairs*, 21: 338–74.

——— (1974), 'The Fiery Chariot: British Prime Ministers and the Search for Love', *British Journal of Political Science*, 4: 345–69.

BIRCH, A. H. (1964), *Representative and Responsible Government* (London: Allen and Unwin).

BLUMLER, J., and MCQUAIL, D. (1968), *Television in Politics* (London: Faber).

——— GUREVITCH, M., and NOSSITER, T. (1989), 'The Earnest versus the Determined: Election Newsmaking at the BBC', in Crewe and Harrop (1989), 157–74.

BOGDANOR, V. (1981), *The People and the Party System* (Cambridge: Cambridge University Press).

——— (1983*a*), *Multi-Party Politics and the Constitution* (Cambridge: Cambridge University Press).

——— (1983*b*) (ed.), *Liberal Party Politics* (Oxford: Oxford University Press).

——— (1994), 'Ministers, Civil Servants and the Constitution', *Government and Opposition*, 29: 676–95.

BRIDGES, E. (1950), *Portrait of a Profession: The Civil Service Tradition* (Cambridge: Cambridge University Press).

BROUGHTON, D. (1995), *Public Opinion Polling and Politics in Britain* (Hemel Hempstead: Harvester Wheatsheaf).

BULPITT, J. (1983), *Territory and Power in the United Kingdom* (Manchester: Manchester University Press).

BURTON, I., and DREWRY, G. (1981), *Legislation and Public Policy* (London: Macmillan).

BUTLER, D. (1986), *Governing without a Majority*, 2nd edn. (London: Macmillan).

——— and KAVANAGH, D. (1980), *The British General Election of 1979* (London: Macmillan).

——— ——— (1988), *The British General Election of 1987* (London: Macmillan).

——— ——— (1992), *The British General Election of 1992* (London: Macmillan).

——— and Stokes, D. (1969), *Political Change in Britain* (London: Macmillan).

——— ——— (1974), *Political Change in Britain*, 2nd edn. (London: Macmillan).

——— ADONIS, A., and TRAVERS, T. (1995), *Failure in British Government* (Oxford: Oxford University Press).

BYRD, P. (1988) (ed.), *British Foreign Policy under Thatcher* (Oxford: Philip Allan).

BYRNE, T. (1994), *Local Government in Britain*, 6th edn. (London: Penguin).

CASTLE, B. (1980), *The Castle Diaries, 1964–70* (London: Weidenfeld & Nicolson).

CHURCHILL, W. (1949), *The Second World War*, ii: *Their Finest Hour* (London: Cassell).

CLARK, A. (1993), *Diaries* (London: Phoenix).

COATES, D. (1975), *The Labour Party and the Struggle for Socialism* (Cambridge: Cambridge University Press).

CREWE, I. (1986), 'On Death and Resurrection of Class Voting', *Political Studies*, 35: 620–38.

——— (1989), 'Values: The Crusade that Failed', in Kavanagh and Seldon (1989), 239–50.

——— (1992) 'Why did Labour Lose (Yet Again)?', *Politics Review*, 2: 2–11.

——— (1993), 'Voting and the Electorate', in Dunleavy *et al.* (1993), 92–122.

——— and HARROP, M. (1986) (eds.), *Political Communications: The General Election Campaign of 1983* (Cambridge: Cambridge University Press).

——— ——— (1989) (eds.), *Political Communications: The General Election Campaign of 1987* (Cambridge: Cambridge University Press).

——— and KING, A. (1994), 'Did Major Win?, Did Kinnock Lose? Leadership Effects in the 1992 Election', in Heath, Jowell, and Curtice (1994), 125–48.

CRIDDLE, B. (1992), 'MPs and Candidates', in Butler and Kavanagh (1992), 211–30.

CROSSMAN, R. (1972), *Inside View* (London: Jonathan Cape).

——— (1975–7), *The Diaries of a Cabinet Minister*, 2 vols. (London: Hamish Hamilton and Jonathan Cape).

CROUCH, C. (1982), 'The Peculiar Relationship: The Party and the Unions', in Kavanagh (1982), 171–90.

——— and MARQUAND, D. (1989) (eds.), *The New Centralism: Britain Out of Step in Europe* (Oxford: Blackwell).

CURTICE, J., and STEED, M. (1980), 'An Analysis of the Voting', in Butler and Kavanagh (1980), 390–431.

——— ——— (1988), 'Analysis', in Butler and Kavanagh (1988), 316–62.

DENVER, D. (1989), *Elections and Voting Behaviour in Britain* (Oxford: Philip Allan).

DICEY, A. (1952), *Introduction to the Law of the Constitution* (London: Macmillan).

DONOUGHUE, B. (1987), *Prime Minister* (London: Cape).

DOWDING, K. (1993), *The Civil Service* (London: Routledge).

DREWRY, G. (1975), 'Judges, and Political Inquiries: Harnessing the Myth', *Political Studies*, 23: 49–61.

——— (1989*a*), *Law, Justice and Politics*, 3rd edn. (London: Longman).

——— (1989*b*) (ed.), *The Select Committees: A Study of the 1979 Reforms*, 2nd edn. (Oxford: Oxford University Press).

——— and BUTCHER, X. (1991), *The Civil Service Today* (Oxford: Blackwell).

DUNLEAVY, P. and HUSBANDS, C. (1985), *British Democracy at the Crossroads* (London: Allen and Unwin).

——— *et al.* (1993) (eds.), *Developments in British Politics 4* (London: Macmillan).

DUNSIRE, A., and HOOD, C. (1989), *Cutback Management and Public Bureaucracies* (Cambridge: Cambridge University Press).

DYNES, M., and WALKER, D. (1995), *The Times Guide to the New British State* (London: Times Books).

ECKSTEIN, H. (1960), *Pressure Group Politics* (London: Allen and Unwin).

FINER, S. (1956), 'The Individual Responsibility of Ministers', *Public Administration*, 34: 377–96.

——— (1958), *Anonymous Empire* (London: Pall Mall).

——— (1975*a*), 'State- and Nation-Building in Europe: The Role of the Military' in C. Tilly (ed.), *The Formation of National States in Western Europe* (Princeton: Princeton University Press), 84-163.

——— (1975*b*) (ed.), *Adversary Politics and Electoral Reform* (London: Wigram).

FOLEY, M. (1993), *The Rise of the British Presidency* (Manchester: Manchester Univesity Press).

FRANKLIN, B. (1994), *Packaging Politics* (London: Arnold).

FRY, G. K. (1975), *The Growth of Government* (London: Allen and Unwin).

GAMBLE, A. (1990), *Britain in Decline*, 3rd edn. (London: Macmillan).

——— (1994), *The Free Economy and the Strong State*, 2nd edn. (London: Macmillan).

GAME, C., and Leach, S. (1995), *The Role of Political Parties in Local Democracy* (London: Commission for Local Democracy).

GEORGE, S. (1990), *An Awkward Partner: Britain in the European Community* (Oxford: Oxford University Press).

——— (1992) (ed.), *Britain and the European Community: The Politics of Semi-Detachment* (Oxford: Oxford University Press).

Glasgow University Media Group (1976), *Bad News* (London: Routledge and Kegan Paul).

——— (1982), *Really Bad News* (London: Readers and Writers).

GOLDTHORPE, J. (1987), *Social Mobility and Class Structure in Modern Britain*, 2nd edn. (Oxford: Clarendon Press).

GORDON WALKER, P. (1970), *The Cabinet* (London: Jonathan Cape).

GRANT, W. (1993), *Business and Politics in Britain*, 2nd edn. (London: Macmillan).

GRIFFITH, J. (1991), *The Politics of the Judiciary*, 4th edn. (Oxford: Clarendon Press).

GUTTSMAN, W. L. (1963), *The British Political Elite* (London: Heinemann).

GWYN, W. (1980), 'Jeremiahs and Pragmatists: Perceptions of British Decline', in Gwyn and Rose (1980), 1–25.

——— and ROSE, R. (1980) (eds.), *Britain, Progress and Decline* (London: Macmillan).

GYFORD, J. (1985), *The Politics of Local Socialism* (London: Allen and Unwin).

HAILSHAM, Lord (1978), *Dilemma of Democracy* (London: Oxford University Press).

HALEVY, E. (1924), *The History of the English People in the 19th Century*, ii (London: Ernest Benn).

Hansard Society (1993), *Making the Law* (London: Hansard Society).

HARDEN, I., and LEWIS, N. (1988), *The Noble Lie: The British Constitution and the Role of Law* (London: Hutchinson).

HARRISON, M. (1985), *TV News: Whose Bias?* (Hermitage, Berks: Policy Journals).

——— (1992), 'Politics on the Air', in Butler and Kavanagh (1992), 155–79.

HARROP, M. (1986), 'Press Coverage of Post-War British Elections' in Crewe and Harrop (1986), 137–49.

——— and SCAMMELL, N. (1992), 'A Tabloid War', in Butler and Kavanagh (1992), 180–210.

HART, J. (1992), *Proportional Representation: Critics of the British Electoral System, 1820–1945* (Oxford: Oxford University Press).

HEADEY, B. (1974), *British Cabinet Ministers* (London: Allen and Unwin).

HEARL, D. (1994), 'Britain in Europe since 1945', *Parliamentary Affairs*, 47: 516–31.

HEATH, A., JOWELL, R., and CURTICE, J. (1985), *How Britain Votes* (Oxford: Pergamon Press).

——— ——— ——— (1987), 'Trendless Fluctuations: A Reply to Crewe', *Political Studies*, 35: 256–77

——— ——— ——— (1991), *Understanding Political Change* (Oxford: Pergamon Press).

——— ——— ——— (1994) (eds.), *Labour's Last Chance* (Aldershot: Dartmouth).

HECLO, H., and WILDAVSKY, A. (1974), *The Private Government of Public Money* (London: Macmillan).

HENNESSY, P. (1986), *Cabinet* (Oxford: Blackwell).

——— (1990), *Whitehall* (London: Secker and Warburg).

——— (1996), *The Good and the Great* (London: London Policy Studies Institute).

HEWART, Lord (1929), *The New Despotism* (London: Ernest Benn).

HOGWOOD, B., and KEATING, M. (1982) (eds.), *Regional Government in England* (Oxford: Oxford University Press).

HOOD, C. (1991), 'A Public Management for all Seasons', *Public Administration*, 69: 3–19.

HOOD PHILLIPS, O. (1978), *Constitutional and Administrative Law*, 6th edn. (London: Sweet and Maxwell).

HOSKYNS, J. (1982), 'Whitehall and Westminster: An Outsider's View', *Fiscal Studies*, 3: 112–72.

—— (1984), 'Conservatism is not Enough', *Political Quarterly*, 55: 3–16.

HUGHES, C., and WINTOUR, P. (1990), *Labour Rebuilt* (London: Fourth Estate).

Institute of Public Policy Research (IPPR) (1991), *The Constitution of the United Kingdom* (London: IPPR).

JAMES, S. (1992), *British Cabinet Government* (London: Routledge).

JENKINS, R. (1975), 'On Being a Minister', in V. Herman and J. Ault (eds.), *Cabinet Studies* (London: Macmillan).

JOHNSON, N. (1977), *In Search of the Constitution* (London: Methuen).

JONES, G. (1973), 'The Prime Minister's Advisers', *Political Studies*, 21: 363–75.

—— (1987), 'The United Kingdom', in W. Plowden (ed.), *Advising the Rulers* (Oxford: Blackwell), 36–66.

JONES, W. (1992), 'Broadcasters, Politicians and the Political Interview', in W. Jones and L. Robinson (eds.), *Two Decades of British Politics* (Manchester: Manchester University Press), 53–78.

—— (1994), '"The Unknown Government": The Conservative Quangocracy', *Talking Politics*, 6/2: 104–5.

KAVANAGH, D. (1980), 'From Gentlemen to Players: Changes in Political Leadership' in Gwyn and Rose (1980), 73–93.

—— (1982) (ed.), *The Politics of the Labour Party* (London: Allen and Unwin).

—— (1985), 'Power in Political Parties: Iron Law or Special Pleading?', *West European Politics*, 8: 5–20.

—— (1990), *Thatcherism and British Politics: The End of Consensus?* (Oxford: Oxford University Press).

—— (1992), 'Changes in the Political Class and its Culture', *Parliamentary Affairs*, 45: 18–32.

—— (1994*a*), 'A Major Agenda', in Kavanagh and Seldon (1994), 3–17.

—— (1994*b*), 'Changes in Electoral Behaviour and the Party System', *Parliamentary Affairs*, 47: 598–612.

—— (1995), *Election Campaigning: The New Politics of Marketing* (Oxford: Blackwell).

—— and Seldon, A. (1989) (eds.), *The Thatcher Effect* (Oxford: Oxford Univesity Press).

—— —— (1994) (eds.), *The Major Effect* (London: Macmillan).

KELLAS, J. (1989), *The Scottish Political System*, 4th edn. (Cambridge: Cambridge University Press).

KELLY, R. (1989), *Conservative Party Conferences* (Manchester: Manchester University Press).

—— and GARNER, R. (1993), *British Political Parties Today* (Manchester: Manchester University Press).

KING, A. (1969 and 1985) (ed.), *The British Prime Minister*, lst and 2nd edns. (London: Macmillan).

—— (1975), 'Overload: Problems of Governing in the 1970s' in *Political Studies*, 23: 284–96.

KING, A. (1976), 'Modes of Executive Relations: Great Britain, France and West Germany', *Legislative Studies Quarterly*, 1: 11–36.

——— (1981), 'The Rise of the Career Politician in Britain–and its Consequences', *British Journal of Political Science*, 11: 249–63.

KOGAN, D., and KOGAN, M. (1982), *The Battle for the Labour Party* (London: Fontana).

LAWSON, N. (1992), *The View from No. 11: Memoirs of a Tory Radical* (London: Bantam Press).

LAYTON-HENRY, Z., and RICH, P. (1986) (eds.), *Race, Government and Politics in Britain* (London: Macmillan).

LEE, S. (1994), 'Law and the Constitution', in Kavanagh and Seldon (1994), 122–44.

MCCORMICK, J. (1991), *British Politics and the Environment* (London: Earthscan).

MCGARRY, J., and O'LEARY, B. (1993), *Politics of Antagonism: Understanding Northern Ireland* (London: Athlone Press).

MCKENZIE, R. (1963), *British Political Parties* (London: Heinemann).

——— (1974), 'Politics, Pressure Groups and the British Political Process', repr. in R. Kimber and J. Richardson (eds.), *Pressure Groups in Great Britain* (London: Dent).

MACKINTOSH, J. (1962), *The British Cabinet* (London: Stevens).

MADGWICK, P., and ROSE, R. (1982) (eds.), *The Territorial Dimension in the United Kingdom Politics* (London: Longman).

MARQUAND, D. (1995), *The State in Context* (London: ESRC Lecture).

MARR, A. (1992), *The Battle for Scotland* (London: Penguin).

MARSH, D. (1992), *The New Politics of British Trade Unions and the Thatcher Legacy* (London: Macmillan).

——— and RHODES, R. (1992), *Implementing Thatcher's Policies: Audit of an Era* (Milton Keynes: Open University).

MARSHALL, G. (1971), *Constitutional Theory* (Oxford: Oxford University Press).

——— (1984), *Constitutional Conventions* (Oxford: Clarendon Press).

MELLORS, C. (1978), *The British MPs* (Farnborough: Saxon House).

MESSINA, A. (1989), *Race and Party Competition in Britain* (Oxford: Oxford University Press).

MIDDLEMAS, K. (1979), *Politics in Industrial Society* (London: Andre Deutsch).

MILIBAND, R. (1961), *Parliamentary Socialism* (London: Merlin).

MINKIN, L. (1980), *The Labour Party Conference* (Manchester: Manchester University Press).

——— (1991), *The Contentious Alliance: Trade Unions and the Labour Party* (Edinburgh: Edinburgh University Press).

MORAN, M. (1981), 'Finance, Capital and Pressure Group Politics in Britain', *British Journal of Political Science*, 11: 381–404.

MOUNT, F. (1992), *The British Constitution Now: Recovery or Decline?* (London: Heinemann).

MULGAN, G. (1994), *Politics in an Anti-political Age* (Oxford: Polity Press).

NEGRINE, R.(1994), *Politics and the Mass Media*, 2nd edn. (London: Routledge).

NEWTON, K. (1982), 'Is Small really so Beautiful?', *Political Studies*, 30: 190–206.

NORRIS, P., and LOVENDUSKI, J. (1994), *Political Recruitment* (Cambridge: Cambridge University Press).

NORTON, P. (1980), *Dissension in the House of Commons, 1974–79* (Oxford: Oxford University Press).

——— (1982), *The Constitution in Flux* (Oxford: Martin Robertson).

——— (1991) (ed.), *New Directions in British Politics* (Aldershot: Edward Elgar).

——— (1993), *Does Parliament Matter?* (Hemel Hempstead: Harvester Wheatsheaf).

——— (1994), 'The Growth of the Constituency Role of the MP', *Parliamentary Affairs*, 47: 705–20.

——— and AUGHEY, A. (1981), *Conservatives and Conservatism* (London: Temple Smith).

NUGENT, N. (1993), 'The European Dimension', in Dunleavy *et al.* (1993), 40–68.

O'CONNOR, J. (1973), *The Fiscal Crisis of the Capitalist State* (New York: St. James' Press).

OLIVER, D., and AUSTIN, R. (1987), 'Political and Constitutional Aspects of the Westland Affair', *Parliamentary Affairs*, 40: 20–40.

OLSON, M. (1982), *The Rise and Fall of Nations* (New Haven: Yale University Press).

OSBORNE, D., and GAEBLER, T. (1992), *Reinventing Government* (Reading, Mass.: Addison-Wesley).

PARRIS, H. (1969), *Constitutional Bureaucracy* (London: Allen and Unwin).

PARRY, G., MOYSER, G., and DAY, M. (1992), *Political Participation and Democracy in Britain* (Cambridge: Cambridge University Press).

PINTO-DUSCHINSKY, M. (1981), *British Political Finance, 1830–1980* (London and Washington, DC: American Enterprise Institute).

PLOWDEN, W. (1994), *Ministers and Mandarins* (London: IPPR).

POLLARD, S. (1981), *The Wasting of the British Economy* (London: Croom Helm).

POLLITT, C. (1993), *Managerialism and the Public Services*, 2nd edn. (Oxford: Blackwell).

PUNNETT, M. (1973), *Front-Bench Opposition* (London: Heinemann).

——— (1992), *Selecting the Party Leaders* (Hemel Hempstead: Harvester Wheatsheaf).

RHODES, R. (1986), *The National World of Local Government* (London: Allen and Unwin).

——— (1988), *Beyond Westminster and Whitehall* (London: Unwin Hyman).

——— (1994), 'The Hollowing-out of the State', *Political Quarterly*, 65: 138–51.

——— and Dunleavy, P. (1995) (eds.), *Prime Minister, Cabinet and Core Executive* (London: Macmillan).

RIDDELL, T. (1993), *Honest Opportunism: The Rise of the Career Politician* (London: Hamish Hamilton).

RIDLEY, F. (1986), 'Political Neutrality and the Civil Service', *Social Studies Review*, 2: 23–8.

——— (1988), 'There is no British Constitution: A Dangerous Case of the Emperor's Clothes', *Parliamentary Affairs*, 41: 339–60.

RIDLEY, N. (1989), *The Local Right* (London: Centre for Policy Studies).

ROSE, R. (1971), *Governing Without Consensus: An Irish Perspective* (London: Faber).

ROSE, R. (1976), 'On the Priorities of Government', *European Journal of Political Research*, 4: 247–89.

——— (1980*a*), 'British Government: The Job at the Top', in R. Rose and E. Suleiman (eds.), *Presidents and Prime Ministers* (Washington, DC: American Enterprise Institute), 1–49.

——— (1980*b*) (ed.), *Challenge to Governance: Studies in Overloaded Politics* (London: Sage).

——— (1982*a*), 'Is the United Kingdom a State? Northern Ireland as a Test Case', in Madgwick and Rose (1982), 100–36.

——— (1982*b*) *Understanding the United Kingdom: The Territorial Dimension in Government* (London: Longman).

——— (1984*a*), *Do Parties Make a Difference?*, 2nd edn. (London: Macmillan).

——— (1984*b*) *Understanding Big Government* (London: Sage).

——— (1987), *Ministers and Ministries* (Oxford: Oxford University Press).

——— (1992), 'Structural Change or Cyclical Fluctuation?', *Parliamentary Affairs*, 45: 451–65.

——— (1995), 'A Crisis of Confidence in British Party Leaders?', *Contemporary Record*, 9 (forthcoming).

——— and Davies, P. (1995), *Inheritance in Public Policy: Change without Choice in Britain* (New Haven: Yale University Press).

——— and Peters, G. (1978), *Can Government Go Bankrupt* (Boston: Basic Books).

RUSH, M. (1994), 'Career Politics in British Politics: First Choose your Party', *Parliamentary Affairs*, 47: 566–82.

SCAMMELL, M. (1995), *Designer Politics* (London: Macmillan).

SCARMAN, Lord (1974), *English Law: The New Dimensions* (London: Stevens).

SCHUMPETER, J. A. (1942), *Capitalism, Socialism and Democracy* (London: Allen and Unwin).

SEARING, D. (1994), *Westminster World: Understanding Political Rules* (Cambridge, Mass.: Harvard University Press).

SEDGEMORE, B. (1980), *The Secret Constitution* (London: Hodder and Stoughton).

SELDON, A. (1990), 'The Cabinet Office and Co-ordination, 1979–87', *Public Administration*, 68: 103–122.

——— (1994), 'Consensus: A Debate too Long?', *Parliamentary Affairs*, 47: 501–15.

——— and BALL, S. (1994) (eds.), *The Conservative Century* (Oxford: Oxford University Press).

SEMETKO, H. A., SCAMMEL, M., and NOSSITER, T. J. (1994), 'The Media's Coverage of the Campaign', in Heath *et al.* (1994), 25–42.

SEYD, P. (1987), *The Rise and Fall of the Labour Left* (London: Macmillan).

——— and Whiteley, P. (1992), *Labour's Grass-Roots: The Politics of Party Membership* (Oxford: Clarendon Press).

——— ——— (1995), 'Labour and Conservative Party Members Compared', *Politics Review*, 4: 2–7.

SEYMOUR-URE, C. (1991), *The British Press and Broadcasting since 1945* (Oxford: Blackwell).

——— (1995), 'Characters and Assassinations: Portrayals of John Major and Neil Kinnock in the *Daily Mirror* and the *Sun*', in I. Crewe and B. Gosschalk (eds.),

Political Communications: The General Election Campaign of 1992 (Cambridge: Cambridge University Press, 1995), 137–59.

SHARPE, J. (1973), 'American Democracy Re-considered, Part II', *British Journal of Political Science*, 3: 129–68.

——— (1982), 'The Labour Party and the Geography of Inequality', in Kavanagh (1982), 135–70.

SHELL, D. (1988), *The House of Lords* (Oxford: Philip Allan).

——— (1994), 'The House of Lords, Time for a Change?', *Parliamentary Affairs*, 47: 721–37.

SMITH, M. (1994), 'Understanding the "Politics of Catch-up": The Modernisation of Labour', *Political Studies*, 42: 700–7.

STEWART, J. (1994), *Accountability to the Public* (London: European Policy Forum).

——— Greer, A., and Hoggett, P. (1995), *The Quango State* (London: Commission for Local Democracy).

THATCHER, M. (1993), *The Downing Street Years* (London: HarperCollins).

THEAKSTON, K., and FRY, G. (1989), 'The British Administrative Elite: Permanent Secretaries 1900–1986', *Public Administration*, 67: 129–47.

TIMMINS, N. (1995), *The Five Giants* (London: HarperCollins).

TIVEY, L. (1988), *Interpretations of British Politics* (Hemel Hempstead: Harvester Wheatsheaf).

VIBERT, F. (1991), *Constitutional Reform in the United Kingdom: An Incremental Agenda* (London: Institute of Economic Affairs).

WAKEHAM, J. (1994), 'Cabinet Government', *Contemporary Record*, 8: 473–83.

WEINER, M. (1981), *English Culture and the Decline of the Industrial Spirit, 1850–1950* (Cambridge: Cambridge University Press; repr. Harmondsworth: Penguin, 1985).

WHITELEY, P., SEYD, P., and RICHARDSON, R. (1994), *True Blues: The Politics of Conservative Party Membership* (Oxford: Oxford University Press).

WHYTE, J. (1990), *Interpreting Northern Ireland* (Oxford: Oxford University Press).

Widdicombe Report (1986), *The Conduct of Local Authority Business*, Cmnd 9797 (London: HMSO).

WILLIAMS, P. (1982), 'Changing Styles of Labour Leadership', in Kavanagh (1982), 50–68.

WILLMAN, J. (1994), 'The Civil Service', in Kavanagh and Seldon (1994), 64–82.

WILSON, D., and GAME, C. (1994), *Local Government in Britain* (London: Macmillan).

WILSON, H. (1977), *The Governance of Britain* (London: Sphere Books).

WINCOTT, D. (1992), 'The Conservative Party and Europe', *Politics Review*, 1: 12–18.

WRIGHT, T. (1994), *Citizens and Subjects* (London: Routledge).

YOUNG, H., and SLOMAN, A. (1982), *No Minister* (London: BBC).

INDEX